P9-DUR-422

Introduction to Autonomous Mobile Robots

Intelligent Robotics and Autonomous Agents
Edited by Ronald C. Arkin

A list of the books published in the Intelligent Robotics and Autonomous Agents series can be found at the back of the book.

Introduction to Autonomous Mobile Robots

second edition

Roland Siegwart, Illah R. Nourbakhsh, and Davide Scaramuzza

The MIT Press
Cambridge, Massachusetts
London, England

© 2011 Massachusetts Institute of Technology
Original edition © 2004

All rights reserved. No part of this book may be reproduced in any form by any electronic or mechanical means (including photocopying, recording, or information storage and retrieval) without permission in writing from the publisher.

For information about special quantity discount, please email special_sales@mitpress.mit.edu

This book was set in Times Roman by the authors using Adobe FrameMaker 9.0.
Printed and bound in the United States of America.

Library of Congress Cataloging-in-Publication Data

Siegwart, Roland.
Introduction to autonomous mobile robots. - 2nd ed. / Roland Siegwart, Illah R. Nourbakhsh, and Davide Scaramuzza.
    p. cm. - (Intelligent robotics and autonomous agents series)
Includes bibliographical references and index.
ISBN 978-0-262-01535-6 (hardcover : alk. paper) 1. Mobile robots. 2. Autonomous robots. I. Nourbakhsh, Illah Reza, 1970- II. Scaramuzza, Davide. III. Title.
TJ211.415.S54 2011
629.8'932-dc22
2010028053

10 9 8 7 6 5 4 3

To Luzia and my children, Janina, Malin, and Yanik, who give me their support and freedom to grow every day — RS

To my parents, Susi and Yvo, who opened my eyes — RS

To Marti, Mitra, and Nikou, who are my love and my inspiration — IRN

To my parents, Fatemeh and Mahmoud, who let me disassemble and investigate everything in our home — IRN

To my parents, Paola and Ermanno, who encouraged and supported my choices every day and introduced me to robotics at the age of three — DS

To my sisters, Lisa and Silvia, for their love — DS

Slides and exercises that go with this book are available at:

http://www.mobilerobots.org

Contents

Acknowledgments

This book is the result of inspirations and contributions from many researchers and students at the Swiss Federal Institutes of Technology Zurich (ETH) and Lausanne (EPFL), Carnegie Mellon University's Robotics Institute, Pittsburgh (CMU), and many others around the globe.

We would like to thank all the researchers in mobile robotics who make this field so rich and stimulating by sharing their goals and visions with the community. It is their work that enabled us to collect the material for this book.

The most valuable and direct support and contribution for this second edition came from our current collaborators at ETH. We would like to thank Friedrich Fraundorfer for his contribution to the section on location recognition; Samir Bouabdallah for his contribution to the section on flying robots; Christian David Remy for his contribution to the section on considerations of dynamics; Martin Rufli for his contribution to path planning; Agostino Martinelli for his careful checking of some of the equations; Deon Sabatta and Jonathan Claassens for their careful review of some sections and their fruitful discussions; and Sarah Bulliard for her useful suggestions. Furthermore, we would like to renew our acknowledgments to the people who contributed to the first edition. In particular Kai Arras for his contribution to uncertainty representation and Kalman filter localization; Matt Mason for his contribution to kinematics; Al Rizzi for his guidance on feedback control; Roland Philippsen and Jan Persson for their contribution to obstacle avoidance; Gilles Caprari and Yves Piguet for their input and suggestions on motion control; Marco Lauria for offering his talent for some of the figures; Marti Louw for her efforts on the cover design; and Nicola Tomatis, Remy Blank, and Marie-Jo Pellaud.

This book was also inspired by other courses, especially by the lecture notes on mobile robotics at the Swiss Federal Institutes of Technology, both in Lausanne (EPFL) and Zurich (ETH). The material for this book has been used for lectures at EPFL, ETH, and CMU since 1997. We thank the hundreds of students who followed the lecture and contributed through their corrections and comments.

It has been a pleasure to work with MIT Press, the publisher of this book. Thanks to Gregory McNamee for his careful and valuable copy-editing, and to Ada Brunstein, Katherine Almeida, Abby Streeter Roake, Marc Lowenthal, and Susan Clark from MIT Press for their help in editing and finalizing the book.

Preface

Mobile robotics is a young field. Its roots include many engineering and science disciplines, from mechanical, electrical, and electronics engineering to computer, cognitive, and social sciences. Each of these parent fields has its share of introductory textbooks that excite and inform prospective students, preparing them for future advanced coursework and research. Our objective in writing this textbook is to provide mobile robotics with such a preparatory guide.

This book presents an introduction to the fundamentals of mobile robotics, spanning the mechanical, motor, sensory, perceptual, and cognitive layers that comprise our field of study. A collection of workshop proceedings and journal publications could present the new student with a snapshot of the state of the art in all aspects of mobile robotics. But here we aim to present a foundation—a formal introduction to the field. The formalism and analysis herein will prove useful even as the frontier of the state-of-the-art advances due to the rapid progress in all of the subdisciplines of mobile robotics.

This second edition largely extends the content of the first edition. In particular, chapters 2, 4, 5, and 6 have been notably expanded and updated to the most recent, state-of-the-art acquisitions in both computer vision and robotics. In particular, we have added in chapter 2 the most recent and popular examples of mobile, legged, and micro aerial robots. In chapter 4, we have added the description of new sensors—such as 3D laser rangefinders, time-of-flight cameras, IMUs, and omnidirectional cameras—and tools—such as image filtering, camera calibration, structure-from-stereo, structure-from-motion, visual odometry, the most popular feature detectors for camera (Harris, FAST, SURF, SIFT) and laser images, and finally bag-of-feature approaches for place recognition and image retrieval. In chapter 5, we have added an introduction to probability theory, and improved and expanded the description of Markov and Kalman filter localization using a better formalism and more examples. Furthermore, we have also added the description of the Simultaneous Localization and Mapping (SLAM) problem along with a description of the most popular approaches to solve it such as extended-Kalman-filter SLAM, graph-based SLAM, particle filter SLAM, and the most recent monocular visual SLAM. Finally, in chapter 6 we have added the description of graph-search algorithms for path planning such as breadth-first, depth first, Dijkstra, A*, D*, and rapidly exploring random trees. Besides these many new additions, we have also provided state-of-the-art references and links to online resources and downloadable software.

We hope that this book will empower both undergraduate and graduate robotics students with the background knowledge and analytical tools they will need to evaluate and even criticize mobile robot proposals and artifacts throughout their careers. This textbook is suitable as a whole for introductory mobile robotics coursework at both the undergraduate and graduate level. Individual chapters such as those on perception or kinematics can be useful as overviews in more focused courses on specific subfields of robotics.

The origins of this book bridge the Atlantic Ocean. The authors have taught courses on mobile robotics at the undergraduate and graduate level at Stanford University, ETH Zurich, Carnegie Mellon University and EPFL. Their combined set of curriculum details and lecture notes formed the earliest versions of this text. We have combined our individual notes, provided overall structure and then test-taught using this textbook for two additional years before settling on the first edition in 2004, and another six years for the current, published text.

For an overview of the organization of the book and summaries of individual chapters, refer to section 1.2.

Finally, for the teacher and the student: we hope that this textbook will prove to be a fruitful launching point for many careers in mobile robotics. That would be the ultimate reward.

1 Introduction

1.1 Introduction

Robotics has achieved its greatest success to date in the world of industrial manufacturing. Robot arms, or *manipulators*, comprise a $ 2 billion industry. Bolted at its shoulder to a specific position in the assembly line, the robot arm can move with great speed and accuracy to perform repetitive tasks such as spot welding and painting (figure 1.1). In the electronics industry, manipulators place surface-mounted components with superhuman precision, making the portable telephone and laptop computer possible.

Yet, for all of their successes, these commercial robots suffer from a fundamental disadvantage: lack of mobility. A fixed manipulator has a limited range of motion that depends

Figure 1.1
Picture of auto assembly plant-spot welding robot of KUKA and a parallel robot Delta of SIG Demaurex SA (invented at EPFL [305]) during packaging of chocolates.

on where it is bolted down. In contrast, a mobile robot would be able to travel throughout the manufacturing plant, flexibly applying its talents wherever it is most effective.

This book focuses on the technology of mobility: how can a mobile robot move unsupervised through real-world environments to fulfill its tasks? The first challenge is locomotion itself. How should a mobile robot move, and what is it about a particular locomotion mechanism that makes it superior to alternative locomotion mechanisms?

Hostile environments such as Mars trigger even more unusual locomotion mechanisms (figure 1.2). In dangerous and inhospitable environments, even on Earth, such *teleoperated* systems have gained popularity (figures 1.3-1.6). In these cases, the low-level complexities of the robot often make it impossible for a human operator to control its motions directly. The human performs localization and cognition activities but relies on the robot's control scheme to provide motion control.

For example, Plustech's walking robot provides automatic leg coordination while the human operator chooses an overall direction of travel (figure 1.3). Figure 1.6 depicts an underwater vehicle that controls three propellers to stabilize the robot submarine autonomously in spite of underwater turbulence and water currents while the operator chooses position goals for the submarine to achieve.

Other commercial robots operate not where humans *cannot* go, but rather share space with humans in human environments (figure 1.7). These robots are compelling not for reasons of mobility but because of their *autonomy*, and so their ability to maintain a sense of position and to navigate without human intervention is paramount.

Figure 1.2
The mobile robot Sojourner was used during the Pathfinder mission to explore Mars in summer 1997. It was almost completely teleoperated from Earth. However, some on-board sensors allowed for obstacle detection (http://ranier.oact.hq.nasa.gov/telerobotics_page/telerobotics.shtm). © NASA/JPL.

Figure 1.3
Plustech developed the first application-driven walking robot. It is designed to move wood out of the forest. The leg coordination is automated, but navigation is still done by the human operator on the robot. (http://www.plustech.fi). © Plustech.

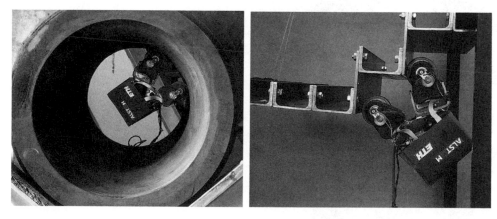

Figure 1.4
The MagneBike robot developed by ASL (ETH Zurich) and ALSTOM. MagneBike is a magnetic wheeled robot with high mobility for inspecting complex shaped structures such as ferromagnetic pipes and turbines (http://www.asl.ethz.ch/). © ALSTOM / ETH Zurich.

Figure 1.5
Picture of Pioneer, a robot designed to explore the Sarcophagus at Chernobyl. © Wide World Photos.

Figure 1.6
The autonomous underwater vehicle (AUV) Sirius being retrieved after a mission aboard the RV Southern Surveyor © Robin Beaman—James Cook University.

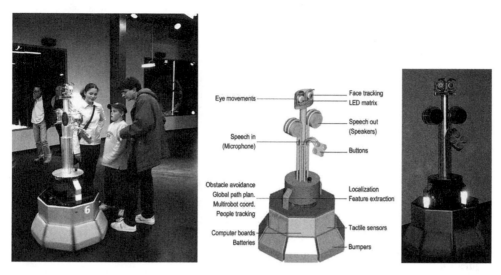

Figure 1.7
Tour-guide robots are able to interact and present exhibitions in an educational way [85, 251, 288, 310,]. Ten Roboxes have operated during five months at the Swiss exhibition EXPO.02, meeting hundreds of thousands of visitors. They were developed by EPFL [288] (http://robotics.epfl.ch) and commercialized by BlueBotics (http://www.bluebotics.com).

For example, AGV (autonomous guided vehicle) robots (figure 1.8) autonomously deliver parts between various assembly stations by following special electrical guidewires installed in the floor (figure 1.8a) or, differently, by using onboard lasers to localize within a user-specified map (figure 1.8b). The Helpmate service robot transports food and medication throughout hospitals by tracking the position of ceiling lights, which are manually specified to the robot beforehand (figure 1.9). Several companies have developed autonomous cleaning robots, mainly for large buildings (figure 1.10). One such cleaning robot is in use at the Paris Metro. Other specialized cleaning robots take advantage of the regular geometric pattern of aisles in supermarkets to facilitate the localization and navigation tasks.

Research into high-level questions of cognition, localization, and navigation can be performed using standard research robot platforms that are tuned to the laboratory environment. This is one of the largest current markets for mobile robots. Various mobile robot platforms are available for programming, ranging in terms of size and terrain capability. Very popular research robots are the Pioneer, BIBA, and the *e-puck* (figures 1.11-1.13) and also very small robots like the Alice from EPFL (Swiss Federal Institute of Technology at Lausanne) (figure 1.14).

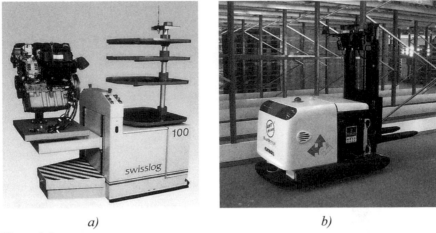

a)   *b)*

Figure 1.8
(a) Autonomous guided vehicle (AGV) by SWISSLOG used to transport motor blocks from one assembly station to another. It is guided by an electrical wire installed in the floor. © Swisslog.
(b) Equipped with the Autonomous Navigation Technology (ANT) from BlueBotics, Paquito, the autonomous forklift by Esatroll, does not rely on electrical wires, magnetic plots, or reflectors, but rather uses the onboard safety lasers to localize itself with respect to the shape of the environment. Image courtesy of BlueBotics (http://www.bluebotics.com).

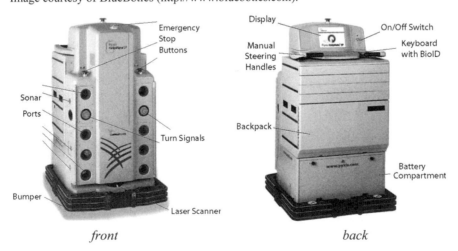

front   *back*

Figure 1.9
HELPMATE is a mobile robot used in hospitals for transportation tasks. It has various on-board sensors for autonomous navigation in the corridors. The main sensor for localization is a camera looking to the ceiling. It can detect the lamps on the ceiling as references, or landmarks (http://www.pyxis.com). © Pyxis Corp.

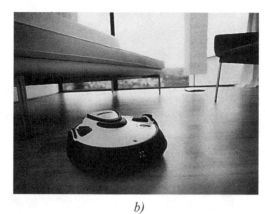

a)   *b)*

Figure 1.10
(a) The Robot40 is a consumer robot developed and sold by Cleanfix for cleaning large gymnasiums. The navigation system of Robo40 is based on a sophisticated sonar and infrared system (http://www.cleanfix.com). © Cleanfix. (b) The RoboCleaner RC 3000 covers badly soiled areas with a special driving strategy until it is really clean. Optical sensors measure the degree of pollution of the aspirated air (http://www.karcher.de). © Alfred Kärcher GmbH & Co.

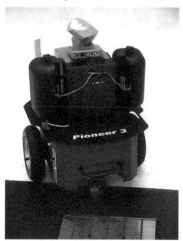
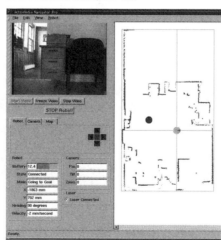

Figure 1.11
PIONEER is a modular mobile robot offering various options like a gripper or an on-board camera. It is equipped with a sophisticated navigation library developed at SRI, Stanford, CA. Reprinted with permission from ActivMedia Robotics, http://www.MobileRobots.com.

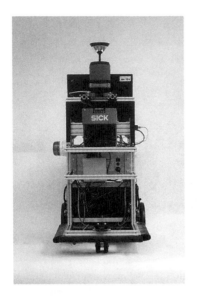

Figure 1.12
BIBA is a very sophisticated mobile robot developed for research purposes and built by BlueBotics (http://www.bluebotics.com/). It has a large variety of sensors for high-performance navigation tasks.

Figure 1.13
The *e-puck* is an educational desktop mobile robot developed at the EPFL [226]. It is only about 70 mm in diameter. As extensions to the basic capabilities, various modules such as additional sensors, actuators, or computational power have been developed. In this picture, two example extensions are shown: (center) an omnidirectional camera and (right) an infrared distance scanner (http://www.e-puck.org/). © Ecole Polytechnique Fédérale de Lausanne (EPFL).

Figure 1.14
Alice is one of the smallest fully autonomous robots. It is approximately $2 \times 2 \times 2$ cm, it has an autonomy of about 8 hours and uses infrared distance sensors, tactile whiskers, or even a small camera for navigation [93].

Although mobile robots have a broad set of applications and markets as summarized above, there is one fact that is true of virtually every successful mobile robot: its design involves the integration of many different bodies of knowledge. No mean feat, this makes mobile robotics as interdisciplinary a field as there can be. To solve locomotion problems, the mobile roboticist must understand mechanism and kinematics, dynamics and control theory. To create robust perceptual systems, the mobile roboticist must leverage the fields of signal analysis and specialized bodies of knowledge such as computer vision to properly employ a multitude of sensor technologies. Localization and navigation demand knowledge of computer algorithms, information theory, artificial intelligence, and probability theory.

Figure 1.15 depicts an abstract control scheme for mobile robot systems that we will use throughout this text. This figure identifies many of the main bodies of knowledge associated with mobile robotics.

This book provides an introduction to all aspects of mobile robotics, including software and hardware design considerations, related technologies, and algorithmic techniques. The intended audience is broad, including both undergraduate and graduate students in introductory mobile robotics courses, as well as individuals fascinated by the field. Although it is not absolutely required, a familiarity with matrix algebra, calculus, probability theory, and computer programming will significantly enhance the reader's experience.

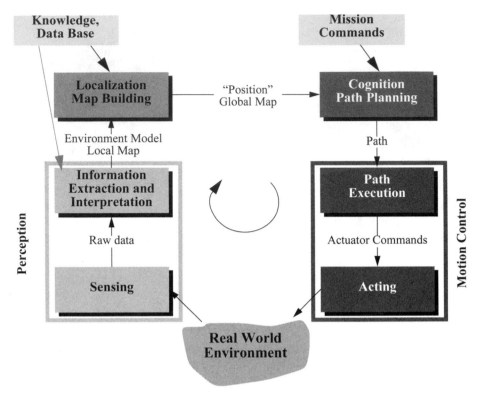

Figure 1.15
Reference control scheme for mobile robot systems used throughout this book.

Mobile robotics is a large field, and this book focuses not on robotics in general, or on mobile robot applications, but rather on mobility itself. From mechanism and perception to localization and navigation, this book focuses on the techniques and technologies that enable robust *mobility*.

Clearly, a useful, commercially viable mobile robot does more than just move. It polishes the supermarket floor, keeps guard in a factory, mows the golf course, provides tours in a museum, or provides guidance in a supermarket. The aspiring mobile roboticist will start with this book but will quickly graduate to coursework and research specific to the desired application, integrating techniques from fields as disparate as human-robot interaction, computer vision, and speech understanding.

1.2   An Overview of the Book

This book introduces the different aspects of a robot in modules, much like the modules shown in figure 1.15. Chapters 2 and 3 focus on the robot's low-level *locomotive ability*. Chapter 4 presents an in-depth view of *perception*. Chapters 5 and 6 take us to the higher-level challenges of *localization* and *mapping* and even higher-level *cognition*, specifically the *ability to navigate robustly*. Each chapter builds upon previous chapters, and so the reader is encouraged to start at the beginning, even if his or her interest is primarily at the high level. Robotics is peculiar in that solutions to high-level challenges are most meaningful only in the context of a solid understanding of the low-level details of the system.

Chapter 2, "Locomotion," begins with a survey of the most important mechanisms that enable locomotion: wheels, legs, and flight. Numerous robotic examples demonstrate the particular talents of each form of locomotion. But designing a robot's locomotive system properly requires the ability to evaluate its overall motion capabilities quantitatively. Chapter 3, "Mobile Robot Kinematics," applies principles of kinematics to the whole robot, beginning with the kinematic contribution of each wheel and graduating to an analysis of robot maneuverability enabled by each mobility mechanism configuration.

The greatest single shortcoming in conventional mobile robotics is, without doubt, perception: mobile robots can travel across much of earth's man-made surfaces, but they cannot perceive the world nearly as well as humans and other animals. Chapter 4, "Perception," begins a discussion of this challenge by presenting a clear language for describing the performance envelope of mobile robot sensors. With this language in hand, chapter 4 goes on to present many of the off-the-shelf sensors available to the mobile roboticist, describing their basic principles of operation as well as their performance limitations. The most promising sensor for the future of mobile robotics is vision, and chapter 4 includes an overview of the theory of camera image formation, omnidirectional vision, camera calibration, structure from stereovision, structure from motion, and visual odometry. But perception is more than sensing. Perception is also the *interpretation* of sensed data in meaningful ways. The second half of chapter 4 describes strategies for feature extraction that have been most useful in both computer vision and mobile robotics applications, including extraction of geometric shapes from range sensing data, as well as point features (such as Harris, SIFT, SURF, FAST, and so on) from camera images. Furthermore, a section is dedicated to the description of the most recent bag-of-feature approach that became popular for place recognition and image retrieval.

Armed with locomotion mechanisms and outfitted with hardware and software for perception, the mobile robot can move and perceive the world. The first point at which mobility and sensing must meet is localization: mobile robots often need to maintain a sense of position. Chapter 5, "Mobile Robot Localization," describes approaches that obviate the need for direct localization, then delves into fundamental ingredients of successful localization strategies: belief representation and map representation. Case studies demonstrate various localization schemes, including both Markov localization and Kalman filter localization. The final part of chapter 5 is devoted to a description of the Simultaneous Localization and Mapping (SLAM) problem along with a description of the most popular approaches to solve it such as extended-Kalman-filter SLAM, graph-based SLAM, particle filter SLAM, and the most recent monocular visual SLAM.

Mobile robotics is so young a discipline that it lacks a standardized architecture. There is as yet no established robot operating system. But the question of architecture is of paramount importance when one chooses to address the higher-level competences of a mobile robot: how does a mobile robot navigate robustly from place to place, interpreting data, and localizing and controlling its motion all the while? For this highest level of robot competence, which we term *navigation competence*, there are numerous mobile robots that showcase particular architectural strategies. Chapter 6, "Planning and Navigation," surveys the state of the art of robot navigation, showing that today's various techniques are quite similar, differing primarily in the manner in which they *decompose* the problem of robot control. But first, chapter 6 addresses two skills that a competent, navigating robot usually must demonstrate: obstacle avoidance and path planning.

There is far more to know about the cross-disciplinary field of mobile robotics than can be contained in a single book. We hope, though, that this broad introduction will place the reader in the context of the collective wisdom of mobile robotics. This is only the beginning. With luck, the first robot you program or build will have only good things to say about you.

2 Locomotion

2.1 Introduction

A mobile robot needs locomotion mechanisms that enable it to move unbounded throughout its environment. But there are a large variety of possible ways to move, and so the selection of a robot's approach to locomotion is an important aspect of mobile robot design. In the laboratory, there are research robots that can walk, jump, run, slide, skate, swim, fly, and, of course, roll. Most of these locomotion mechanisms have been inspired by their biological counterparts (see figure 2.1).

There is, however, one exception: the actively powered wheel is a human invention that achieves extremely high efficiency on flat ground. This mechanism is not completely foreign to biological systems. Our bipedal walking system can be approximated by a rolling polygon, with sides equal in length d to the span of the step (figure 2.2). As the step size decreases, the polygon approaches a circle or wheel. But nature did not develop a fully rotating, actively powered joint, which is the technology necessary for wheeled locomotion.

Biological systems succeed in moving through a wide variety of harsh environments. Therefore, it can be desirable to copy their selection of locomotion mechanisms. However, replicating nature in this regard is extremely difficult for several reasons. To begin with, mechanical complexity is easily achieved in biological systems through structural replication. Cell division, in combination with specialization, can readily produce a millipede with several hundred legs and several tens of thousands of individually sensed cilia. In man-made structures, each part must be fabricated individually, and so no such economies of scale exist. Additionally, the cell is a microscopic building block that enables extreme miniaturization. With very small size and weight, insects achieve a level of robustness that we have not been able to match with human fabrication techniques. Finally, the biological energy storage system and the muscular and hydraulic activation systems used by large animals and insects achieve torque, response time, and conversion efficiencies that far exceed similarly scaled man-made systems.

Type of motion	Resistance to motion	Basic kinematics of motion
Flow in a Channel 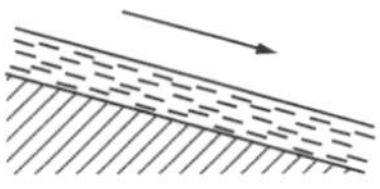	Hydrodynamic forces	Eddies 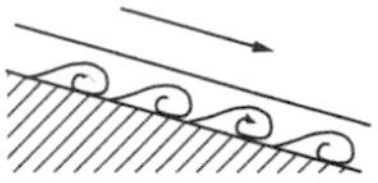
Crawl 	Friction forces	 Longitudinal vibration
Sliding 	Friction forces	 Transverse vibration
Running 	Loss of kinetic energy	Periodic bouncing on a spring 
Walking 	Loss of kinetic energy	Rolling of a polygon (see figure 2.2) 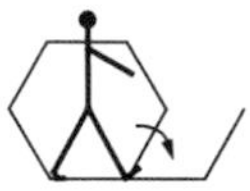

Figure 2.1
Locomotion mechanisms used in biological systems.

Owing to these limitations, mobile robots generally locomote either using wheeled mechanisms, a well-known human technology for vehicles, or using a small number of articulated legs, the simplest of the biological approaches to locomotion (see figure 2.2).

In general, legged locomotion requires higher degrees of freedom and therefore greater mechanical complexity than wheeled locomotion. Wheels, in addition to being simple, are extremely well suited to flat ground. As figure 2.3 depicts, on flat surfaces wheeled locomotion is one to two orders of magnitude more efficient than legged locomotion. The railway is ideally engineered for wheeled locomotion because rolling friction is minimized on a hard and flat steel surface. But as the surface becomes soft, wheeled locomotion accumulates inefficiencies due to rolling friction, whereas legged locomotion suffers much less because it consists only of point contacts with the ground. This is demonstrated in figure 2.3 by the dramatic loss of efficiency in the case of a tire on soft ground.

In effect, the efficiency of wheeled locomotion depends greatly on environmental qualities, particularly the flatness and hardness of the ground, while the efficiency of legged

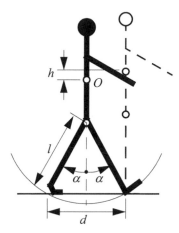

Figure 2.2
A biped walking system can be approximated by a rolling polygon, with sides equal in length d to the span of the step. As the step size decreases, the polygon approaches a circle or wheel with the radius l.

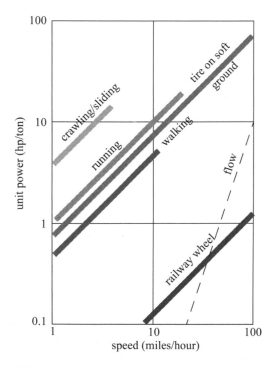

Figure 2.3
Specific power versus attainable speed of various locomotion mechanisms [52].

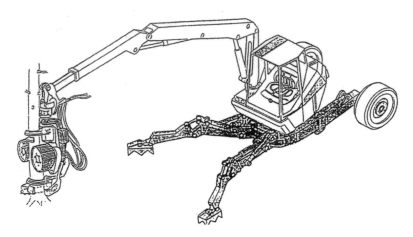

Figure 2.4
RoboTrac, a hybrid wheel-leg vehicle for rough terrain [282].

locomotion depends on the leg mass and body mass, both of which the robot must support at various points in a legged gait.

It is understandable, therefore, that nature favors legged locomotion, since locomotion systems in nature must operate on rough and unstructured terrain. For example, in the case of insects in a forest, the vertical variation in ground height is often an order of magnitude greater than the total height of the insect. By the same token, the human environment frequently consists of engineered, smooth surfaces, both indoors and outdoors. Therefore, it is also understandable that virtually all industrial applications of mobile robotics utilize some form of wheeled locomotion. Recently, for more natural outdoor environments, there has been some progress toward hybrid and legged industrial robots such as the forestry robot shown in figure 2.4.

In section 2.1.1, we present general considerations that concern all forms of mobile robot locomotion. Following this, in sections 2.2. 2.3, and 2.4 we present overviews of legged locomotion, wheeled locomotion, and aerial locomotion techniques for mobile robots.

2.1.1 Key issues for locomotion

Locomotion is the complement of manipulation. In manipulation, the robot arm is fixed but moves objects in the workspace by imparting force to them. In locomotion, the environment is fixed and the robot moves by imparting force to the environment. In both cases, the scientific basis is the study of actuators that generate interaction forces and mechanisms that implement desired kinematic and dynamic properties. Locomotion and manipulation thus share the same core issues of stability, contact characteristics, and environmental type:

- stability
  - number and geometry of contact points
  - center of gravity
  - static/dynamic stability
  - inclination of terrain

- characteristics of contact
  - contact point/path size and shape
  - angle of contact
  - friction

- type of environment
  - structure
  - medium (e.g., water, air, soft or hard ground)

A theoretical analysis of locomotion begins with mechanics and physics. From this starting point, we can formally define and analyze all manner of mobile robot locomotion systems. However, this book focuses on the mobile robot *navigation* problem, particularly stressing perception, localization, and cognition. Thus, we will not delve deeply into the physical basis of locomotion. Nevertheless, the three remaining sections in this chapter present overviews of issues in legged locomotion [52], wheeled locomotion, and aerial locomotion. Chapter 3 presents a more detailed analysis of the kinematics and control of wheeled mobile robots.

2.2   Legged Mobile Robots

Legged locomotion is characterized by a series of point contacts between the robot and the ground. The key advantages include adaptability and maneuverability in rough terrain (figure 2.5). Because only a set of point contacts is required, the quality of the ground between those points does not matter as long as the robot can maintain adequate ground clearance. In addition, a walking robot is capable of crossing a hole or chasm so long as its reach exceeds the width of the hole. A final advantage of legged locomotion is the potential to manipulate objects in the environment with great skill. An excellent insect example, the dung beetle, is capable of rolling a ball while locomoting by way of its dexterous front legs.

The main disadvantages of legged locomotion include power and mechanical complexity. The leg, which may include several degrees of freedom, must be capable of sustaining part of the robot's total weight, and in many robots it must be capable of lifting and lowering the robot. Additionally, high maneuverability will only be achieved if the legs have a

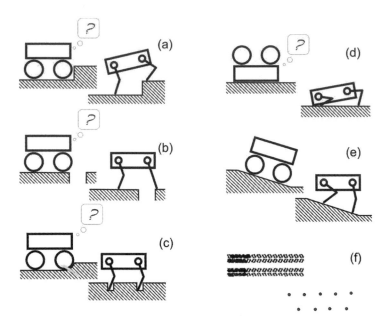

Figure 2.5 Legged robots are particularly suited for rough terrain, where they are able to traverse obstacles such as steps (a), gaps (b), or sandy patches (c) that are impassable for wheeled systems. Additionally, the high number of degrees of freedom allows the robot to stand up when fallen (d) and keep its payload leveled (e). Because legged systems do not require a continuous path for support, they can rely on a few selected footholds, which also reduces the environmental impact (f). Image courtesy of D. Remy.

sufficient number of degrees of freedom to impart forces in a number of different directions.

2.2.1 Leg configurations and stability

Because legged robots are biologically inspired, it is instructive to examine biologically successful legged systems. A number of different leg configurations have been successful in a variety of organisms (figure 2.6). Large animals, such as mammals and reptiles, have four legs, whereas insects have six or more legs. In some mammals, the ability to walk on only two legs has been perfected. Especially in the case of humans, balance has progressed to the point that we can even jump with one leg.[1] This exceptional maneuverability comes at a price: much more complex active control to maintain balance.

---

1. In child development, one of the tests used to determine if the child is acquiring advanced locomotion skills is the ability to jump on one leg.

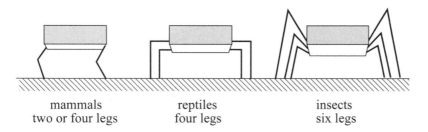

mammals
two or four legs

reptiles
four legs

insects
six legs

Figure 2.6
Arrangement of the legs of various animals.

In contrast, a creature with three legs can exhibit a static, stable pose provided that it can ensure that its center of gravity is within the tripod of ground contact. Static stability, demonstrated by a three-legged stool, means that balance is maintained with no need for motion. A small deviation from stability (e.g., gently pushing the stool) is passively corrected toward the stable pose when the upsetting force stops.

But a robot must be able to lift its legs in order to walk. In order to achieve static walking, a robot must have at least four legs, moving one of it at a time. For six legs, it is possible to design a gait in which a statically stable tripod of legs is in contact with the ground at all times (figure 2.9).

Insects and spiders are immediately able to walk when born. For them, the problem of balance during walking is relatively simple. Mammals, with four legs, can achieve static walking, which, however, is less stable due to the high center of gravity than, for example, reptile walking. Fawns, for example, spend several minutes attempting to stand before they are able to do so, then spend several more minutes learning to walk without falling. Humans, with two legs, can also stand statically stable due to their large feet. Infants require months to stand and walk, and even longer to learn to jump, run, and stand on one leg.

There is also the potential for great variety in the complexity of each individual leg. Once again, the biological world provides ample examples at both extremes. For instance, in the case of the caterpillar, each leg is extended using hydraulic pressure by constricting the body cavity and forcing an increase in pressure, and each leg is retracted longitudinally by relaxing the hydraulic pressure, then activating a single tensile muscle that pulls the leg in toward the body. Each leg has only a single degree of freedom, which is oriented longitudinally along the leg. Forward locomotion depends on the hydraulic pressure in the body, which extends the distance between pairs of legs. The caterpillar leg is therefore mechanically very simple, using a minimal number of extrinsic muscles to achieve complex overall locomotion.

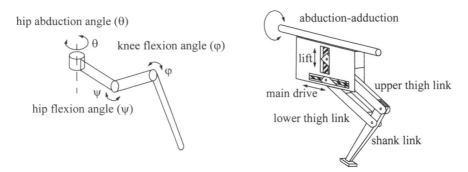

Figure 2.7
Two examples of legs with three degrees of freedom.

At the other extreme, the human leg has more than seven major degrees of freedom, combined with further actuation at the toes. More than fifteen muscle groups actuate eight complex joints.

In the case of legged mobile robots, a minimum of two degrees of freedom is generally required to move a leg forward by lifting the leg and swinging it forward. More common is the addition of a third degree of freedom for more complex maneuvers, resulting in legs such as those shown in figure 2.7. Recent successes in the creation of bipedal walking robots have added a fourth degree of freedom at the ankle joint. The ankle enables the robot to shift the resulting force vector of the ground contact by actuating the pose of the sole of the foot.

In general, adding degrees of freedom to a robot leg increases the maneuverability of the robot, both augmenting the range of terrains on which it can travel and the ability of the robot to travel with a variety of gaits. The primary disadvantages of additional joints and actuators are, of course, energy, control, and mass. Additional actuators require energy and control, and they also add to leg mass, further increasing power and load requirements on existing actuators.

In the case of a multilegged mobile robot, there is the issue of leg coordination for loco-motion, or gait control. The number of possible gaits depends on the number of legs [52]. The gait is a sequence of lift and release events for the individual legs. For a mobile robot with k legs, the total number of distinct event sequences N for a walking machine is:

$$N = (2k - 1)!$$ (2.1)

For a biped walker $k = 2$ legs, the number of distinct event sequences N is:

$$N = (2k - 1)! = 3! = 3 \cdot 2 \cdot 1 = 6 \tag{2.2}$$

The six distinct event sequences that can be combined for more complex sequences are:

1. both legs down – right down / left up – both legs down;

2. both legs down – right leg up / left leg down – both legs down;

3. both legs down – both legs up – both legs down;

4. right leg down / left leg up – right leg up / left leg down – right leg down / left leg up;

5. right leg down / left leg up – both legs up – right leg down / left leg up;

6. right leg up / left leg down – both legs up – right leg up / left leg down.

Of course, this quickly grows quite large. For example, a robot with six legs has far more events:

$$N = 11! = 39916800 , \tag{2.3}$$

with an even higher number of theoretically possible gaits.

Figures 2.8 and 2.9 depict several four-legged gaits and the static six-legged tripod gait.

2.2.2 Consideration of dynamics

The *cost of transportation* expresses how much energy a robot uses to travel a certain distance. To better compare differently sized systems, this value is usually normalized by the robot's weight and expressed in $J/(N \cdot m)$ —a dimensionless quantity—where J stands for joule, N for newton, and m for meter. When a robot moves with constant speed on a level surface, its potential and kinetic energy remain constant. In theory, no physical work is necessary to keep it moving, which makes it possible to get from one place to another with zero cost of transportation. In reality, however, some energy is always dissipated, and robots have to be equipped with actuators and batteries to compensate for the losses. For a wheeled robot, the main causes for such losses are the friction in the drive train and the rolling resistance of the wheels on the ground. Similarly, friction is present in the joints of legged systems and energy is dissipated by the foot-ground interaction. However, these effects cannot explain why legged systems usually consume considerably more energy than their wheeled counterparts. The bulk part of energy loss actually originates in the fact that legs—in contrast to wheels or tracks—are not performing a continuous motion, but are periodically moving back and forth. Joints have to undergo alternating phases of acceleration and deceleration, and, as we have only a very limited ability to recuperate the negative work of deceleration, energy is irrecoverably lost in the process. Because of the segmented structure of

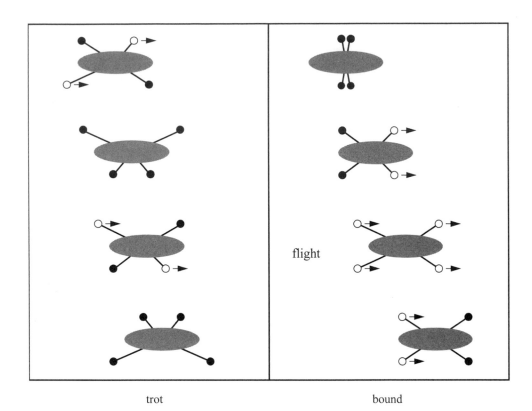

Figure 2.8
Two gaits with four legs.

the legs, it can even happen that energy that is fed into one joint (e.g., the knee) is simultaneously dissipated in another joint (e.g., the hip) without creating any net work at the feet. Therefore, actuators are working against each other [180].

A solution to this problem is a better exploitation of the dynamics of the mechanical structure. The natural oscillations of pendula and springs, can—if they are well designed—automatically create the required periodic motions. For example, the motion of a swinging leg can be grasped by the dynamics of a simple double pendulum. If the lengths and the inertial properties of the leg segments are correctly selected, such a pendulum will automatically swing forward, clear the ground, and extend the leg to touch the ground in front of the main body. If, on the other side, a foot is on the ground and the leg is kept stiff, an inverted pendulum motion will efficiently propel the main body forward. During running, these inverted pendulum dynamics are additionally enhanced by springs, which store

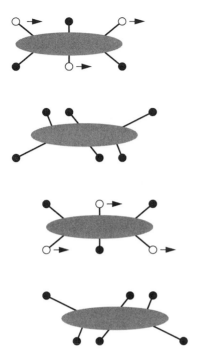

Figure 2.9
Static walking with six legs. A tripod formed by three legs always exists.

energy during the ground phase and allow the main body to take off for the subsequent flight phase (figure 2.10).

With this approach it is, in fact, possible to build legged robots that do not have actuation of any kind. Such passive dynamic walkers [211,344] walk down a shallow incline (which compensates for frictional losses), but, because no actuators are present, no negative work is performed and energetic losses due to braking are eliminated. In addition to creating a periodic motion, the dynamics of such walkers must be designed to ensure dynamic stability. The mechanical structure must passively reject small disturbances which would otherwise accumulate over time and eventually cause the robot to fall. Actuated robots built according to these principles can walk with a remarkable efficiency [104] and one of them, the Cornell Ranger, currently holds the distance record for autonomous legged robots [345] (figure 2.11).

Passive dynamic walkers also have a striking similarity to the physique and motion patters of human gait. During the evolution, humans and animals have become quite efficient walkers, and a look at electromyography recordings shows that during walking our muscles

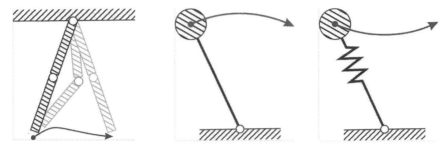

Figure 2.10 Dynamic elements that are exploited in energy efficient walking include the double-pendulum for leg swing, the inverted pendulum for the stance phase of walking, and springy legs for running gaits. Image courtesy of D. Remy.

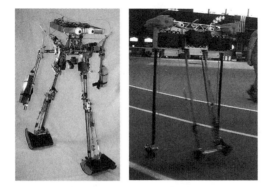

Figure 2.11 The Cornell powered two-legged and four-legged biped robots. In April 2008, the four-legged bipedal robot Ranger walked a distance of 9.07 km without being touched by a person. Image courtesy of the Biorobotics and Locomotion Lab—Cornell University.

are far less active than one would expect for a task in which most of our limbs are in constant motion. To some degree, humans are passive dynamic walkers.

It is obvious that such an exploitation of the mechanical dynamics can only work at specific velocities. When the locomotion speed changes, characteristic properties such as stride length or stride frequency change as well, and—because these have to be matched with the spring and pendulum oscillations of the mechanical structure—more and more actuator effort is needed to force the joints to follow their required trajectories. For human walking, the optimal walking speed is approximately 1 m/s, which is also the range that subjectively feels most comfortable. For both higher and lower speeds, the cost of transportation will increase, and more energy is needed to travel the same distance. For this reason, humans change their gait from walking to running when they want to travel at higher speeds, which is more efficient than just performing the same motion faster and faster.

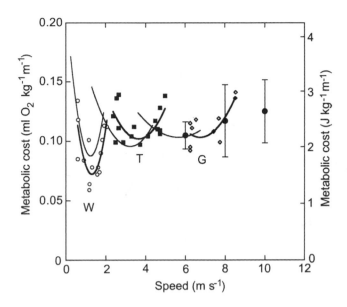

Figure 2.12 Metabolic cost of transportation (here normalized by body mass) for different gaits of horses: walking (W), running (R), and galloping (G). Each gait has a specific velocity that minimizes energy expenditure. This explains why animals and humans change their gait when traveling at different speeds. Image courtesy of A. E. Minetti [221].

Changing the gait allows us to use a different set of natural dynamics, which better matches the stride frequency and step length that are needed for higher velocities. Likewise, the wide variety of gaits found in animals can be explained by the use of different sets of dynamic elements, which minimize the energy necessary for transportation (figure 2.12).

2.2.3 Examples of legged robot locomotion
Although there are no high-volume industrial applications to date, legged locomotion is an important area of long-term research. Several interesting designs are presented here, beginning with the one-legged robot and finishing with six-legged robots.

2.2.3.1 One leg
The minimum number of legs a legged robot can have is, of course, one. Minimizing the number of legs is beneficial for several reasons. Body mass is particularly important to walking machines, and the single leg minimizes cumulative leg mass. Leg coordination is required when a robot has several legs, but with one leg no such coordination is needed. Perhaps most important, the one-legged robot maximizes the basic advantage of legged locomotion: legs have single points of contact with the ground in lieu of an entire track, as

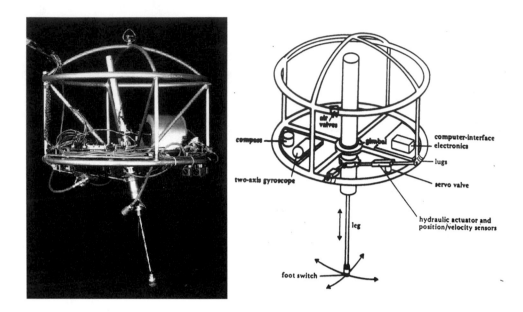

Figure 2.13
The Raibert hopper [42, 264]. Image courtesy of the LegLab and Marc Raibert. © 1983.

with wheels. A single-legged robot requires only a sequence of single contacts, making it amenable to the roughest terrain. Furthermore, a hopping robot can dynamically cross a gap that is larger than its stride by taking a running start, whereas a multilegged walking robot that cannot run is limited to crossing gaps that are as large as its reach.

The major challenge in creating a single-legged robot is balance. For a robot with one leg, static walking is not only impossible, but static stability when stationary is also impossible. The robot must actively balance itself by either changing its center of gravity or by imparting corrective forces. Thus, the successful single-legged robot must be dynamically stable.

Figure 2.13 shows the Raibert hopper [42, 264], one of the best-known single-legged hopping robots created. This robot makes continuous corrections to body attitude and to robot velocity by adjusting the leg angle with respect to the body. The actuation is hydraulic, including high-power longitudinal extension of the leg during stance to hop back into the air. Although powerful, these actuators require a large, off-board hydraulic pump to be connected to the robot at all times.

Figure 2.14 shows a more energy-efficient design that takes advantage of well-designed mechanical dynamics [83]. Instead of supplying power by means of an off-board hydraulic pump, the bow leg hopper is designed to capture the kinetic energy of the robot as it lands,

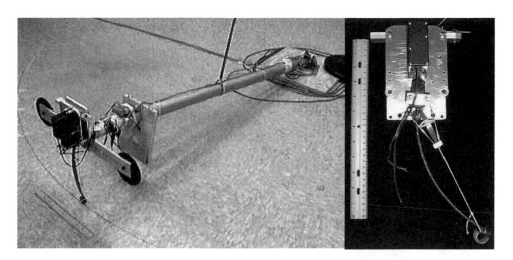

Figure 2.14
The 2D single bow leg hopper [83]. Image courtesy of H. Benjamin Brown and Garth Zeglin, CMU.

using an efficient bow spring leg. This spring returns approximately 85% of the energy, meaning that stable hopping requires only the addition of 15% of the required energy on each hop. This robot, which is constrained along one axis by a boom, has demonstrated continuous hopping for 20 minutes using a single set of batteries carried on board the robot. As with the Raibert hopper, the bow leg hopper controls velocity by changing the angle of the leg to the body at the hip joint.

Ringrose [266] demonstrates the very important duality of mechanics and controls as applied to a single-legged hopping machine. Often clever mechanical design can perform the same operations as complex active control circuitry. In this robot, the physical shape of the foot is exactly the right curve so that when the robot lands without being perfectly vertical, the proper corrective force is provided from the impact, making the robot vertical by the next landing. This robot is dynamically stable, and is furthermore passive. The correction is provided by physical interactions between the robot and its environment, with no computer or any active control in the loop.

2.2.3.2 Two legs (biped)

A variety of successful bipedal robots have been demonstrated over the past ten years. Two legged robots have been shown to run, jump, travel up and down stairways, and even do aerial tricks such as somersaults. In the commercial sector, both Honda and Sony have made significant advances over the past decade that have enabled highly capable bipedal robots. Both companies designed small, powered joints that achieve power-to-weight per-

Specifications:

Weight:	7 kg
Height:	58 cm
Neck DOF:	4
Body DOF:	2
Arm DOF:	2×5
Legs DOF:	2×6
Five-finger Hands	

Figure 2.15
The Sony SDR-4X II. © 2003 Sony Corporation.

formance unheard of in commercially available servomotors. These new "intelligent" servos provide not only strong actuation but also compliant actuation by means of torque sensing and closed-loop control.

The Sony Dream Robot, model SDR-4X II, is shown in figure 2.15. This current model is the result of research begun in 1997 with the basic objective of motion entertainment and communication entertainment (i.e., dancing and singing). This robot with thirty-eight degrees of freedom has seven microphones for fine localization of sound, image person recognition, on-board miniature stereo depth-map reconstruction, and limited speech recognition. Given the goal of fluid and entertaining motion, Sony spent considerable effort designing a motion prototyping application system to enable their engineers to script dances in a straightforward manner. Note that the SDR-4X II is relatively small, standing at 58 cm and weighing only 7 kg.

The Honda humanoid project has a significant history, but, again, it has tackled the very important engineering challenge of actuation. Figure 2.16 shows model *P2*, which is an immediate predecessor to the most recent Asimo (advanced step in innovative mobility) model. Note that the latest Honda Asimo model is still much larger than the SDR-4X at 120 cm tall and 52 kg. This enables practical mobility in the human world of stairs and ledges while maintaining a nonthreatening size and posture. Perhaps the first robot to demonstrate

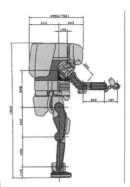

Specifications:

Maximum speed:	2 km/h
Autonomy:	15 min
Weight:	210 kg
Height:	1.82 m
Leg DOF:	2×6
Arm DOF:	2×7

Figure 2.16
The humanoid robot P2 from Honda, Japan. © Honda Motor Cooperation.

biomimetic bipedal stair climbing and descending, these Honda humanoid series robots are being designed not for entertainment purposes but as human aids throughout society. Honda refers, for instance, to the height of Asimo as the minimum height that enables it to manage nonetheless operation of the human world, for instance, control of light switches.

An important feature of bipedal robots is their anthropomorphic shape. They can be built to have the same approximate dimensions as humans, and this makes them excellent vehicles for research in human-robot interaction. WABIAN-2R is a robot built at Waseda University, Japan (figure 2.17) for just such research [255]. WABIAN-2R is designed to emulate human motion, and it is even designed to dance like a human.

Bipedal robots can only be statically stable within some limits, and so robots such as P2 and WABIAN-2R generally must perform continuous balance-correcting servoing even when standing still. Furthermore, each leg must have sufficient capacity to support the full weight of the robot. In the case of four-legged robots, the balance problem is facilitated along with the load requirements of each leg. An elegant design of a biped robot is the Spring Flamingo of MIT (figure 2.18). This robot inserts springs in series with the leg actuators to achieve a more elastic gait. Combined with "kneecaps" that limit knee joint angles, the Flamingo achieves surprisingly biomimetic motion.

Specifications:

Weight: 64 [kg] (with batteries)
Height: 1.55 [m]

DOF:
    Leg: 6×2
    Foot: 1×2 (passive)
    Waist: 2
    Trunk: 2
    Arm: 7×2
    Hand: 3×2
    Neck: 3

Figure 2.17
The humanoid robot WABIAN-2R developed at Waseda University in Japan [255] (http://www.takanishi.mech.waseda.ac.jp/). © Atsuo Takanishi Lab, Waseda University.

Figure 2.18
The Spring Flamingo developed at MIT [262]. Image courtesy of Jerry Pratt, MIT Leg Laboratory.

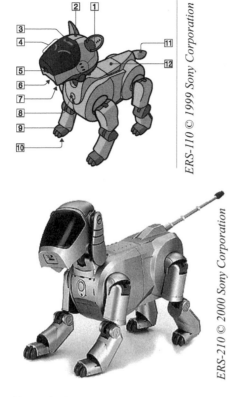

1 Stereo microphone: Allows AIBO to pick up surrounding sounds.
2 Head sensor: Senses when a person taps or pets AIBO on the head.
3 Mode indicator: Shows AIBO's operation mode.
4 Eye lights: These light up in blue-green or red to indicate AIBO's emotional state.
5 Color camera: Allows AIBO to search for objects and recognize them by color and movement.
6 Speaker: Emits various musical tones and sound effects.
7 Chin sensor: Senses when a person touches AIBO on the chin.
8 Pause button: Press to activate AIBO or to pause AIBO.
9 Chest light: Gives information about the status of the robot.
10 Paw sensors: Located on the bottom of each paw.
11 Tail light: Lights up blue or orange to show AIBO's emotional state.
12 Back sensor: Senses when a person touches AIBO on the back.

Figure 2.19
AIBO, the artificial dog from Sony, Japan.

2.2.3.3 Four legs (quadruped)

Although standing still on four legs is passively stable, walking remains challenging because to remain stable, the robot's center of gravity must be actively shifted during the gait. Sony invested several million dollars to develop a four-legged robot called AIBO (figure 2.19). To create this robot, Sony produced both a new robot operating system that is near real-time and new geared servomotors that are of sufficiently high torque to support the robot, yet are back-drivable for safety. In addition to developing custom motors and software, Sony incorporated a color vision system that enables AIBO to chase a brightly colored ball. The robot is able to function for at most one hour before requiring recharging. Early sales of the robot have been very strong, with more than 60,000 units sold in the first year. Nevertheless, the number of motors and the technology investment behind this robot dog resulted in a very high price of approximately $1,500.

Specifications:

Weight: 19 kg
Height: 0.25 m
DOF: 4 × 3

Figure 2.20
Titan VIII, a quadruped robot developed at Tokyo Institute of Technology
(http://www-robot.mes.titech.ac.jp). © Tokyo Institute of Technology.

Four-legged robots have the potential to serve as effective artifacts for research in
human-robot interaction (figure 2.20). Humans can treat the Sony robot, for example, as a
pet and might develop an emotional relationship similar to that between man and dog. Furthermore, Sony has designed AIBO's walking style and general behavior to emulate learning and maturation, resulting in dynamic behavior over time that is more interesting for the
owner who can track the changing behavior. As the challenges of high energy storage and
motor technology are solved, it is likely that quadruped robots much more capable than
AIBO will become common throughout the human environment.

BigDog and LittleDog (figure 2.21) are two recent examples of quadruped robots developed by Boston Dynamics and commissioned by the American Defense Advanced
Research Projects Agency (DARPA). BigDog is a rough-terrain robot that walks, runs,
climbs, and carries heavy loads. It is powered by an engine that drives a hydraulic actuation
system. Its legs are articulated like an animal's, with compliant elements to absorb shock
and recycle energy between two steps. The goal of this project is to make it able go anywhere people and animals can go. The program is funded by the Tactical Technology
Office at DARPA. Conversely, LittleDog is a small-size robot designed for research on
learning locomotion. Each leg has a large range of motion and is powered by three electric
motors. Therefore, the robot is strong enough for climbing and dynamic locomotion gaits.

Another example four-legged robot is ALoF, the quadruped developed at the ASL (ETH
Zurich) (figure 2.22).This robot serves as a platform to study energy-efficient locomotion.
This is done by exploiting passive dynamic in ways that have shown to be effective in
bipedal robots, as has been explained in section 2.2.2.

Figure 2.21 LittleDog and BigDog quadruped robots developed by Boston Dynamics. Image courtesy of Boston Dynamics (http://www.bostondynamics.com).

Figure 2.22 ALoF, the quadruped robot developed at the ASL (ETH Zurich), has been built to investigate energy efficient locomotion. This is done by exploiting passive dynamics (http://www.asl.ethz.ch/). © ASL-ETH Zurich.

2.2.3.4 Six legs (hexapod)

Six-legged configurations have been extremely popular in mobile robotics because of their static stability during walking, thus reducing the control complexity (figures 2.23 and 1.3). In most cases, each leg has three degrees of freedom, including hip flexion, knee flexion,

Specifications:

Maximum speed:	0.5 m/s
Weight:	16 kg
Height:	0.3 m
Length:	0.7 m
No. of legs:	6
DOF in total:	6×3
Power consumption:	10 W

Figure 2.23
Lauron II, a hexapod platform developed at the University of Karlsruhe, Germany.
© University of Karlsruhe.

Figure 2.24
Genghis, one of the most famous walking robots from MIT, uses hobby servomotors as its actuators
(http://www.ai.mit.edu/projects/genghis). © MIT AI Lab.

and hip abduction (see figure 2.7). Genghis is a commercially available hobby robot that
has six legs, each of which has two degrees of freedom provided by hobby servos (figure
2.24). Such a robot, which consists only of hip flexion and hip abduction, has less maneuverability in rough terrain but performs quite well on flat ground. Because it consists of a
straightforward arrangement of servomotors and straight legs, such robots can be readily
built by a robot hobbyist.

Insects, which are arguably the most successful locomoting creatures on earth, excel at
traversing all forms of terrain with six legs, even upside down. Currently, the gap between

the capabilities of six-legged insects and artificial six-legged robots is still quite large. Interestingly, this is not due to a lack of sufficient numbers of degrees of freedom on the robots. Rather, insects combine a small number of active degrees of freedom with passive structures, such as microscopic barbs and textured pads, that increase the gripping strength of each leg significantly. Robotic research into such passive tip structures has only recently begun. For example, a research group is attempting to re-create the complete mechanical function of the cockroach leg [124].

It is clear from these examples that legged robots have much progress to make before they are competitive with their biological equivalents. Nevertheless, significant gains have been realized recently, primarily due to advances in motor design. Creating actuation systems that approach the efficiency of animal muscles remains far from the reach of robotics, as does energy storage with the energy densities found in organic life forms.

2.3   Wheeled Mobile Robots

The wheel has been by far the most popular locomotion mechanism in mobile robotics and in man-made vehicles in general. It can achieve very good efficiencies, as demonstrated in figure 2.3, and it does so with a relatively simple mechanical implementation.

In addition, balance is not usually a research problem in wheeled robot designs, because wheeled robots are almost always designed so that all wheels are in ground contact at all times. Thus, three wheels are sufficient to guarantee stable balance, although, as we shall see below, two-wheeled robots can also be stable. When more than three wheels are used, a suspension system is required to allow all wheels to maintain ground contact when the robot encounters uneven terrain.

Instead of worrying about balance, wheeled robot research tends to focus on the problems of traction and stability, maneuverability, and control: can the robot wheels provide sufficient traction and stability for the robot to cover all of the desired terrain, and does the robot's wheeled configuration enable sufficient control over the velocity of the robot?

2.3.1 Wheeled locomotion: The design space

As we shall see, there is a very large space of possible wheel configurations when one considers possible techniques for mobile robot locomotion. We begin by discussing the wheel in detail, since there are a number of different wheel types with specific strengths and weaknesses. We then examine complete wheel configurations that deliver particular forms of locomotion for a mobile robot.

2.3.1.1 Wheel design

There are four major wheel classes, as shown in figure 2.25. They differ widely in their kinematics, and therefore the choice of wheel type has a large effect on the overall kinemat-

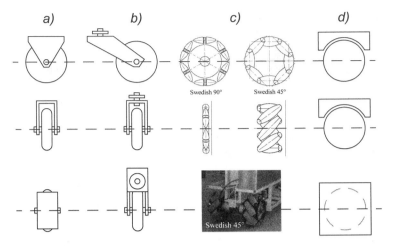

Figure 2.25
The four basic wheel types. (a) Standard wheel: two degrees of freedom; rotation around the (motorized) wheel axle and the contact point.(b) castor wheel: two degrees of freedom; rotation around an offset steering joint. (c) Swedish wheel: three degrees of freedom; rotation around the (motorized) wheel axle, around the rollers, and around the contact point. (d) Ball or spherical wheel: realization technically difficult.

ics of the mobile robot. The standard wheel and the castor wheel have a primary axis of rotation and are thus highly directional. To move in a different direction, the wheel must be steered first along a vertical axis. The key difference between these two wheels is that the standard wheel can accomplish this steering motion with no side effects, since the center of rotation passes through the contact patch with the ground, whereas the castor wheel rotates around an offset axis, causing a force to be imparted to the robot chassis during steering.

The Swedish wheel and the spherical wheel are both designs that are less constrained by directionality than the conventional standard wheel. The Swedish wheel functions as a normal wheel but provides low resistance in another direction as well, sometimes perpendicular to the conventional direction, as in the Swedish 90, and sometimes at an intermediate angle, as in the Swedish 45. The small rollers attached around the circumference of the wheel are passive and the wheel's primary axis serves as the only actively powered joint. The key advantage of this design is that, although the wheel rotation is powered only along the one principal axis (through the axle), the wheel can kinematically move with very little friction along many possible trajectories, not just forward and backward.

The spherical wheel is a truly omnidirectional wheel, often designed so that it may be actively powered to spin along any direction. One mechanism for implementing this spherical design imitates the computer mouse, providing actively powered rollers that rest against the top surface of the sphere and impart rotational force.

Figure 2.26
The Tartan Racing self-driving vehicle developed at CMU, which won the 2007 DARPA Urban Challenge. Image courtesy of the Tartan Racing Team—http://www.tartanracing.org.

Regardless of what wheel is used, in robots designed for all-terrain environments and in robots with more than three wheels, a suspension system is normally required to maintain wheel contact with the ground. One of the simplest approaches to suspension is to design flexibility into the wheel itself. For instance, in the case of some four-wheeled indoor robots that use castor wheels, manufacturers have applied a deformable tire of soft rubber to the wheel to create a primitive suspension. Of course, this limited solution cannot compete with a sophisticated suspension system in applications where the robot needs a more dynamic suspension for significantly nonflat terrain.

2.3.1.2 Wheel geometry
The choice of wheel types for a mobile robot is strongly linked to the choice of wheel arrangement, or wheel geometry. The mobile robot designer must consider these two issues simultaneously when designing the locomoting mechanism of a wheeled robot. Why do wheel type and wheel geometry matter? Three fundamental characteristics of a robot are governed by these choices: maneuverability, controllability, and stability.

Unlike automobiles, which are largely designed for a highly standardized environment (the road network), mobile robots are designed for applications in a wide variety of situations. Automobiles all share similar wheel configurations because there is one region in the design space that maximizes maneuverability, controllability, and stability for their standard environment: the paved roadway. However, there is no single wheel configuration that maximizes these qualities for the variety of environments faced by different mobile robots. So you will see great variety in the wheel configurations of mobile robots. In fact, few robots use the Ackerman wheel configuration of the automobile because of its poor maneuverability, with the exception of mobile robots designed for the road system (figure 2.26).

Table 2.1 gives an overview of wheel configurations ordered by the number of wheels. This table shows both the selection of particular wheel types and their geometric configuration on the robot chassis. Note that some of the configurations shown are of little use in mobile robot applications. For instance, the two-wheeled bicycle arrangement has moderate maneuverability and poor controllability. Like a single-legged hopping machine, it can never stand still. Nevertheless, this table provides an indication of the large variety of wheel configurations that are possible in mobile robot design.

The number of variations in table 2.1 is quite large. However, there are important trends and groupings that can aid in comprehending the advantages and disadvantages of each configuration. We next identify some of the key trade-offs in terms of the three issues we identified earlier: stability, maneuverability, and controllability.

2.3.1.3 Stability

Surprisingly, the minimum number of wheels required for static stability is two. As shown above, a two-wheel differential-drive robot can achieve static stability if the center of mass is below the wheel axle. Cye is a commercial mobile robot that uses this wheel configuration (figure 2.27).

However, under ordinary circumstances such a solution requires wheel diameters that are impractically large. Dynamics can also cause a two-wheeled robot to strike the floor with a third point of contact, for instance, with sufficiently high motor torques from standstill. Conventionally, static stability requires a minimum of three wheels, with the additional caveat that the center of gravity must be contained within the triangle formed by the ground contact points of the wheels. Stability can be further improved by adding more wheels, although once the number of contact points exceeds three, the hyperstatic nature of the geometry will require some form of flexible suspension on uneven terrain.

2.3.1.4 Maneuverability

Some robots are omnidirectional, meaning that they can move at any time in any direction along the ground plane (x, y) regardless of the orientation of the robot around its vertical axis. This level of maneuverability requires wheels that can move in more than just one direction, and so omnidirectional robots usually employ Swedish or spherical wheels that are powered. A good example is Uranus (figure 2.30), a robot that uses four Swedish wheels to rotate and translate independently and without constraints.

In general, the ground clearance of robots with Swedish and spherical wheels is somewhat limited due to the mechanical constraints of constructing omnidirectional wheels. An interesting recent solution to the problem of omnidirectional navigation while solving this ground-clearance problem is the four-castor wheel configuration in which each castor wheel is actively steered and actively translated. In this configuration, the robot is truly omnidirectional because, even if the castor wheels are facing a direction perpendicular to

Table 2.1
Wheel configurations for rolling vehicles

# of wheels	Arrangement	Description	Typical examples
2		One steering wheel in the front, one traction wheel in the rear	Bicycle, motorcycle
		Two-wheel differential drive with the center of mass (COM) below the axle	Cye personal robot
3		Two-wheel centered differential drive with a third point of contact	Nomad Scout, smartRob EPFL
		Two independently driven wheels in the rear/front, one unpowered omnidirectional wheel in the front/rear	Many indoor robots, including the EPFL robots Pygmalion and Alice
		Two connected traction wheels (differential) in rear, one steered free wheel in front	Piaggio minitrucks
		Two free wheels in rear, one steered traction wheel in front	Neptune (Carnegie Mellon University), Hero-1
		Three motorized Swedish or spherical wheels arranged in a triangle; omnidirectional movement is possible	Stanford wheel Tribolo EPFL, Palm Pilot Robot Kit (CMU)
		Three synchronously motorized and steered wheels; the orientation is not controllable	"Synchro drive" Denning MRV-2, Georgia Institute of Technology, I-Robot B24, Nomad 200

Table 2.1
Wheel configurations for rolling vehicles

# of wheels	Arrangement	Description	Typical examples
4		Two motorized wheels in the rear, two steered wheels in the front; steering has to be different for the two wheels to avoid slipping/skidding.	Car with rear-wheel drive
		Two motorized and steered wheels in the front, two free wheels in the rear; steering has to be different for the two wheels to avoid slipping/skidding.	Car with front-wheel drive
		Four steered and motorized wheels	Four-wheel drive, four-wheel steering Hyperion (CMU)
		Two traction wheels (differential) in rear/front, two omnidirectional wheels in the front/rear	Charlie (DMT-EPFL)
		Four omnidirectional wheels	Carnegie Mellon Uranus
		Two-wheel differential drive with two additional points of contact	EPFL Khepera, Hyperbot Chip
		Four motorized and steered castor wheels	Nomad XR4000

Table 2.1
Wheel configurations for rolling vehicles

# of wheels	Arrangement	Description	Typical examples
6		Two motorized and steered wheels aligned in center, one omnidirectional wheel at each corner	First
		Two traction wheels (differential) in center, one omnidirectional wheel at each corner	Terregator (Carnegie Mellon University)

Icons for the each wheel type are as follows:	
○	unpowered omnidirectional wheel (spherical, castor, Swedish)
▨	motorized Swedish wheel (Stanford wheel)
▭	unpowered standard wheel
▨	motorized standard wheel
▨○	motorized and steered castor wheel
⊤	steered standard wheel
⊥	connected wheels

the desired direction of travel, the robot can still move in the desired direction by steering these wheels. Because the vertical axis is offset from the ground-contact path, the result of this steering motion is robot motion.

In the research community, other classes of mobile robots are popular that achieve high maneuverability, only slightly inferior to that of the omnidirectional configurations. In such robots, motion in a particular direction may initially require a rotational motion. With a circular chassis and an axis of rotation at the center of the robot, such a robot can spin without

Figure 2.27
Cye, a domestic robot, was designed to vacuum floors and make domestic deliveries.

changing its ground footprint. The most popular such robot is the two-wheel differential-drive robot where the two wheels rotate around the center point of the robot. One or two additional ground contact points may be used for stability, based on the application specifics.

In contrast to these configurations, consider the Ackerman steering configuration common in automobiles. Such a vehicle typically has a turning diameter that is larger than the car. Furthermore, for such a vehicle to move sideways requires a parking maneuver consisting of repeated changes in direction forward and backward. Nevertheless, Ackerman steering geometries have been especially popular in the hobby robotics market, where a robot can be built by starting with a remote control racecar kit and adding sensing and autonomy to the existing mechanism. In addition, the limited maneuverability of Ackerman steering has an important advantage: its directionality and steering geometry provide it with very good lateral stability in high-speed turns.

2.3.1.5 Controllability

There is generally an inverse correlation between controllability and maneuverability. For example, the omnidirectional designs such as the four-castor wheel configuration require significant processing to convert desired rotational and translational velocities to individual wheel commands. Furthermore, such omnidirectional designs often have greater degrees of freedom at the wheel. For instance, the Swedish wheel has a set of free rollers along the wheel perimeter. These degrees of freedom cause an accumulation of slippage, tend to reduce dead-reckoning accuracy, and increase the design complexity.

Controlling an omnidirectional robot for a specific direction of travel is also more difficult and often less accurate when compared to less maneuverable designs. For example, an Ackerman steering vehicle can go straight simply by locking the steerable wheels and driving the drive wheels. In a differential-drive vehicle, the two motors attached to the two

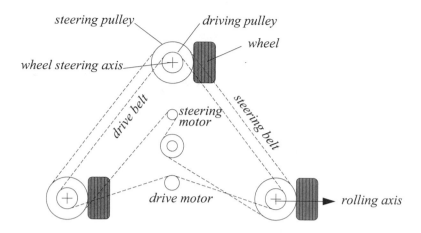

Figure 2.28
Synchro drive: The robot can move in any direction; however, the orientation of the chassis is not controllable.

wheels must be driven along exactly the same velocity profile, which can be challenging considering variations between wheels, motors, and environmental differences. With four-wheel omnidrive, such as the Uranus robot, which has four Swedish wheels, the problem is even harder because all four wheels must be driven at exactly the same speed for the robot to travel in a perfectly straight line.

In summary, there is no "ideal" drive configuration that simultaneously maximizes stability, maneuverability, and controllability. Each mobile robot application places unique constraints on the robot design problem, and the designer's task is to choose the most appropriate drive configuration possible from among this space of compromises.

2.3.2 Wheeled locomotion: Case studies

We next describe four specific wheel configurations, in order to demonstrate concrete applications of the concepts discussed above to mobile robots built for real-world activities.

2.3.2.1 Synchro drive

The synchro drive configuration (figure 2.28) is a popular arrangement of wheels in indoor mobile robot applications. It is an interesting configuration because, although there are three driven and steered wheels, only two motors are used in total. The one translation motor sets the speed of all three wheels together, and the one steering motor spins all the wheels together about each of their individual vertical steering axes. But note that the wheels are being steered with respect to the robot chassis, and therefore there is no direct way of reorienting the robot chassis. In fact, the chassis orientation does drift over time due to uneven tire slippage, causing rotational dead-reckoning error.

Synchro drive is particularly advantageous in cases where omnidirectionality is sought. So long as each vertical steering axis is aligned with the contact path of each tire, the robot can always reorient its wheels and move along a new trajectory without changing its footprint. Of course, if the robot chassis has directionality and the designers intend to reorient the chassis purposefully, then synchro drive is appropriate only when combined with an independently rotating turret that attaches to the wheel chassis. Commercial research robots such as the Nomadics 150 or the RWI B21r have been sold with this configuration (figure 1.12).

In terms of dead reckoning, synchro drive systems are generally superior to true omnidirectional configurations but inferior to differential-drive and Ackerman steering systems. There are two main reasons for this. First and foremost, the translation motor generally drives the three wheels using a single belt. Because of slop and backlash in the drive train, whenever the drive motor engages, the closest wheel begins spinning before the furthest wheel, causing a small change in the orientation of the chassis. With additional changes in motor speed, these small angular shifts accumulate to create a large error in orientation during dead reckoning. Second, the mobile robot has no direct control over the orientation of the chassis. Depending on the orientation of the chassis, the wheel thrust can be highly asymmetric, with two wheels on one side and the third wheel alone, or symmetric, with one wheel on each side and one wheel straight ahead or behind, as shown in figure 2.22. The asymmetric cases result in a variety of errors when tire-ground slippage can occur, again causing errors in dead reckoning of robot orientation.

2.3.2.2 Omnidirectional drive

As we will see later in section 3.4.2, omnidirectional movement is of great interest for complete maneuverability. Omnidirectional robots that are able to move in any direction (x, y, θ) at any time are also holonomic (see section 3.4.2). They can be realized by using spherical, castor, or Swedish wheels. Three examples of such holonomic robots are presented here.

Omnidirectional locomotion with three spherical wheels. The omnidirectional robot depicted in figure 2.29 is based on three spherical wheels, each actuated by one motor. In this design, the spherical wheels are suspended by three contact points, two given by spherical bearings and one by a wheel connected to the motor axle. This concept provides excellent maneuverability and is simple in design. However, it is limited to flat surfaces and small loads, and it is quite difficult to find round wheels with high friction coefficients.

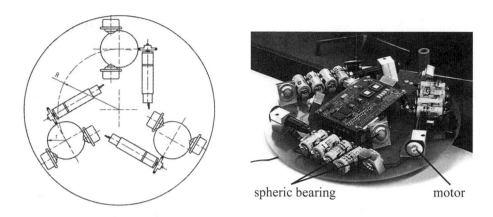

Figure 2.29
The Tribolo designed at EPFL (Swiss Federal Institute of Technology, Lausanne, Switzerland). Left: arrangement of spheric bearings and motors (bottom view). Right: Picture of the robot without the spherical wheels (bottom view).

Omnidirectional locomotion with four Swedish wheels. The omnidirectional arrangement depicted in figure 2.30 has been used successfully on several research robots, including the Carnegie Mellon Uranus. This configuration consists of four Swedish 45-degree wheels, each driven by a separate motor. By varying the direction of rotation and relative speeds of the four wheels, the robot can be moved along any trajectory in the plane and, even more impressively, can simultaneously spin around its vertical axis.

For example, when all four wheels spin "forward" or "backward," the robot as a whole moves in a straight line forward or backward, respectively. However, when one diagonal pair of wheels is spun in the same direction and the other diagonal pair is spun in the opposite direction, the robot moves laterally.

This four-wheel arrangement of Swedish wheels is not minimal in terms of control motors. Because there are only three degrees of freedom in the plane, one can build a three-wheel omnidirectional robot chassis using three Swedish 90-degree wheels as shown in table 2.1. However, existing examples such as Uranus have been designed with four wheels owing to capacity and stability considerations.

One application for which such omnidirectional designs are particularly amenable is mobile manipulation. In this case, it is desirable to reduce the degrees of freedom of the manipulator arm to save arm mass by using the mobile robot chassis motion for gross motion. As with humans, it would be ideal if the base could move omnidirectionally with-

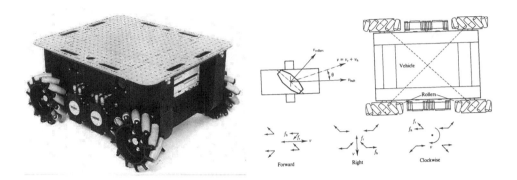

Figure 2.30
The Carnegie Mellon Uranus robot, an omnidirectional robot with four powered Swedish 45-wheels.

Figure 2.31
The Nomad XR4000 from Nomadic Technologies had an arrangement of four castor wheels for holonomic motion. All the castor wheels are driven and steered, thus requiring a precise synchronization and coordination to obtain a precise movement in x, y, and θ.

out greatly impacting the position of the manipulator tip, and a base such as Uranus can afford precisely such capabilities.

Omnidirectional locomotion with four castor wheels and eight motors. Another solution for omnidirectionality is to use castor wheels. This is done for the Nomad XR4000 from Nomadic Technologies (figure 2.31), giving it excellent maneuverability. Unfortunately, Nomadic has ceased production of mobile robots.

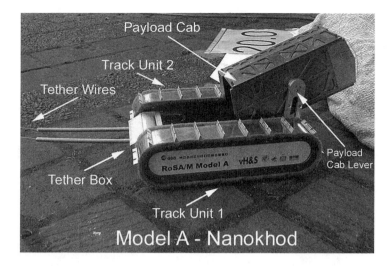

Figure 2.32
The microrover Nanokhod, developed by von Hoerner & Sulger GmbH and the Max Planck Institute, Mainz, for the European Space Agency (ESA), will probably go to Mars [302, 327].

The preceding three examples are drawn from table 2.1, but this is not an exhaustive list of all wheeled locomotion techniques. Hybrid approaches that combine legged and wheeled locomotion, or tracked and wheeled locomotion, can also offer particular advantages. Following are two unique designs created for specialized applications.

2.3.2.3 Tracked slip/skid locomotion

In the wheel configurations discussed earlier, we have made the assumption that wheels are not allowed to skid against the surface. An alternative form of steering, termed slip/skid, may be used to reorient the robot by spinning wheels that are facing the same direction at different speeds or in opposite directions. The army tank operates this way, and the Nanokhod (figure 2.32) is an example of a mobile robot based on the same concept.

Robots that make use of tread have much larger ground contact patches, and this can significantly improve their maneuverability in loose terrain compared to conventional wheeled designs. However, due to this large ground contact patch, changing the orientation of the robot usually requires a skidding turn, wherein a large portion of the track must slide against the terrain.

The disadvantage of such configurations is coupled to the slip/skid steering. Because of the large amount of skidding during a turn, the exact center of rotation of the robot is hard to predict and the exact change in position and orientation is also subject to variations depending on the ground friction. Therefore, dead reckoning on such robots is highly inac-

Figure 2.33
Shrimp, an all-terrain robot with outstanding passive climbing abilities (EPFL [184, 289]).

curate. This is the trade-off that is made in return for extremely good maneuverability and traction over rough and loose terrain. Furthermore, a slip/skid approach on a high-friction surface can quickly overcome the torque capabilities of the motors being used. In terms of power efficiency, this approach is reasonably efficient on loose terrain but extremely inefficient otherwise.

2.3.2.4 Walking wheels

Walking robots might offer the best maneuverability in rough terrain. However, they are inefficient on flat ground and need sophisticated control. Hybrid solutions, combining the adaptability of legs with the efficiency of wheels, offer an interesting compromise. Solutions that passively adapt to the terrain are of particular interest for field and space robotics. The Sojourner robot of NASA/JPL (see figure 1.2) represents such a hybrid solution, able to overcome objects up to the size of the wheels. A more recent mobile robot design for similar applications has been produced by EPFL (figure 2.33). This robot, called Shrimp, has six motorized wheels and is capable of climbing objects up to two times its wheel diameter [184, 289]. This enables it to climb regular stairs even though the robot is even smaller than the Sojourner. Using a rhombus configuration, the Shrimp has a steering wheel in the front and the rear and two wheels arranged on a bogie on each side. The front wheel has a spring suspension to guarantee optimal ground contact of all wheels at any time. The steer-

Figure 2.34
The Personal Rover, demonstrating ledge climbing using active center-of-mass shifting.

ing of the rover is realized by synchronizing the steering of the front and rear wheels and
the speed difference of the bogie wheels. This allows for high-precision maneuvers and
turning on the spot with minimum slip/skid of the four center wheels. The use of parallel
articulations for the front wheel and the bogies creates a virtual center of rotation at the
level of the wheel axis. This ensures maximum stability and climbing abilities even for very
low friction coefficients between the wheel and the ground.

The climbing ability of the Shrimp is extraordinary in comparison to most robots of similar mechanical complexity, owing much to the specific geometry and thereby the manner
in which the center of mass (COM) of the robot shifts with respect to the wheels over time.
In contrast, the Personal Rover demonstrates active COM shifting to climb ledges that are
also several times the diameter of its wheels, as demonstrated in figure 2.34. A majority of
the weight of the Personal Rover is borne at the upper end of its swinging boom. A dedicated motor drives the boom to change the front/rear weight distribution in order to facilitate step-climbing. Because this COM-shifting scheme is active, a control loop must
explicitly decide how to move the boom during a climbing scenario. In this case, the Personal Rover accomplished this closed-loop control by inferring terrain based on measurements of current flowing to each independently driven wheel [125].

As mobile robotics research matures, we find ourselves able to design more intricate
mechanical systems. At the same time, the control problems of inverse kinematics and
dynamics are now so readily conquered that these complex mechanics can in general be
controlled. So, in the near future, we can expect to see a great number of unique, hybrid
mobile robots that draw together advantages from several of the underlying locomotion

mechanisms that we have discussed in this chapter. They will be technologically impressive, and each will be designed as the expert robot for its particular environmental niche.

2.4   Aerial Mobile Robots

2.4.1   Introduction

Flying objects have always exerted a great fascination on humans, encouraging all kinds of research and development. This introduction is written in a time at which the robotics community is showing a growing interest in micro aerial vehicle (MAV) development. The scientific challenge in MAV design, control, and navigation in cluttered environments and the lack of existing solutions is the main leitmotiv. On the other hand, the broad field of applications in both military and civilian markets is encouraging the funding of MAV-related projects. However, the task is not trivial due to several open challenges.

In the field of sensing technologies, industry can currently provide a new generation of integrated micro inertial measurement units (IMU, section 4.1.7) composed generally of micro electro-mechanical systems (MEMS) technology, inertial and magneto-resistive sensors. The latest technology in high density power storage offers about 230Wh/kg (Li-Ion technology in 2009), which is a real jump ahead, especially for micro aerial robotics. This technology was originally developed for handheld applications and is now widely used in aerial robotics. The cost and size reduction of such systems makes it very interesting for the civilian market. Simultaneously, this reduction of cost and size implies performance limitations and thus a more challenging control problem. Moreover, the miniaturization of inertial sensors imposes the use of MEMS technology, which is still much less accurate than the conventional sensors because of noise and drift. The use of low-cost IMUs demands less effective data processing and thus a bad orientation data prediction in addition to a weak drift rejection. On the other hand, and in spite of the latest progress in miniature actuators, the scaling laws are still unfavorable and one has to face the problem of actuator saturation. That is to say, even though the design of micro aerial robots is possible, the control is still a challenging goal.

Investigating relations between size and weight of flying objects yields some interesting findings. Tennekes's Great Flight Diagram [50] (figure 2.35) plots weight versus wing loading for all sizes comprising insects, birds, and sailplanes all the way up to the Boeing 747. It illustrates the fundamental simplified assumption that the weight W scales with the wingspan b to the power of three (b^3), while the wing surface S may be seen as scaling with b^2. Figure 2.35 shows Tennekes's Great Flight Diagram augmented with some unmanned solar airplanes and radio controlled airplanes [248] that are comparable to small robotic unmanned aerial systems. The curve W/S represents an average of the shown data points. Notice that different constructions still yield different results: the extremely lightweight solar airplane *Helios* by NASA, for example, with its 75 meters of wingspan and a surface

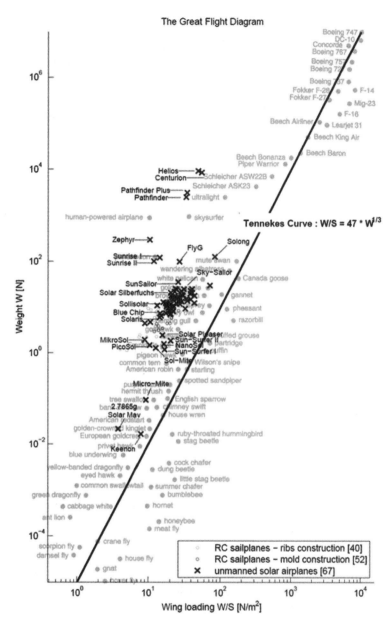

Figure 2.35 Tennekes's Great Flight Diagram [50] augmented with RC sailplanes and unmanned solar airplanes. Image courtesy of A. Noth [248].

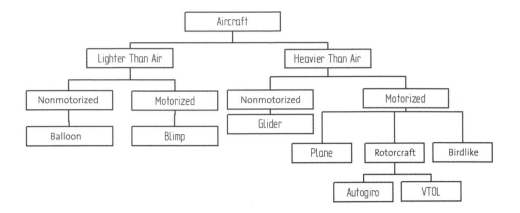

Figure 2.36 General classification of aircrafts.

area of $184\,m^2$, has approximately the same wing loading as a pelican but it is heavier by a factor of 1,000 and of course is much larger.

2.4.2 Aircraft configurations

In general, aerial vehicles can be divided into two categories: Lighter Than Air (LTA) and Heavier Than Air (HTA). Figure 2.36 presents a general classification of aircraft depending on the flying principle and the propulsion mode. Table 2.2 gives a nonexhaustive comparison between different flying principles from the miniaturization point of view. From this table, one can easily conclude that Vertical Take-Off and Landing (VTOL) systems such as helicopters or blimps have an unquestionable advantage compared with the other concepts. This superiority is owed to their unique ability for vertical, stationary, and low-speed flight. The key advantage of blimps is the *autolift* and simplicity of control, which can be essential for critical applications, such as aerial surveillance and space exploration. However, VTOL vehicles in different configurations represent today one of the most promising flying concepts seen in terms of miniaturization. Figure 2.37 lists different configurations commonly used in MAV research and industry.

2.4.3 State of the art in autonomous VTOL

The state of the art in MAV research has dramatically changed in the last few years. The number of projects tackling this problem has considerably and suddenly increased. Until 2006, the main research problem was MAV stabilization, especially for mini quadrotors. Since 2007, the research community shifted its interest toward autonomous navigation, first outdoor and more recently even indoor.

Table 2.2 Flying principle comparison (1 = Bad, 3 = Good)

	Airplane	Helicopter	Bird	Autogiro	Blimp
Power cost	2	1	2	2	3
Control cost	2	1	1	2	3
Payload/volume	3	2	2	2	1
Maneuverability	2	3	3	2	1
Stationary flight	1	3	2	1	3
Low speed fly	1	3	2	2	3
Vulnerability	2	2	3	2	2
VTOL	1	3	2	1	3
Endurance	2	1	2	1	3
Miniaturization	2	3	3	2	1
Indoor usage	1	3	2	1	2
Total	19	25	24	18	25

The CSAIL laboratory at MIT is presently one of the leaders in terms of MAV navigation in GPS-denied environments. The quadrotor that it used in the 2009 edition of the AUVSI competition uses laser scanners to localize and navigate autonomously inside buildings. The quadrotor from ALU Freiburg [142] is also equipped with a laser scanner; it achieves global localization using a particle filter and a graph-based SLAM algorithm (both these algorithms will be treated in section 5.8.2). It is thus able to navigate autonomously indoors while avoiding obstacles. STARMAC, from Stanford University, targets the demonstration of multiagent control of quadrotors of about 1 kg, outdoors, using GPS. ETH Zurich is also participating in this endeavor with different projects. The European project sFly (www.sfly.org) targets outdoor autonomous navigation of a swarm of small quadrotors using monocular vision as the main sensor (no laser, no GPS). To the best of our knowledge, the smallest existing autonomous helicopter is the muFly helicopter, developed at ETH Zurich within the European project muFly (www.mufly.org): it weighs 80 g and has an overall span of 17.5 cm. In addition to an IMU, muFly is equipped with a 360-degree laser scanner, a down-looking micro camera, and a miniature omnidirectional camera that weighs less than 5 g. These projects are listed in figure 2.38.   .

Configuration *e.g.*	Advantages	Drawbacks	Picture
Fixed-wing (AeroVironment)	Simple mechanics, silent operation	No hovering	
Single rotor (A. V de Rostyne)	Good controllability, good maneuverability	Complex mechanics, large rotor, long tail boom	
Axial rotor (Univ. of Maryland)	Simple mechanics, compactness	Complex aerodynamics	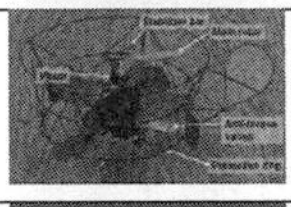
Coaxial rotors (EPSON)	Simple mechanics, compactness	Complex aerodynamics	
Tandem rotors (Heudiasyc)	Good controllability, simple aerodynamics	Complex mechanics, large size	
Quadrotor (EPFL-ETHZ)	Good maneuverability, simple mechanics, increased payload	High energy consumption, large size	
Blimp (EPFL)	Low power, long flight operation, auto-lift	Large size, weak maneuverability	
Hybrid quadrotor-blimp (MIT)	Good maneuverability, good survivability	Large size, weak maneuverability	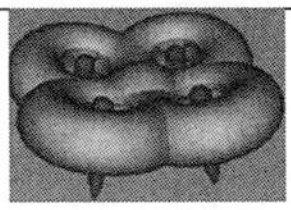
Birdlike (Caltech)	Good maneuverability, compactness	Complex mechanics, complex control	
Insectlike (UC Berkeley)	Good maneuverability, compactness	Complex mechanics, complex control	
Fishlike (US Naval Lab)	Multimode mobility, efficient aerodynamics	Complex control, weak maneuverability	

Figure 2.37 Common MAV configurations.

Projects	University	Status	Picture
MIT-MAV	MIT	Ended	
Freiburg MAV	ALU Freiburg	In progress	
Starmac	Stanford	In progress	
sFly	ETH Zürich	In progress	
muFly	ETH Zürich	Ended	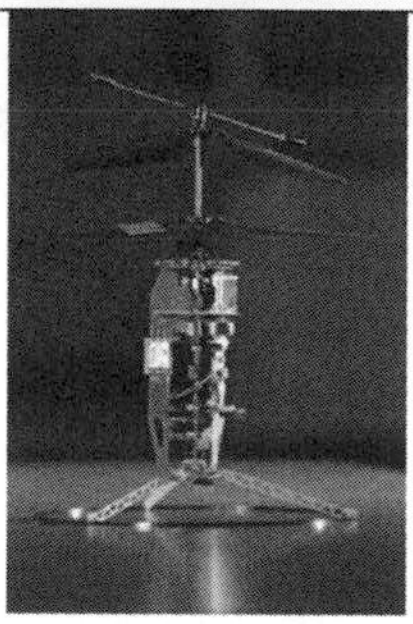

Figure 2.38 Progress in autonomous VTOL systems.

2.5   Problems

1. Consider an eight-legged walking robot. Consider gaits in terms of lift/release events as in this chapter. (a) How many possible events exist for this eight-legged machine? (b) Specify two different statically stable walking gaits using the notation of figure 2.8.

2. Describe two wheel configurations that enable omnidirectional motion that are not identified in section 2.3.2.2. Note that you may use any type of wheel in these two designs. Draw the configurations using the notation of table 2.1.

3. You wish to build a dynamically stable robot with a single wheel only. For each of the four basic wheel types, explain whether or not it may be used for such a robot.

4. **Challenge Question.**
   Four-legged machines are normally not statically stable. Design a four-legged locomotion machine that is statically stable. Draw it and describe the gait used.

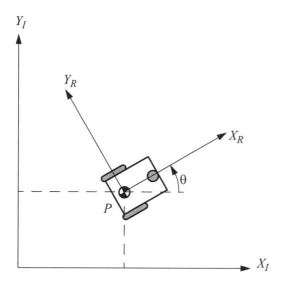

Figure 3.1
The global reference frame and the robot local reference frame.

In order to specify the position of the robot on the plane, we establish a relationship between the global reference frame of the plane and the local reference frame of the robot, as in figure 3.1. The axes X_I and Y_I define an arbitrary inertial basis on the plane as the global reference frame from some origin O: $\{X_I, Y_I\}$. To specify the position of the robot, choose a point P on the robot chassis as its position reference point. The basis $\{X_R, Y_R\}$ defines two axes relative to P on the robot chassis and is thus the robot's local reference frame. The position of P in the global reference frame is specified by coordinates x and y, and the angular difference between the global and local reference frames is given by θ . We can describe the pose of the robot as a vector with these three elements. Note the use of the subscript I to clarify the basis of this pose as the global reference frame:

$$\xi_I = \begin{bmatrix} x \\ y \\ \theta \end{bmatrix} \tag{3.1}$$

To describe robot motion in terms of component motions, it will be necessary to map motion along the axes of the global reference frame to motion along the axes of the robot's local reference frame. Of course, the mapping is a function of the current pose of the robot. This mapping is accomplished using the *orthogonal rotation matrix*:

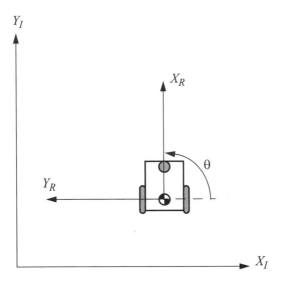

Figure 3.2
The mobile robot aligned with a global axis.

$$R(\theta) = \begin{bmatrix} \cos\theta & \sin\theta & 0 \\ -\sin\theta & \cos\theta & 0 \\ 0 & 0 & 1 \end{bmatrix}. \tag{3.2}$$

This matrix can be used to map motion in the global reference frame $\{X_I, Y_I\}$ to motion in terms of the local reference frame $\{X_R, Y_R\}$. This operation is denoted by $R(\theta)\dot{\xi}_I$ because the computation of this operation depends on the value of θ:

$$\dot{\xi}_R = R(\frac{\pi}{2})\dot{\xi}_I \quad . \tag{3.3}$$

For example, consider the robot in figure 3.2. For this robot, because $\theta = \frac{\pi}{2}$ we can easily compute the instantaneous rotation matrix R:

$$R(\frac{\pi}{2}) = \begin{bmatrix} 0 & 1 & 0 \\ -1 & 0 & 0 \\ 0 & 0 & 1 \end{bmatrix}. \tag{3.4}$$

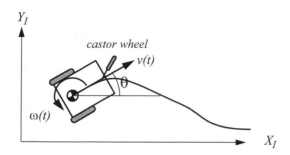

Figure 3.3
A differential-drive robot in its global reference frame.

Given some velocity $(\dot{x}, \dot{y}, \dot{\theta})$ in the global reference frame we can compute the components of motion along this robot's local axes X_R and Y_R. In this case, due to the specific angle of the robot, motion along X_R is equal to $\dot{y}$, and motion along Y_R is $-\dot{x}$:

$$\dot{\xi}_R = R(\frac{\pi}{2})\dot{\xi}_I = \begin{bmatrix} 0 & 1 & 0 \\ -1 & 0 & 0 \\ 0 & 0 & 1 \end{bmatrix}\begin{bmatrix} \dot{x} \\ \dot{y} \\ \dot{\theta} \end{bmatrix} = \begin{bmatrix} \dot{y} \\ -\dot{x} \\ \dot{\theta} \end{bmatrix}. \tag{3.5}$$

3.2.2 Forward kinematic models

In the simplest cases, the mapping described by equation (3.3) is sufficient to generate a formula that captures the forward kinematics of the mobile robot: how does the robot move, given its geometry and the speeds of its wheels? More formally, consider the example shown in figure 3.3.

This differential drive robot has two wheels, each with diameter r. Given a point P centered between the two drive wheels, each wheel is a distance l from P. Given r, l, θ, and the spinning speed of each wheel, $\dot{\varphi}_1$ and $\dot{\varphi}_2$, a forward kinematic model would predict the robot's overall speed in the global reference frame:

$$\dot{\xi}_I = \begin{bmatrix} \dot{x} \\ \dot{y} \\ \dot{\theta} \end{bmatrix} = f(l, r, \theta, \dot{\varphi}_1, \dot{\varphi}_2). \tag{3.6}$$

From equation (3.3) we know that we can compute the robot's motion in the global reference frame from motion in its local reference frame: $\dot{\xi}_I = R(\theta)^{-1}\dot{\xi}_R$. Therefore, the strategy will be to first compute the contribution of each of the two wheels in the local reference frame, $\dot{\xi}_R$. For this example of a differential-drive chassis, this problem is particularly straightforward.

Suppose that the robot's local reference frame is aligned such that the robot moves forward along $+X_R$, as shown in figure 3.1. First, consider the contribution of each wheel's spinning speed to the translation speed at P in the direction of $+X_R$. If one wheel spins while the other wheel contributes nothing and is stationary, since P is halfway between the two wheels, it will move instantaneously with half the speed: $\dot{x}_{r1} = (1/2)r\dot{\varphi}_1$ and $\dot{x}_{r2} = (1/2)r\dot{\varphi}_2$. In a differential-drive robot, these two contributions can simply be added to calculate the $\dot{x}_R$ component of $\dot{\xi}_R$. Consider, for example, a differential robot in which each wheel spins with equal speed but in opposite directions. The result is a stationary, spinning robot. As expected, $\dot{x}_R$ will be zero in this case. The value of $\dot{y}_R$ is even simpler to calculate. Neither wheel can contribute to sideways motion in the robot's reference frame, and so $\dot{y}_R$ is always zero. Finally, we must compute the rotational component $\dot{\theta}_R$ of $\dot{\xi}_R$. Once again, the contributions of each wheel can be computed independently and just added. Consider the right wheel (we will call this wheel 1). Forward spin of this wheel results in *counterclockwise* rotation at point P. Recall that if wheel 1 spins alone, the robot pivots around wheel 2. The rotation velocity ω_1 at P can be computed because the wheel is instantaneously moving along the arc of a circle of radius $2l$:

$$\omega_1 = \frac{r\dot{\varphi}_1}{2l}.$$ (3.7)

The same calculation applies to the left wheel, with the exception that forward spin results in *clockwise* rotation at point P:

$$\omega_2 = \frac{-r\dot{\varphi}_2}{2l}.$$ (3.8)

Combining these individual formulas yields a kinematic model for the differential-drive example robot:

$$\dot{\xi}_I = R(\theta)^{-1} \begin{bmatrix} \dfrac{r\dot{\varphi}_1}{2} + \dfrac{r\dot{\varphi}_2}{2} \\ 0 \\ \dfrac{r\dot{\varphi}_1}{2l} + \dfrac{-r\dot{\varphi}_2}{2l} \end{bmatrix}.$$ (3.9)

We can now use this kinematic model in an example. However, we must first compute $R(\theta)^{-1}$. In general, calculating the inverse of a matrix may be challenging. In this case, however, it is easy because it is simply a transform from $\dot{\xi}_R$ to $\dot{\xi}_I$ rather than vice versa:

$$R(\theta)^{-1} = \begin{bmatrix} \cos\theta & -\sin\theta & 0 \\ \sin\theta & \cos\theta & 0 \\ 0 & 0 & 1 \end{bmatrix}. \tag{3.10}$$

Suppose that the robot is positioned such that $\theta = \pi/2$, $r = 1$, and $l = 1$. If the robot engages its wheels unevenly, with speeds $\dot{\varphi}_1 = 4$ and $\dot{\varphi}_2 = 2$, we can compute its velocity in the global reference frame:

$$\dot{\xi}_I = \begin{bmatrix} \dot{x} \\ \dot{y} \\ \dot{\theta} \end{bmatrix} = \begin{bmatrix} 0 & -1 & 0 \\ 1 & 0 & 0 \\ 0 & 0 & 1 \end{bmatrix} \begin{bmatrix} 3 \\ 0 \\ 1 \end{bmatrix} = \begin{bmatrix} 0 \\ 3 \\ 1 \end{bmatrix}. \tag{3.11}$$

So this robot will move instantaneously along the y-axis of the global reference frame with speed 3 while rotating with speed 1. This approach to kinematic modeling can provide information about the motion of a robot given its component wheel speeds in straightforward cases. However, we wish to determine the space of possible motions for each robot chassis design. To do this, we must go further, describing formally the constraints on robot motion imposed by each wheel. Section 3.2.3 begins this process by describing constraints for various wheel types; the rest of this chapter provides tools for analyzing the characteristics and workspace of a robot given these constraints.

3.2.3 Wheel kinematic constraints

The first step to a kinematic model of the robot is to express constraints on the motions of individual wheels. Just as shown in section 3.2.2, the motions of individual wheels can later be combined to compute the motion of the robot as a whole. As discussed in chapter 2, there are four basic wheel types with widely varying kinematic properties. Therefore, we begin by presenting sets of constraints specific to each wheel type.

However, several important assumptions will simplify this presentation. We assume that the plane of the wheel always remains vertical and that there is in all cases one single point of contact between the wheel and the ground plane. Furthermore, we assume that there is no sliding at this single point of contact. That is, the wheel undergoes motion only under conditions of pure rolling and rotation about the vertical axis through the contact point. For a more thorough treatment of kinematics, including sliding contact, refer to [38].

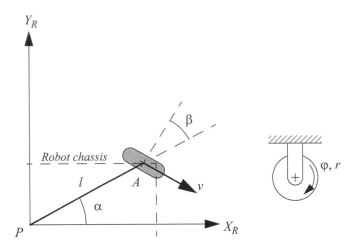

Figure 3.4
A fixed standard wheel and its parameters.

Under these assumptions, we present two constraints for every wheel type. The first constraint enforces the concept of rolling contact—that the wheel must roll when motion takes place in the appropriate direction. The second constraint enforces the concept of no lateral slippage—that the wheel must not slide orthogonal to the wheel plane.

3.2.3.1 Fixed standard wheel

The fixed standard wheel has no vertical axis of rotation for steering. Its angle to the chassis is thus fixed, and it is limited to motion back and forth along the wheel plane and rotation around its contact point with the ground plane. Figure 3.4 depicts a fixed standard wheel A and indicates its position pose relative to the robot's local reference frame $\{X_R, Y_R\}$. The position of A is expressed in polar coordinates by distance l and angle α. The angle of the wheel plane relative to the chassis is denoted by β, which is fixed since the fixed standard wheel is not steerable. The wheel, which has radius r, can spin over time, and so its rotational position around its horizontal axle is a function of time t: $\varphi(t)$.

The rolling constraint for this wheel enforces that all motion along the direction of the wheel plane must be accompanied by the appropriate amount of wheel spin so that there is pure rolling at the contact point:

$$\left[\sin(\alpha + \beta) \ -\cos(\alpha + \beta) \ (-l)\cos\beta\right] R(\theta)\dot{\xi}_I - r\dot{\varphi} = 0. \tag{3.12}$$

The first term of the sum denotes the total motion along the wheel plane. The three elements of the vector on the left represent mappings from each of $\dot{x}, \dot{y}, \dot{\theta}$ to their contributions for motion along the wheel plane. Note that the $R(\theta)\dot{\xi}_I$ term is used to transform the motion parameters $\dot{\xi}_I$ that are in the global reference frame $\{X_I, Y_I\}$ into motion parameters in the local reference frame $\{X_R, Y_R\}$ as shown in example equation (3.5). This is necessary because all other parameters in the equation, α, β, l, are in terms of the robot's local reference frame. This motion along the wheel plane must be equal, according to this constraint, to the motion accomplished by spinning the wheel, $r\dot{\varphi}$.

The sliding constraint for this wheel enforces that the component of the wheel's motion orthogonal to the wheel plane must be zero:

$$\begin{bmatrix} \cos(\alpha + \beta) & \sin(\alpha + \beta) & l\sin\beta \end{bmatrix} R(\theta)\dot{\xi}_I = 0. \tag{3.13}$$

For example, suppose that wheel A is in a position such that $\{(\alpha = 0), (\beta = 0)\}$. This would place the contact point of the wheel on X_I with the plane of the wheel oriented parallel to Y_I. If $\theta = 0$, then the sliding constraint [equation (3.13)] reduces to

$$\begin{bmatrix} 1 & 0 & 0 \end{bmatrix} \begin{bmatrix} 1 & 0 & 0 \\ 0 & 1 & 0 \\ 0 & 0 & 1 \end{bmatrix} \begin{bmatrix} \dot{x} \\ \dot{y} \\ \dot{\theta} \end{bmatrix} = \begin{bmatrix} 1 & 0 & 0 \end{bmatrix} \begin{bmatrix} \dot{x} \\ \dot{y} \\ \dot{\theta} \end{bmatrix} = 0. \tag{3.14}$$

This constrains the component of motion along X_I to be zero, and since X_I and X_R are parallel in this example, the wheel is constrained from sliding sideways, as expected.

3.2.3.2 Steered standard wheel

The steered standard wheel differs from the fixed standard wheel only in that there is an additional degree of freedom: the wheel may rotate around a vertical axis passing through the center of the wheel and the ground contact point. The equations of position for the steered standard wheel (figure 3.5) are identical to that of the fixed standard wheel shown in figure 3.4 with one exception. The orientation of the wheel to the robot chassis is no longer a single fixed value, β, but instead varies as a function of time: $\beta(t)$. The rolling and sliding constraints are

$$\begin{bmatrix} \sin(\alpha + \beta) & -\cos(\alpha + \beta) & (-l)\cos\beta \end{bmatrix} R(\theta)\dot{\xi}_I - r\dot{\varphi} = 0. \tag{3.15}$$

$$\begin{bmatrix} \cos(\alpha + \beta) & \sin(\alpha + \beta) & l\sin\beta \end{bmatrix} R(\theta)\dot{\xi}_I = 0. \tag{3.16}$$

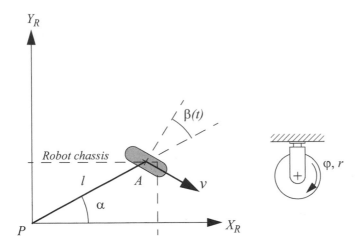

Figure 3.5
A steered standard wheel and its parameters.

These constraints are identical to those of the fixed standard wheel because, unlike $\dot{\varphi}$, $\dot{\beta}$ does not have a direct impact on the instantaneous motion constraints of a robot. It is only by integrating over time that changes in steering angle can affect the mobility of a vehicle. This may seem subtle, but is a very important distinction between change in steering position, $\dot{\beta}$, and change in wheel spin, $\dot{\varphi}$.

3.2.3.3 Castor wheel

Castor wheels are able to steer around a vertical axis. However, unlike the steered standard wheel, the vertical axis of rotation in a castor wheel does not pass through the ground contact point. Figure 3.6 depicts a castor wheel, demonstrating that formal specification of the castor wheel's position requires an additional parameter.

The wheel contact point is now at position B, which is connected by a rigid rod AB of fixed length d to point A, fixes the location of the vertical axis about which B steers, and this point A has a position specified in the robot's reference frame, as in figure 3.6. We assume that the plane of the wheel is aligned with AB at all times. Similar to the steered standard wheel, the castor wheel has two parameters that vary as a function of time. $\varphi(t)$ represents the wheel spin over time as before. $\beta(t)$ denotes the steering angle and orientation of AB over time.

For the castor wheel, the rolling constraint is identical to equation (3.15) because the offset axis plays no role during motion that is aligned with the wheel plane:

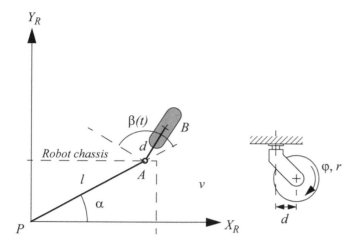

Figure 3.6
A castor wheel and its parameters.

$$\left[\sin(\alpha+\beta) \ -\cos(\alpha+\beta) \ (-l)\cos\beta\right] R(\theta)\dot{\xi}_I - r\dot{\varphi} \ = \ 0 \, . \tag{3.17}$$

The castor geometry does, however, have significant impact on the sliding constraint. The critical issue is that the lateral force on the wheel occurs at point A because this is the attachment point of the wheel to the chassis. Because of the offset ground contact point relative to A, the constraint that there be zero lateral movement would be wrong. Instead, the constraint is much like a rolling constraint, in that appropriate rotation of the vertical axis must take place:

$$\left[\cos(\alpha+\beta) \ \sin(\alpha+\beta) \ d+l\sin\beta\right] R(\theta)\dot{\xi}_I + d\dot{\beta} \ = \ 0 \, . \tag{3.18}$$

In equation (3.18), any motion orthogonal to the wheel plane must be balanced by an equivalent and opposite amount of castor steering motion. This result is critical to the success of castor wheels because by setting the value of $\dot{\beta}$ any arbitrary lateral motion can be acceptable. In a steered standard wheel, the steering action does not by itself cause a movement of the robot chassis. But in a castor wheel, the steering action itself moves the robot chassis because of the offset between the ground contact point and the vertical axis of rotation.

Figure 3.7
Office chair with five castor wheels.

More concisely, it can be surmised from equations (3.17) and (3.18) that, given *any* robot chassis motion $\dot{\xi}_I$, there exists some value for spin speed $\dot{\varphi}$ and steering speed $\dot{\beta}$ such that the constraints are met. Therefore, a robot with only castor wheels can move with any velocity in the space of possible robot motions. We term such systems *omnidirectional*.

A real-world example of such a system is the five-castor wheel office chair shown in figure 3.7. Assuming that all joints are able to move freely, you may select any motion vector on the plane for the chair and push it by hand. Its castor wheels will spin and steer as needed to achieve that motion without contact point sliding. By the same token, if each of the chair's castor wheels housed two motors, one for spinning and one for steering, then a control system would be able to move the chair along any trajectory in the plane. Thus, although the kinematics of castor wheels is somewhat complex, such wheels do not impose any real constraints on the kinematics of a robot chassis.

3.2.3.4 Swedish wheel

Swedish wheels have no vertical axis of rotation, yet are able to move *omnidirectionally* like the castor wheel. This is possible by adding a degree of freedom to the fixed standard wheel. Swedish wheels consist of a fixed standard wheel with rollers attached to the wheel perimeter with axes that are antiparallel to the main axis of the fixed wheel component. The exact angle γ between the roller axes and the wheel plane can vary, as shown in figure 3.8.

For example, given a Swedish 45-degree wheel, the motion vectors of the principal axis and the roller axes can be drawn as in figure 3.8. Since each axis can spin clockwise or counterclockwise, one can combine any vector along one axis with any vector along the other axis. These two axes are not necessarily independent (except in the case of the Swedish 90-degree wheel); however, it is visually clear that any desired direction of motion is achievable by choosing the appropriate two vectors.

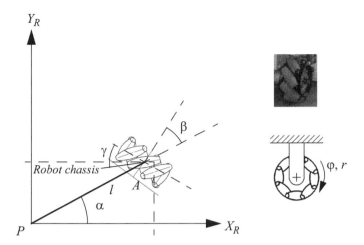

Figure 3.8
A Swedish wheel and its parameters.

The pose of a Swedish wheel is expressed exactly as in a fixed standard wheel, with the addition of a term, γ, representing the angle between the main wheel plane and the axis of rotation of the small circumferential rollers. This is depicted in figure 3.8 within the robot's reference frame.

Formulating the constraint for a Swedish wheel requires some subtlety. The instantaneous constraint is due to the specific orientation of the small rollers. The axis around which these rollers spin is a zero component of velocity at the contact point. That is, moving in that direction without spinning the main axis is not possible without sliding. The motion constraint that is derived looks identical to the rolling constraint for the fixed standard wheel in equation (3.12), except that the formula is modified by adding γ such that the effective direction along which the rolling constraint holds is along this zero component rather than along the wheel plane:

$$\left[\sin(\alpha+\beta+\gamma) \ -\cos(\alpha+\beta+\gamma) \ (-l)\cos(\beta+\gamma)\right]R(\theta)\dot{\xi}_I - r\dot{\varphi}\cos\gamma \ = \ 0 \ . \qquad (3.19)$$

Orthogonal to this direction the motion is not constrained because of the free rotation $\dot{\varphi}_{sw}$ of the small rollers.

$$\left[\cos(\alpha+\beta+\gamma) \ \sin(\alpha+\beta+\gamma) \ l\sin(\beta+\gamma)\right]R(\theta)\dot{\xi}_I - r\dot{\varphi}\sin\gamma - r_{sw}\dot{\varphi}_{sw} \ = \ 0 \ . \qquad (3.20)$$

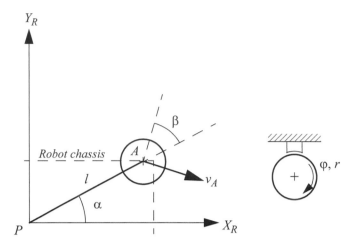

Figure 3.9
A spherical wheel and its parameters.

The behavior of this constraint and thereby the Swedish wheel changes dramatically as the value γ varies. Consider $\gamma = 0$. This represents the Swedish 90-degree wheel. In this case, the zero component of velocity is in line with the wheel plane, and so equation (3.19) reduces exactly to equation (3.12), the fixed standard wheel rolling constraint. But because of the rollers, there is no sliding constraint orthogonal to the wheel plane (see equation [3.20]). By varying the value of $\dot{\varphi}$, any desired motion vector can be made to satisfy equation (3.19), and therefore the wheel is omnidirectional. In fact, this special case of the Swedish design results in fully decoupled motion, in that the rollers and the main wheel provide orthogonal directions of motion.

At the other extreme, consider $\gamma = \pi/2$. In this case, the rollers have axes of rotation that are parallel to the main wheel axis of rotation. Interestingly, if this value is substituted for γ in equation (3.19) the result is the fixed standard wheel sliding constraint, equation (3.13). In other words, the rollers provide no benefit in terms of lateral freedom of motion since they are simply aligned with the main wheel. However, in this case the main wheel never needs to spin, and therefore the rolling constraint disappears. This is a degenerate form of the Swedish wheel, and therefore we assume in the remainder of this chapter that $\gamma \neq \pi/2$.

3.2.3.5 Spherical wheel
The final wheel type, a ball or spherical wheel, places no direct constraints on motion (figure 3.9). Such a mechanism has no principal axis of rotation, and therefore no appropriate rolling or sliding constraints exist. As with castor wheels and Swedish wheels, the spherical

wheel is clearly omnidirectional and places no constraints on the robot chassis kinematics. Therefore, equation (3.21) simply describes the roll rate of the ball in the direction of motion v_A of point A of the robot.

$$\left[\sin(\alpha + \beta) \quad -\cos(\alpha + \beta) \quad (-l)\cos\beta \right] R(\theta)\dot{\xi}_I - r\dot{\varphi} = 0. \tag{3.21}$$

By definition, the wheel rotation orthogonal to this direction is zero.

$$\left[\cos(\alpha + \beta) \quad \sin(\alpha + \beta) \quad l\sin\beta \right] R(\theta)\dot{\xi}_I = 0. \tag{3.22}$$

As can be seen, the equations for the spherical wheel are exactly the same as for the fixed standard wheel. However, the interpretation of equation (3.22) is different. The omnidirectional spherical wheel can have any arbitrary direction of movement, where the motion direction given by β is a free variable deduced from equation (3.22). Consider the case that the robot is in pure translation in the direction of Y_R. Then equation (3.22) reduces to $\sin(\alpha + \beta) = 0$, thus $\beta = -\alpha$, which makes sense for this special case.

3.2.4 Robot kinematic constraints

Given a mobile robot with M wheels we can now compute the kinematic constraints of the robot chassis. The key idea is that each wheel imposes zero or more constraints on robot motion, and so the process is simply one of appropriately combining all of the kinematic constraints arising from all of the wheels based on the placement of those wheels on the robot chassis.

We have categorized all wheels into five categories: (1) fixed and (2) steerable standard wheels, (3) castor wheels, (4) Swedish wheels, and (5) spherical wheels. But note from the wheel kinematic constraints in equations (3.17), (3.18), and (3.19) that the castor wheel, Swedish wheel, and spherical wheel impose *no* kinematic constraints on the robot chassis, since $\dot{\xi}_I$ can range freely in all of these cases owing to the internal wheel degrees of freedom.

Therefore, only fixed standard wheels and steerable standard wheels have impact on robot chassis kinematics and therefore require consideration when computing the robot's kinematic constraints. Suppose that the robot has a total of N standard wheels, comprising N_f fixed standard wheels and N_s steerable standard wheels. We use $\beta_s(t)$ to denote the variable steering angles of the N_s steerable standard wheels. In contrast, β_f refers to the orientation of the N_f fixed standard wheels as depicted in figure 3.4. In the case of wheel spin, both the fixed and steerable wheels have rotational positions around the horizontal axle that vary as a function of time. We denote the fixed and steerable cases separately as $\varphi_f(t)$ and $\varphi_s(t)$ and use $\varphi(t)$ as an aggregate matrix that combines both values:

$$\varphi(t) = \begin{bmatrix} \varphi_f(t) \\ \varphi_s(t) \end{bmatrix}. \tag{3.23}$$

The rolling constraints of all wheels can now be collected in a single expression:

$$J_1(\beta_s)R(\theta)\dot{\xi}_I - J_2\dot{\varphi} = 0. \tag{3.24}$$

This expression bears a strong resemblance to the rolling constraint of a single wheel, but substitutes matrices in lieu of single values, thus taking into account all wheels. J_2 is a constant diagonal $N \times N$ matrix whose entries are radii r of all standard wheels. $J_1(\beta_s)$ denotes a matrix with projections for all wheels to their motions along their individual wheel planes:

$$J_1(\beta_s) = \begin{bmatrix} J_{1f} \\ J_{1s}(\beta_s) \end{bmatrix}. \tag{3.25}$$

Note that $J_1(\beta_s)$ is only a function of β_s and not β_f. This is because the orientations of steerable standard wheels vary as a function of time, whereas the orientations of fixed standard wheels are constant. J_{1f} is therefore a constant matrix of projections for all fixed standard wheels. It has size ($N_f \times 3$), with each row consisting of the three terms in the three-matrix from equation (3.12) for each fixed standard wheel. $J_{1s}(\beta_s)$ is a matrix of size ($N_s \times 3$), with each row consisting of the three terms in the three-matrix from equation (3.15) for each steerable standard wheel.

In summary, equation (3.24) represents the constraint that all standard wheels must spin around their horizontal axis an appropriate amount based on their motions along the wheel plane so that rolling occurs at the ground contact point.

We use the same technique to collect the sliding constraints of all standard wheels into a single expression with the same structure as equations (3.13) and (3.16):

$$C_1(\beta_s)R(\theta)\dot{\xi}_I = 0. \tag{3.26}$$

$$C_1(\beta_s) = \begin{bmatrix} C_{1f} \\ C_{1s}(\beta_s) \end{bmatrix}. \tag{3.27}$$

C_{1f} and C_{1s} are ($N_f \times 3$) and ($N_s \times 3$) matrices whose rows are the three terms in the three-matrix of equations (3.13) and (3.16) for all fixed and steerable standard wheels. Thus equation (3.26) is a constraint over all standard wheels that their components of motion orthogonal to their wheel planes must be zero. This sliding constraint over all standard wheels has the most significant impact on defining the overall maneuverability of the robot chassis, as explained in the next section.

3.2.5 Examples: Robot kinematic models and constraints

In section 3.2.2 we presented a forward kinematic solution for $\dot{\xi}_I$ in the case of a simple differential-drive robot by combining each wheel's contribution to robot motion. We can now use the tools presented above to construct the same kinematic expression by direct application of the rolling constraints for every wheel type. We proceed with this technique applied again to the differential drive robot, enabling verification of the method as compared to the results of section 3.2.2. Then we proceed to the case of the three-wheeled omnidirectional robot.

3.2.5.1 A differential-drive robot example

First, refer to equations (3.24) and (3.26). These equations relate robot motion to the rolling and sliding constraints $J_1(\beta_s)$ and $C_1(\beta_s)$, and the wheel spin speed of the robot's wheels, $\dot{\varphi}$. Fusing these two equations yields the following expression:

$$\begin{bmatrix} J_1(\beta_s) \\ C_1(\beta_s) \end{bmatrix} R(\theta)\dot{\xi}_I = \begin{bmatrix} J_2\dot{\varphi} \\ 0 \end{bmatrix}. \tag{3.28}$$

Once again, consider the differential drive robot in figure 3.3. We will construct $J_1(\beta_s)$ and $C_1(\beta_s)$ directly from the rolling constraints of each wheel. The castor is unpowered and is free to move in any direction, so we ignore this third point of contact altogether. The two remaining drive wheels are not steerable, and therefore $J_1(\beta_s)$ and $C_1(\beta_s)$ simplify to J_{1f} and C_{1f} respectively. To employ the fixed standard wheel's rolling constraint formula, equation (3.12), we must first identify each wheel's values for α and β. Suppose that the robot's local reference frame is aligned such that the robot moves forward along $+X_R$, as shown in figure 3.1. In this case, for the right wheel $\alpha = -\pi/2$, $\beta = \pi$, and for the left wheel, $\alpha = \pi/2$, $\beta = 0$. Note the value of β for the right wheel is necessary to ensure that positive spin causes motion in the $+X_R$ direction (figure 3.4). Now we can compute the J_{1f} and C_{1f} matrix using the matrix terms from equations (3.12) and (3.13). Because the two fixed standard wheels are parallel, equation (3.13) results in only one independent equation, and equation (3.28) gives:

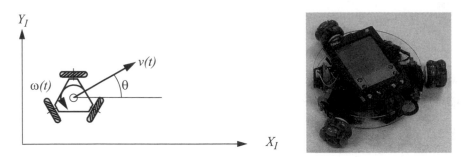

Figure 3.10
A three-wheel omnidrive robot developed by Carnegie Mellon University (www.cs.cmu.edu/~pprk).

$$
\begin{bmatrix} \begin{bmatrix} 1 & 0 & l \\ 1 & 0 & -l \end{bmatrix} \\ \begin{bmatrix} 0 & 1 & 0 \end{bmatrix} \end{bmatrix} R(\theta)\dot{\xi}_I = \begin{bmatrix} J_2\dot{\varphi} \\ 0 \end{bmatrix}. \tag{3.29}
$$

Inverting equation (3.29) yields the kinematic equation specific to our differential drive robot:

$$
\dot{\xi}_I = R(\theta)^{-1} \begin{bmatrix} 1 & 0 & l \\ 1 & 0 & -l \\ 0 & 1 & 0 \end{bmatrix}^{-1} \begin{bmatrix} J_2\dot{\varphi} \\ 0 \end{bmatrix} = R(\theta)^{-1} \begin{bmatrix} \dfrac{1}{2} & \dfrac{1}{2} & 0 \\ 0 & 0 & 1 \\ \dfrac{1}{2l} & -\dfrac{1}{2l} & 0 \end{bmatrix} \begin{bmatrix} J_2\dot{\varphi} \\ 0 \end{bmatrix}. \tag{3.30}
$$

This demonstrates that, for the simple differential-drive case, the combination of wheel rolling and sliding constraints describes the kinematic behavior, based on our manual calculation in section 3.2.2.

3.2.5.2 An omnidirectional robot example

Consider the omniwheel robot shown in figure 3.10. This robot has three Swedish 90-degree wheels, arranged radially symmetrically, with the rollers perpendicular to each main wheel.

First we must impose a specific local reference frame upon the robot. We do so by choosing point P at the center of the robot, then aligning the robot with the local reference

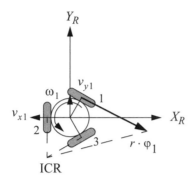

Figure 3.11
The local reference frame plus detailed parameters for wheel 1.

frame such that X_R is colinear with the axis of wheel 2. Figure 3.11 shows the robot and its local reference frame arranged in this manner.

We assume that the distance between each wheel and P is l, and that all three wheels have the same radius, r. Once again, the value of $\dot{\xi}_I$ can be computed as a combination of the rolling constraints of the robot's three omnidirectional wheels, as in equation (3.28). As with the differential-drive robot, since this robot has no steerable wheels, $J_1(\beta_s)$ simplifies to J_{1f}:

$$\dot{\xi}_I = R(\theta)^{-1} J_{1f}^{-1} J_2 \dot{\varphi} . \tag{3.31}$$

We calculate J_{1f} using the matrix elements of the rolling constraints for the Swedish wheel, given by equation (3.19). But to use these values, we must establish the values α, β, γ for each wheel. Referring to figure (3.8), we can see that $\gamma = 0$ for the Swedish 90-degree wheel. Note that this immediately simplifies equation (3.19) to equation (3.12), the rolling constraints of a fixed standard wheel. Given our particular placement of the local reference frame, the value of α for each wheel is easily computed: $(\alpha_1 = \pi/3)$, $(\alpha_2 = \pi)$, $(\alpha_3 = -\pi/3)$. Furthermore, $\beta = 0$ for all wheels because the wheels are tangent to the robot's circular body. Constructing and simplifying J_{1f} using equation (3.12) yields:

$$
J_{1f} = \begin{bmatrix} \sin\dfrac{\pi}{3} & -\cos\dfrac{\pi}{3} & -l \\[2mm] 0 & -\cos\pi & -l \\[2mm] \sin-\dfrac{\pi}{3} & -\cos-\dfrac{\pi}{3} & -l \end{bmatrix} = \begin{bmatrix} \dfrac{\sqrt{3}}{2} & -\dfrac{1}{2} & -l \\[2mm] 0 & 1 & -l \\[2mm] -\dfrac{\sqrt{3}}{2} & -\dfrac{1}{2} & -l \end{bmatrix}. \tag{3.32}
$$

Once again, computing the value of $\dot{\xi}_I$ requires calculating the inverse, J_{1f}^{-1}, as needed in equation (3.31). One approach would be to apply rote methods for calculating the inverse of a 3×3 square matrix. A second approach would be to compute the contribution of each Swedish wheel to chassis motion, as shown in section 3.2.2. We leave this process as an exercise for the enthusiast. Once the inverse is obtained, $\dot{\xi}_I$ can be isolated:

$$
\dot{\xi}_I = R(\theta)^{-1} \begin{bmatrix} \dfrac{1}{\sqrt{3}} & 0 & -\dfrac{1}{\sqrt{3}} \\[2mm] -\dfrac{1}{3} & \dfrac{2}{3} & -\dfrac{1}{3} \\[2mm] -\dfrac{1}{3l} & -\dfrac{1}{3l} & -\dfrac{1}{3l} \end{bmatrix} J_2 \dot{\varphi}. \tag{3.33}
$$

Consider a specific omnidrive chassis with $l = 1$ and $r = 1$ for all wheels. The robot's local reference frame and global reference frame are aligned, so that $\theta = 0$. If wheels 1, 2, and 3 spin at speeds $(\varphi_1 = 4)$, $(\varphi_2 = 1)$, $(\varphi_3 = 2)$, what is the resulting motion of the whole robot? Using the equation above, the answer can be calculated readily:

$$
\dot{\xi}_I = \begin{bmatrix} \dot{x} \\ \dot{y} \\ \dot{\theta} \end{bmatrix} = \begin{bmatrix} 1 & 0 & 0 \\ 0 & 1 & 0 \\ 0 & 0 & 1 \end{bmatrix} \begin{bmatrix} \dfrac{1}{\sqrt{3}} & 0 & -\dfrac{1}{\sqrt{3}} \\[2mm] -\dfrac{1}{3} & \dfrac{2}{3} & -\dfrac{1}{3} \\[2mm] -\dfrac{1}{3} & \dfrac{1}{3} & -\dfrac{1}{3} \end{bmatrix} \begin{bmatrix} 1 & 0 & 0 \\ 0 & 1 & 0 \\ 0 & 0 & 1 \end{bmatrix} \begin{bmatrix} 4 \\ 1 \\ 2 \end{bmatrix} = \begin{bmatrix} \dfrac{2}{\sqrt{3}} \\[2mm] -\dfrac{4}{3} \\[2mm] -\dfrac{7}{3} \end{bmatrix}. \tag{3.34}
$$

So this robot will move instantaneously along the x-axis with positive speed and along the y axis with negative speed while rotating clockwise. We can see from the preceding examples that robot motion can be predicted by combining the rolling constraints of individual wheels.

The sliding constraints comprising $C_1(\beta_s)$ can be used to go even further, enabling us to evaluate the maneuverability and workspace of the robot rather than just its predicted motion. Next, we examine methods for using the sliding constraints, sometimes in conjunction with rolling constraints, to generate powerful analyses of the maneuverability of a robot chassis.

3.3   Mobile Robot Maneuverability

The kinematic mobility of a robot chassis is its ability to directly move in the environment. The basic constraint limiting mobility is the rule that every wheel must satisfy its sliding constraint. Therefore, we can formally derive robot mobility by starting from equation (3.26).

In addition to instantaneous kinematic motion, a mobile robot is able to further manipulate its position, over time, by steering steerable wheels. As we will see in section 3.3.3, the overall maneuverability of a robot is thus a combination of the mobility available based on the kinematic sliding constraints of the standard wheels, plus the additional freedom contributed by steering and spinning the steerable standard wheels.

3.3.1   Degree of mobility

Equation (3.26) imposes the constraint that every wheel must avoid any lateral slip. Of course, this holds separately for each and every wheel, and so it is possible to specify this constraint separately for fixed and for steerable standard wheels:

$$C_{1f}R(\theta)\dot{\xi}_I = 0 .$$
(3.35)

$$C_{1s}(\beta_s)R(\theta)\dot{\xi}_I = 0 .$$
(3.36)

For both of these constraints to be satisfied, the motion vector $R(\theta)\dot{\xi}_I$ must belong to the *null space* of the projection matrix $C_1(\beta_s)$, which is simply a combination of C_{1f} and C_{1s}. Mathematically, the null space of $C_1(\beta_s)$ is the space N such that for any vector n in N, $C_1(\beta_s)n = 0$. If the kinematic constraints are to be honored, then the motion of the robot must always be within this space N. The kinematic constraints [equations (3.35) and (3.36)] can also be demonstrated geometrically using the concept of a robot's *instantaneous center of rotation* (ICR).

Consider a single standard wheel. It is forced by the sliding constraint to have zero lateral motion. This can be shown geometrically by drawing a *zero motion line* through its horizontal axis, perpendicular to the wheel plane (figure 3.12). At any given instant, wheel motion along the zero motion line must be zero. In other words, the wheel must be moving

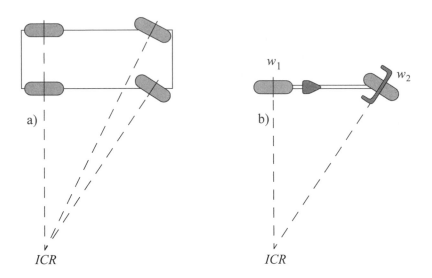

Figure 3.12
(a) Four-wheel with carlike Ackerman steering. (b) Bicycle.

instantaneously along some circle of radius R such that the center of that circle is located on the zero motion line. This center point, called the instantaneous center of rotation, may lie anywhere along the zero motion line. When R is at infinity, the wheel moves in a straight line.

A robot such as the Ackerman vehicle in figure 3.12a can have several wheels, but it must always have a single ICR. Because all of its zero motion lines meet at a single point, there is a single solution for robot motion, placing the ICR at this meet point.

This ICR geometric construction demonstrates how robot mobility is a function of the number of constraints on the robot's motion, not the number of wheels. In figure 3.12b, the bicycle shown has two wheels, w_1 and w_2. Each wheel contributes a constraint, or a zero motion line. Taken together, the two constraints result in a single point as the only remaining solution for the ICR. This is because the two constraints are independent, and thus each further constrains overall robot motion.

But in the case of the differential drive robot in figure 3.13a, the two wheels are aligned along the same horizontal axis. Therefore, the ICR is constrained to lie along a line, not at a specific point. In fact, the second wheel imposes no additional kinematic constraints on robot motion since its zero motion line is identical to that of the first wheel. Thus, although the bicycle and differential-drive chassis have the same number of nonomnidirectional wheels, the former has two independent kinematic constraints while the latter has only one.

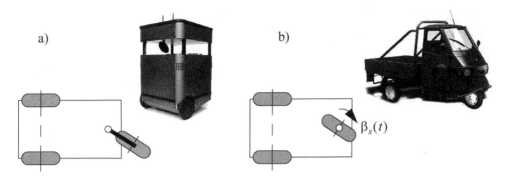

Figure 3.13
(a) Differential-drive robot with two individually motorized wheels and a castor wheel, e.g., the Pygmalion robot at EPFL. (b) Tricycle with two fixed standard wheels and one steered standard wheel, e.g., Piaggio minitransporter.

The Ackerman vehicle of figure 3.12a demonstrates another way in which a wheel may be unable to contribute an independent constraint to the robot kinematics. This vehicle has two steerable standard wheels. Given the instantaneous position of just one of these steerable wheels and the position of the fixed rear wheels, there is only a single solution for the ICR. The position of the second steerable wheel is absolutely constrained by the ICR. Therefore, it offers no independent constraints to robot motion.

Robot chassis kinematics is therefore a function of the set of *independent* constraints arising from all standard wheels. The mathematical interpretation of independence is related to the *rank* of a matrix. Recall that the rank of a matrix is the largest number of linearly independent rows or columns. Equation (3.26) represents all sliding constraints imposed by the wheels of the mobile robot. Therefore, $rank \left[C_1(\beta_s) \right]$ is the number of independent constraints.

The greater the number of independent constraints, and therefore the greater the rank of $C_1(\beta_s)$, the more constrained is the mobility of the robot. For example, consider a robot with a single fixed standard wheel. Remember that we consider only standard wheels. This robot may be a unicycle or it may have several Swedish wheels; however, it has exactly one fixed standard wheel. The wheel is at a position specified by parameters α, β, l relative to the robot's local reference frame. $C_1(\beta_s)$ comprises C_{1f} and C_{1s}. However, since there are no steerable standard wheels, C_{1s} is empty, and therefore $C_1(\beta_s)$ contains only C_{1f}. Because there is one fixed standard wheel, this matrix has a rank of one, and therefore this robot has a single independent constrain on mobility:

$$C_1(\beta_s) = C_{1f} = \left[\cos(\alpha + \beta) \ \sin(\alpha + \beta) \ l\sin\beta \right].$$ (3.37)

Now let us add an additional fixed standard wheel to create a differential-drive robot by constraining the second wheel to be aligned with the same horizontal axis as the original wheel. Without loss of generality, we can place point P at the midpoint between the centers of the two wheels. Given α_1, β_1, l_1 for wheel w_1 and α_2, β_2, l_2 for wheel w_2, it holds geometrically that $\{(l_1 = l_2), (\beta_1 = \beta_2 = 0), (\alpha_1 + \pi = \alpha_2)\}$. Therefore, in this case, the matrix $C_1(\beta_s)$ has two constraints but a rank of one:

$$C_1(\beta_s) = C_{1f} = \begin{bmatrix} \cos(\alpha_1) & \sin(\alpha_1) & 0 \\ \cos(\alpha_1 + \pi) & \sin(\alpha_1 + \pi) & 0 \end{bmatrix}. \tag{3.38}$$

Alternatively, consider the case when w_2 is placed in the wheel plane of w_1 but with the same orientation, as in a bicycle with the steering locked in the forward position. We again place point P between the two wheel centers and orient the wheels such that they lie on axis x_1. This geometry implies that $\{(l_1 = l_2), (\beta_1 = \beta_2 = \pi/2), (\alpha_1 = 0), (\alpha_2 = \pi)\}$ and, therefore, the matrix $C_1(\beta_s)$ retains two independent constraints and has a rank of two:

$$C_1(\beta_s) = C_{1f} = \begin{bmatrix} \cos(\pi/2) & \sin(\pi/2) & l_1\sin(\pi/2) \\ \cos(3\pi/2) & \sin(3\pi/2) & l_1\sin(\pi/2) \end{bmatrix} = \begin{bmatrix} 0 & 1 & l_1 \\ 0 & -1 & l_1 \end{bmatrix}. \tag{3.39}$$

In general, if $rank\begin{bmatrix} C_{1f} \end{bmatrix} > 1$ then the vehicle can, at best, only travel along a circle or along a straight line. This configuration means that the robot has two or more independent constraints due to fixed standard wheels that do not share the same horizontal axis of rotation. Because such configurations have only a degenerate form of mobility in the plane, we do not consider them in the remainder of this chapter. Note, however, that some degenerate configurations such as the four-wheeled slip/skid steering system are useful in certain environments, such as on loose soil and sand, even though they fail to satisfy sliding constraints. Not surprisingly, the price that must be paid for such violations of the sliding constraints is that dead reckoning based on odometry becomes less accurate and power efficiency is reduced dramatically.

In general, a robot will have zero or more fixed standard wheels and zero or more steerable standard wheels. We can therefore identify the possible range of rank values for any robot: $0 \leq rank\begin{bmatrix} C_1(\beta_s) \end{bmatrix} \leq 3$. Consider the case $rank\begin{bmatrix} C_1(\beta_s) \end{bmatrix} = 0$. This is possible only if there are zero independent kinematic constraints in $C_1(\beta_s)$. In this case there are neither fixed nor steerable standard wheels attached to the robot frame: $N_f = N_s = 0$.

Consider the other extreme, $rank\begin{bmatrix} C_1(\beta_s) \end{bmatrix} = 3$. This is the maximum possible rank since the kinematic constraints are specified along three degrees of freedom (i.e., the constraint matrix is three columns wide). Therefore, there cannot be more than three independent constraints. In fact, when $rank\left[C_1(\beta_s)\right] = 3$, then the robot is completely constrained in all directions and is therefore degenerate, since motion in the plane is totally impossible.

Now we are ready to formally define a robot's *degree of mobility* δ_m:

$$\delta_m = dimN\left[C_1(\beta_s)\right] = 3 - rank\left[C_1(\beta_s)\right]. \tag{3.40}$$

The dimensionality of the null space ($dimN$) of matrix $C_1(\beta_s)$ is a measure of the number of degrees of freedom of the robot chassis that can be immediately manipulated through changes in wheel velocity. It is logical, therefore, that δ_m must range between 0 and 3.

Consider an ordinary differential-drive chassis. On such a robot there are two fixed standard wheels sharing a common horizontal axis. As discussed earlier, the second wheel adds no independent kinematic constraints to the system. Therefore, $rank\left[C_1(\beta_s)\right] = 1$ and $\delta_m = 2$. This fits with intuition: a differential drive robot can control both the rate of its change in orientation and its forward/reverse speed, *simply by manipulating wheel velocities*. In other words, its *ICR* is constrained to lie on the infinite line extending from its wheels' horizontal axles.

In contrast, consider a bicycle chassis. This configuration consists of one fixed standard wheel and one steerable standard wheel. In this case, each wheel contributes an independent sliding constraint to $C_1(\beta_s)$. Therefore, $\delta_m = 1$. Note that the bicycle has the same total number of nonomnidirectional wheels as the differential-drive chassis, and indeed one of its wheels is steerable. Yet it has one less degree of mobility. Upon reflection, this is appropriate. A bicycle only has control over its forward/reverse speed by direct manipulation of wheel velocities. Only by steering can the bicycle change its *ICR*.

As expected, based on equation (3.40), any robot consisting only of omnidirectional wheels such as Swedish or spherical wheels will have the maximum mobility, $\delta_m = 3$. Such a robot can directly manipulate all three degrees of freedom.

3.3.2 Degree of steerability

The degree of mobility defined above quantifies the degrees of controllable freedom based on changes to wheel velocity. Steering can also have an eventual impact on a robot chassis pose ξ, although the impact is indirect because after changing the angle of a steerable standard wheel, the robot must move for the change in steering angle to have impact on pose.

As with mobility, we care about the number of independently controllable steering parameters when defining the *degree of steerability* δ_s:

$$\delta_s = rank\left[C_{1s}(\beta_s)\right]. \tag{3.41}$$

Recall that in the case of mobility, an increase in the rank of $C_1(\beta_s)$ implied more kinematic constraints and thus a less mobile system. In the case of steerability, an increase in the rank of $C_{1s}(\beta_s)$ implies more degrees of steering freedom and thus greater eventual maneuverability. Since $C_1(\beta_s)$ includes $C_{1s}(\beta_s)$, this means that a steered standard wheel can both decrease mobility and increase steerability: its particular orientation at any instant imposes a kinematic constraint, but its ability to change that orientation can lead to additional trajectories.

The range of δ_s can be specified: $0 \le \delta_s \le 2$. The case $\delta_s = 0$ implies that the robot has no steerable standard wheels, $N_s = 0$. The case $\delta_s = 1$ is most common when a robot configuration includes one or more steerable standard wheels.

For example, consider an ordinary automobile. In this case $N_f = 2$ and $N_s = 2$. But the fixed wheels share a common axle and so $rank\left[C_{1f}\right] = 1$. The fixed wheels and any one of the steerable wheels constrain the ICR to be a point along the line extending from the rear axle. Therefore, the second steerable wheel cannot impose any independent kinematic constraint and so $rank\left[C_{1s}(\beta_s)\right] = 1$. In this case $\delta_m = 1$ and $\delta_s = 1$.

The case $\delta_s = 2$ is possible only in robots with no fixed standard wheels: $N_f = 0$. Under these circumstances, it is possible to create a chassis with two separate steerable standard wheels, like a pseudobicycle (or the two-steer) in which both wheels are steerable. Then, orienting one wheel constrains the ICR to a line, while the second wheel can constrain the ICR to any point along that line. Interestingly, this means that the $\delta_s = 2$ implies that the robot can place its ICR anywhere on the ground plane.

3.3.3 Robot maneuverability

The overall degrees of freedom that a robot can manipulate, called the *degree of maneuverability* δ_M, can be readily defined in terms of mobility and steerability:

$$\delta_M = \delta_m + \delta_s. \tag{3.42}$$

Therefore, maneuverability includes both the degrees of freedom that the robot manipulates directly through wheel velocity and the degrees of freedom that it indirectly manipulates by changing the steering configuration and moving. Based on the investigations of the previous sections, one can draw the basic types of wheel configurations. They are depicted in figure 3.14.

Note that two robots with the same δ_M are not necessarily equivalent. For example, differential drive and tricycle geometries (figure 3.13) have equal maneuverability $\delta_M = 2$. In differential drive all maneuverability is the result of direct mobility because $\delta_m = 2$ and $\delta_s = 0$. In the case of a tricycle the maneuverability results from steering also: $\delta_m = 1$ and $\delta_s = 1$. Neither of these configurations allows the ICR to range anywhere on the plane. In both cases, the ICR must lie on a predefined line with respect to the robot refer-

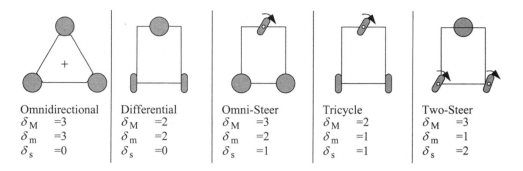

Figure 3.14
The five basic types of three-wheel configurations. The spherical wheels can be replaced by castor or Swedish wheels without influencing maneuverability. More configurations with various numbers of wheels are found in chapter 2.

ence frame. In the case of differential drive, this line extends from the common axle of the two fixed standard wheels, with the differential wheel velocities setting the ICR point on this line. In a tricycle, this line extends from the shared common axle of the fixed wheels, with the steerable wheel setting the ICR point along this line.

More generally, for any robot with $\delta_M = 2$ the ICR is always constrained to lie on a line and for any robot with $\delta_M = 3$ the ICR can be set to any point on the plane.

One final example will demonstrate the use of the tools we have developed here. One common robot configuration for indoor mobile robotics research is the *synchro drive* configuration (figure 2.28). Such a robot has two motors and three wheels that are locked together. One motor provides power for spinning all three wheels, while the second motor provides power for steering all three wheels. In a three-wheeled synchro drive robot $N_f = 0$ and $N_s = 3$. Therefore, $rank\left[C_{1s}(\beta_s)\right]$ can be used to determine both δ_m and δ_s. The three wheels do not share a common axle; therefore, two of the three contribute independent sliding constraints. The third must be dependent on these two constraints for motion to be possible. Therefore, $rank\left[C_{1s}(\beta_s)\right] = 2$ and $\delta_m = 1$. This is intuitively correct. A synchro drive robot with the steering frozen manipulates only one degree of freedom, consisting of traveling back and forth on a straight line.

However, an interesting complication occurs when considering δ_s. Based on equation (3.41) the robot should have $\delta_s = 2$. Indeed, for a three-wheel-steering robot with the geometric configuration of a synchro drive robot this would be correct. However, we have additional information: in a synchro drive configuration a single motor steers all three wheels using a belt drive. Therefore, although ideally, if the wheels were independently steerable, then the system would achieve $\delta_s = 2$; in the case of synchro drive the drive system further constrains the kinematics such that in reality $\delta_s = 1$. Finally, we can compute maneuverability based on these values: $\delta_M = 2$ for a synchro drive robot.

This result implies that a synchro drive robot can only manipulate, in total, two degrees of freedom. In fact, if the reader reflects on the wheel configuration of a synchro drive robot, it will become apparent that there is no way for the chassis orientation to change. Only the $x - y$ position of the chassis can be manipulated, and so, indeed, a synchro drive robot has only two degrees of freedom, in agreement with our mathematical conclusion.

3.4   Mobile Robot Workspace

For a robot, maneuverability is equivalent to its control degrees of freedom. But the robot is situated in some environment, and the next question is to situate our analysis in the environment. We care about the ways in which the robot can use its control degrees of freedom to position itself in the environment. For instance, consider the Ackerman vehicle, or automobile. The total number of control degrees of freedom for such a vehicle is $\delta_M = 2$, one for steering and the second for actuation of the drive wheels. But what are the total degrees of freedom of the vehicle in its environment? In fact, it is three: the car can position itself on the plane at any x, y point and with any angle θ.

Thus, identifying a robot's space of possible configurations is important because, surprisingly, it can exceed δ_M. In addition to *workspace*, we care about how the robot is able to move between various configurations: What are the types of paths that it can follow and, furthermore, what are its possible trajectories through this configuration space? In the remainder of this discussion, we move away from inner kinematic details such as wheels and focus instead on the robot chassis pose and the chassis degrees of freedom. With this in mind, let us place the robot in the context of its workspace now.

3.4.1   Degrees of freedom

In defining the workspace of a robot, it is useful to first examine its *admissible velocity space*. Given the kinematic constraints of the robot, its velocity space describes the independent components of robot motion that the robot can control. For example, the velocity space of a unicycle can be represented with two axes, one representing the instantaneous forward speed of the unicycle and the second representing the instantaneous change in orientation, $\dot{\theta}$, of the unicycle.

The number of dimensions in the velocity space of a robot is the number of independently achievable velocities. This is also called the *differential degrees of freedom (DDOF)*. A robot's *DDOF* is *always* equal to its degree of mobility δ_m. For example, a bicycle has the following degree of maneuverability: $\delta_M = \delta_m + \delta_s = 1 + 1 = 2$. The *DDOF* of a bicycle is indeed 1.

In contrast to a bicycle, consider an *omnibot*, a robot with three Swedish wheels. We know that in this case there are zero standard wheels, and therefore $\delta_M = \delta_m + \delta_s = 3 + 0 = 3$. So, the omnibot has three differential degrees of freedom. This is appropriate, given that because such a robot has no kinematic motion constraints, it is able to independently set all three pose variables: $\dot{x}, \dot{y}, \dot{\theta}$.

Given the difference in *DDOF* between a bicycle and an omnibot, consider the overall degrees of freedom in the workspace of each configuration. The omnibot can achieve any pose (x, y, θ) in its environment and can do so by directly achieving the goal positions of all three axes simultaneously because *DDOF* = 3. Clearly, it has a workspace with *DOF* = 3.

Can a bicycle achieve any pose (x, y, θ) in its environment? It can do so, but achieving some goal points may require more time and energy than an equivalent omnibot. For example, if a bicycle configuration must move laterally 1 m, the simplest successful maneuver would involve either a spiral or a back-and-forth motion similar to *parallel parking* of automobiles. Nevertheless, a bicycle can achieve any (x, y, θ), and therefore the workspace of a bicycle has *DOF* = 3 as well.

Clearly, there is an inequality relation at work: $DDOF \le \delta_M \le DOF$. Although the dimensionality of a robot's workspace is an important attribute, it is clear from the preceding example that the particular paths available to a robot matter as well. Just as workspace *DOF* governs the robot's ability to achieve various poses, so the robot's *DDOF* governs its ability to achieve various paths.

3.4.2 Holonomic robots

In the robotics community, when describing the path space of a mobile robot, often the concept of holonomy is used. The term *holonomic* has broad applicability to several mathematical areas, including differential equations, functions, and constraint expressions. In mobile robotics, the term refers specifically to the kinematic constraints of the robot chassis. A *holonomic robot* is a robot that has zero nonholonomic kinematic constraints. Conversely, a *nonholonomic robot* is a robot with one or more nonholonomic kinematic constraints.

A *holonomic kinematic constraint* can be expressed as an explicit function of position variables only. For example, in the case of a mobile robot with a single fixed standard wheel, a holonomic kinematic constraint would be expressible using $\alpha_1, \beta_1, l_1, r_1, \varphi_1,$ x, y, θ only. Such a constraint may not use derivatives of these values, such as $\dot{\varphi}$ or $\dot{\xi}$. A *nonholonomic kinematic constraint* requires a differential relationship, such as the derivative of a position variable. Furthermore, it cannot be integrated to provide a constraint in terms of the position variables only. Because of this latter point of view, nonholonomic systems are often called *nonintegrable* systems.

Consider the fixed standard wheel sliding constraint:

$$\left[\cos(\alpha + \beta)\ \sin(\alpha + \beta)\ l\sin\beta\right] R(\theta)\dot{\xi}_I = 0.\tag{3.43}$$

This constraint must use robot motion $\dot{\xi}$ rather than pose ξ because the point is to constrain robot motion perpendicular to the wheel plane to be zero. The constraint is nonintegrable, depending explicitly on robot motion. Therefore, the sliding constraint is a nonholonomic constraint. Consider a bicycle configuration, with one fixed standard wheel and one steerable standard wheel. Because the fixed wheel sliding constraint will be in force for such a robot, we can conclude that the bicycle is a nonholonomic robot.

But suppose that one locks the bicycle steering system, so that it becomes two fixed standard wheels with separate but parallel axes. We know that $\delta_M = 1$ for such a configuration. Is it nonholonomic? Although it may not appear so because of the sliding and rolling constraints, the locked bicycle is actually holonomic. Consider the workspace of this locked bicycle. It consists of a single infinite line along which the bicycle can move (assuming the steering was frozen straight ahead). For formulaic simplicity, assume that this infinite line is aligned with X_I in the global reference frame and that $\{\beta_{1,2} = \pi/2,\ \alpha_1 = 0,\ \alpha_2 = \pi\}$. In this case the sliding constraints of both wheels can be replaced with an equally complete set of constraints on the robot pose: $\{y = 0,\ \theta = 0\}$. This eliminates two nonholonomic constraints, corresponding to the sliding constraints of the two wheels.

The only remaining nonholonomic kinematic constraints are the rolling constraints for each wheel:

$$\left[-\sin(\alpha + \beta)\ \cos(\alpha + \beta)\ l\cos\beta\right] R(\theta)\dot{\xi}_I + r\dot{\varphi} = 0.\tag{3.44}$$

This constraint is required for each wheel to relate the speed of wheel spin to the speed of motion projected along the wheel plane. But in the case of our locked bicycle, given the initial rotational position of a wheel at the origin, φ_o, we can replace this constraint with one that directly relates position on the line, x, with wheel rotation angle, φ: $\varphi = (x/r) + \varphi_o$.

The locked bicycle is an example of the first type of holonomic robot—where constraints do exist but are all holonomic kinematic constraints. This is the case for all holonomic robots with $\delta_M < 3$. The second type of holonomic robot exists when there are no kinematic constraints, that is, $N_f = 0$ and $N_s = 0$. Since there are no kinematic constraints, there are also no nonholonomic kinematic constraints, and so such a robot is always holonomic. This is the case for all holonomic robots with $\delta_M = 3$.

An alternative way to describe a holonomic robot is based on the relationship between the differential degrees of freedom of a robot and the degrees of freedom of its workspace: *a robot is holonomic if and only if* $DDOF = DOF$. Intuitively, this is because it is only through nonholonomic constraints (imposed by steerable or fixed standard wheels) that a robot can achieve a workspace with degrees of freedom exceeding its differential degrees of freedom, $DOF > DDOF$. Examples include differential drive and bicycle/tricycle configurations.

In mobile robotics, useful chassis generally must achieve poses in a workspace with dimensionality 3, so in general we require $DOF = 3$ for the chassis. But the "holonomic" abilities to maneuver around obstacles without affecting orientation and to track at a target while following an arbitrary path are important additional considerations. For these reasons, the particular form of holonomy most relevant to mobile robotics is that of $DDOF = DOF = 3$. We define this class of robot configurations as omnidirectional: an *omnidirectional robot* is a holonomic robot with $DDOF = 3$.

3.4.3 Path and trajectory considerations

In mobile robotics, we care not only about the robot's ability to reach the required final configurations but also about *how* it gets there. Consider the issue of a robot's ability to follow paths: in the best case, a robot should be able to trace any path through its workspace of poses. Clearly, any omnidirectional robot can do this because it is holonomic in a three-dimensional workspace. Unfortunately, omnidirectional robots must use unconstrained wheels, limiting the choice of wheels to Swedish wheels, castor wheels, and spherical wheels. These wheels have not yet been incorporated into designs that allow far larger amounts of ground clearance and suspensions. Although powerful from a path space point of view, they are thus much less common than fixed and steerable standard wheels, mainly because their design and fabrication are somewhat complex and expensive.

Additionally, nonholonomic constraints might drastically improve stability of movements. Consider an omnidirectional vehicle driving at high speed on a curve with constant diameter. During such a movement the vehicle will be exposed to a non-negligible centripetal force. This lateral force pushing the vehicle out of the curve has to be counteracted by the motor torque of the omnidirectional wheels. In case of motor or control failure, the vehicle will be thrown out of the curve. However, for a carlike robot with kinematic constraints, the lateral forces are passively counteracted through the sliding constraints, mitigating the demands on motor torque.

But recall an earlier example of high maneuverability using standard wheels: the bicycle on which both wheels are steerable, often called the *two-steer*. This vehicle achieves a degree of steerability of 2, resulting in a high degree of maneuverability: $\delta_M = \delta_m + \delta_s = 1 + 2 = 3$. Interestingly, this configuration is not holonomic, yet it has a high degree of maneuverability in a workspace with $DOF = 3$.

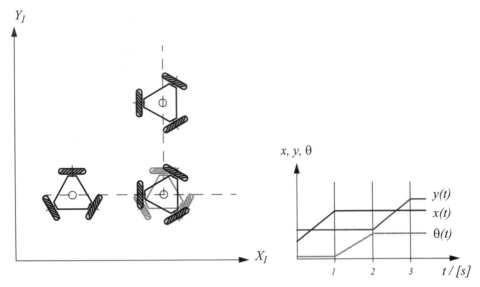

Figure 3.15
Example of robot trajectory with omnidirectional robot: move for 1 second with constant speed of 1 m/s along axis X_I; change orientation counterclockwise 90 degree, in 1 second; move for 1 second with constant speed of 1 m/s along axis Y_I.

The maneuverability result, $\delta_M = 3$, means that the two-steer can select any *ICR* by appropriately steering its two wheels. So, how does this compare to an omnidirectional robot? The ability to manipulate its *ICR* in the plane means that the two-steer can follow *any* path in its workspace. More generally, any robot with $\delta_M = 3$ can follow any path in its workspace from its initial pose to its final pose. An omnidirectional robot can also follow any path in its workspace and, not surprisingly, since $\delta_m = 3$ in an omnidirectional robot, then it must follow that $\delta_M = 3$.

But there is still a difference between a degree of freedom granted by steering versus by direct control of wheel velocity. This difference is clear in the context of *trajectories* rather than paths. A trajectory is like a path, except that it occupies an additional dimension: time. Therefore, for an omnidirectional robot on the ground plane, a path generally denotes a trace through a 3D space of pose; for the same robot, a trajectory denotes a trace through the 4D space of pose plus time.

For example, consider a goal trajectory in which the robot moves along axis X_I at a constant speed of 1 m/s for 1 second, then changes orientation counterclockwise 90 degrees also in 1 second, then moves parallel to axis Y_I for 1 final second. The desired 3-second trajectory is shown in figure 3.15, using plots of x, y and θ in relation to time.

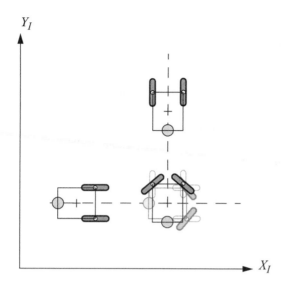

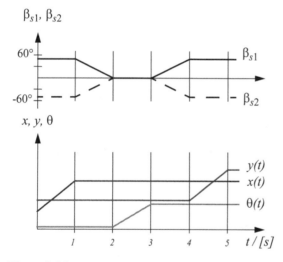

Figure 3.16
Example of robot trajectory similar to figure 3.15 with two steered wheels: move for 1 second with constant speed of 1 m/s along axis X_I; rotate steered wheels -50 / 50 degree respectively; change orientation counterclockwise 90 degree in 1 second; rotate steered wheels 50 / -50 degree respectively; move for 1 second with constant speed of 1 m/s along axis Y_I.

Can the omnidirectional robot accomplish this trajectory? We assume that the robot can achieve some arbitrary, finite velocity at each wheel. For simplicity, we further assume that acceleration is infinite; that is, it takes zero time to reach any desired velocity. Under these assumptions, the omnidirectional robot can indeed follow the trajectory of figure 3.15. The transition between the motion of second 1 and second 2, for example, involves only changes to the wheel velocities.

Because the two-steer has $\delta_M = 3$, it must be able to follow the path that would result from projecting this trajectory into timeless workspace. However, it cannot follow this 4D trajectory. Even if steering velocity is finite and arbitrary, although the two-steer would be able to change steering speed instantly, it would have to wait for the angle of the steerable wheels to change to the desired position before initiating a change in the robot chassis orientation. In short, the two-steer requires changes to internal degrees of freedom and because these changes take time, arbitrary trajectories are not attainable. Figure 3.16 depicts the most similar trajectory that a two-steer can achieve. In contrast to the desired three phases of motion, this trajectory has five phases.

3.5   Beyond Basic Kinematics

This discussion of mobile robot kinematics is only an introduction to a far richer topic. When speed and force are also considered, as is particularly necessary in the case of high-speed mobile robots, dynamic constraints must be expressed in addition to kinematic constraints. Furthermore, many mobile robots such as tank-type chassis and four-wheel slip/skid systems violate the preceding kinematic models. When analyzing such systems, it is often necessary to explicitly model the dynamics of viscous friction between the robot and the ground plane.

More significantly, the kinematic analysis of a mobile robot system provides results concerning the theoretical workspace of that mobile robot. However, to move effectively in this workspace a mobile robot must have appropriate actuation of its degrees of freedom. This problem, called motorization, requires further analysis of the forces that must be actively supplied to realize the kinematic range of motion available to the robot.

In addition to motorization, there is the question of controllability: under what conditions can a mobile robot travel from the initial pose to the goal pose in bounded time? Answering this question requires knowledge, both knowledge of the robot kinematics and knowledge of the control systems that can be used to actuate the mobile robot. Mobile robot control is therefore a return to the practical question of designing a real-world control algorithm that can drive the robot from pose to pose using the trajectories demanded for the application.

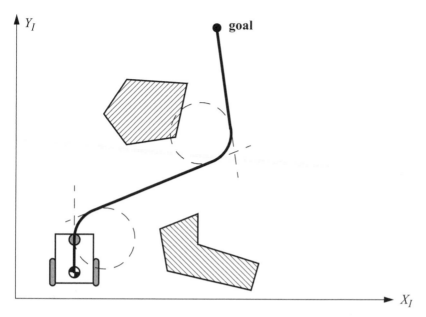

Figure 3.17
Open-loop control of a mobile robot based on straight lines and circular trajectory segments.

3.6   Motion Control (Kinematic Control)

As we have seen, motion control might not be an easy task for nonholonomic systems. However, it has been studied by various research groups, for example, [10, 61, 90, 92, 300], and some adequate solutions for motion control of a mobile robot system are available.

3.6.1   Open loop control (trajectory-following)
The objective of a kinematic controller is to follow a trajectory described by its position or velocity profile as a function of time. This is often done by dividing the trajectory (path) in motion segments of clearly defined shape, for example, straight *lines* and segments of a *circle*. The control problem is thus to precompute a smooth trajectory based on line and circle segments that drives the robot from the initial position to the final position (figure 3.17). This approach can be regarded as open-loop motion control, because the measured robot position is not fed back for velocity or position control. It has several disadvantages:

- It is not at all an easy task to precompute a feasible trajectory if all limitations and constraints of the robot's velocities and accelerations have to be considered.

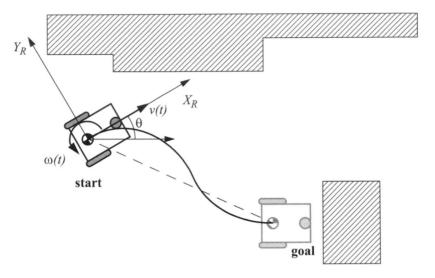

Figure 3.18
Typical situation for feedback control of a mobile robot

- The robot will not automatically adapt or correct the trajectory if dynamic changes of the environment occur.

- The resulting trajectories are usually not smooth, because the transitions from one trajectory segment to another are, for most of the commonly used segments (e.g., lines and part of circles), not smooth. This means there is a discontinuity in the robot's acceleration.

3.6.2 Feedback control
A more appropriate approach in motion control of a mobile robot is to use a real-state feedback controller. With such a controller the robot's path-planning task is reduced to setting intermediate positions (subgoals) lying on the requested path. One useful solution for a stabilizing feedback control of differential-drive mobile robots is presented in section 3.6.2.1. It is very similar to the controllers presented in [61, 189]. Others can be found in [10, 90, 92, 300].

3.6.2.1 Problem statement
Consider the situation shown in figure 3.18, with an arbitrary position and orientation of the robot and a predefined goal position and orientation. The actual pose error vector given in

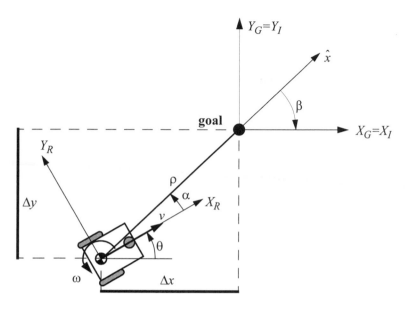

Figure 3.19
Robot kinematics and its frames of interest.

the robot reference frame $\{X_R, Y_R, \theta\}$ is $e = {}^R[x, y, \theta]^T$ with x, y, and θ being the goal coordinates of the robot.

The task of the controller layout is to find a control matrix K, if it exists

$$K = \begin{bmatrix} k_{11} & k_{12} & k_{13} \\ k_{21} & k_{22} & k_{23} \end{bmatrix} \qquad \text{with } k_{ij} = k(t, e), \tag{3.45}$$

such that the control of $v(t)$ and $\omega(t)$

$$\begin{bmatrix} v(t) \\ \omega(t) \end{bmatrix} = K \cdot e = K \; {}^R\!\begin{bmatrix} x \\ y \\ \theta \end{bmatrix} \tag{3.46}$$

drives the error e toward zero.[2]

---

2. Remember that $v(t)$ is always heading in the X_R direction of the robot's reference frame due to the nonholonomic constraint.

$$\lim_{t \to \infty} e(t) = 0 . \tag{3.47}$$

3.6.2.2 Kinematic model

We assume, without loss of generality, that the goal is at the origin of the inertial frame (figure 3.19). In the following the position vector $[x, y, \theta]^T$ is always represented in the inertial frame.

The kinematics of a differential-drive mobile robot described in the inertial frame $\{X_I, Y_I, \theta\}$ is given by

$$^I\begin{bmatrix} \dot{x} \\ \dot{y} \\ \dot{\theta} \end{bmatrix} = \begin{bmatrix} \cos\theta & 0 \\ \sin\theta & 0 \\ 0 & 1 \end{bmatrix} \begin{bmatrix} v \\ \omega \end{bmatrix} . \tag{3.48}$$

where $\dot{x}$ and $\dot{y}$ are the linear velocities in the direction of the X_I and Y_I of the inertial frame.

Let α denote the angle between the x_R axis of the robot's reference frame and the vector $\hat{x}$ connecting the center of the axle of the wheels with the final position. If $\alpha \in I_1$, where

$$I_1 = \left(-\frac{\pi}{2}, \frac{\pi}{2} \right] , \tag{3.49}$$

then consider the coordinate transformation into polar coordinates with its origin at the goal position.

$$\rho = \sqrt{\Delta x^2 + \Delta y^2} . \tag{3.50}$$

$$\alpha = -\theta + \text{atan}2(\Delta y, \Delta x) . \tag{3.51}$$

$$\beta = -\theta - \alpha . \tag{3.52}$$

This yields a system description, in the new polar coordinates, using a matrix equation

$$
\begin{bmatrix} \dot{\rho} \\ \dot{\alpha} \\ \dot{\beta} \end{bmatrix} = \begin{bmatrix} -\cos\alpha & 0 \\ \dfrac{\sin\alpha}{\rho} & -1 \\ -\dfrac{\sin\alpha}{\rho} & 0 \end{bmatrix} \begin{bmatrix} v \\ \omega \end{bmatrix}, \tag{3.53}
$$

where ρ is the distance between the center of the robot's wheel axle and the goal position, θ denotes the angle between the X_R axis of the robot reference frame and the X_I axis associated with the final position, and v and ω are the tangent and the angular velocity respectively.

On the other hand, if $\alpha \in I_2$, where

$$
I_2 = (-\pi, -\pi/2] \cup (\pi/2, \pi], \tag{3.54}
$$

redefining the forward direction of the robot by setting $v = -v$, we obtain a system described by a matrix equation of the form

$$
\begin{bmatrix} \dot{\rho} \\ \dot{\alpha} \\ \dot{\beta} \end{bmatrix} = \begin{bmatrix} \cos\alpha & 0 \\ -\dfrac{\sin\alpha}{\rho} & 1 \\ \dfrac{\sin\alpha}{\rho} & 0 \end{bmatrix} \begin{bmatrix} v \\ \omega \end{bmatrix}. \tag{3.55}
$$

3.6.2.3 Remarks on the kinematic model in polar coordinates [eq. (3.53) and (3.55)]

- The coordinate transformation is not defined at $x = y = 0$; in such a point the determinant of the Jacobian matrix of the transformation is not defined, that is unbounded.

- For $\alpha \in I_1$ the forward direction of the robot points toward the goal; for $\alpha \in I_2$ it is the reverse direction.

- By properly defining the forward direction of the robot at its initial configuration, it is always possible to have $\alpha \in I_1$ at $t = 0$. However, this does not mean that α remains in I_1 for all time t. Hence, to avoid that the robot changes direction during approaching the goal, it is necessary to determine, if possible, the controller in such a way that $\alpha \in I_1$ for all t, whenever $\alpha(0) \in I_1$. The same applies for the reverse direction (see the following stability issues).

3.6.2.4 The control law

The control signals v and ω must now be designed to drive the robot from its actual configuration, say $(\rho_0, \alpha_0, \beta_0)$, to the goal position. It is obvious that equation (3.53) presents a discontinuity at $\rho = 0$; thus, the theorem of Brockett does not obstruct smooth stabilizability.

If we consider now the linear control law

$$v = k_\rho \rho, \tag{3.56}$$

$$\omega = k_\alpha \alpha + k_\beta \beta, \tag{3.57}$$

we get with equation (3.53) a closed-loop system described by

$$\begin{bmatrix} \dot{\rho} \\ \dot{\alpha} \\ \dot{\beta} \end{bmatrix} = \begin{bmatrix} -k_\rho \rho \cos\alpha \\ k_\rho \sin\alpha - k_\alpha \alpha - k_\beta \beta \\ -k_\rho \sin\alpha \end{bmatrix}. \tag{3.58}$$

The system does not have any singularity at $\rho = 0$ and has a unique equilibrium point at $(\rho, \alpha, \beta) = (0, 0, 0)$. Thus, it will drive the robot to this point, which is the goal position.

- In the Cartesian coordinate system the control law (equation [3.57]) leads to equations that are not defined at $x = y = 0$.

- Be aware of the fact that the angles α and β have always to be expressed in the range $(-\pi, \pi)$.

- Observe that the control signal v always has a constant sign—that is, it is positive whenever $\alpha(0) \in I_1$ and it is always negative otherwise. This implies that the robot performs its parking maneuver always in a single direction and without reversing its motion.

In figure 3.20 you find the resulting paths when the robot is initially on a circle in the xy plane. All movements have smooth trajectories toward the goal in the center. The control parameters for this simulation were set to

$$k = (k_\rho, k_\alpha, k_\beta) = (3, 8, -1.5). \tag{3.59}$$

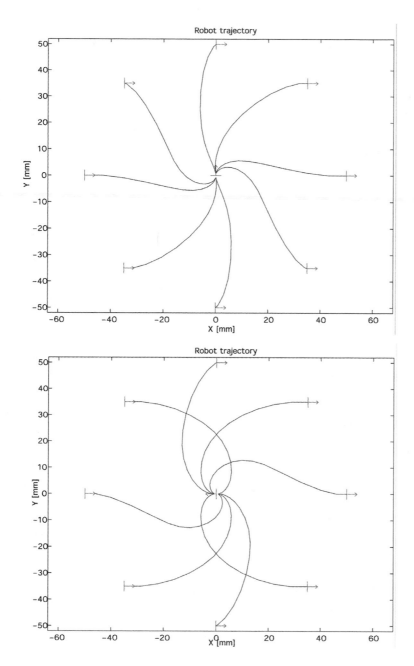

Figure 3.20
Resulting paths when the robot is initially on the unit circle in the x,y plane.

3.6.2.5 Local stability issue

It can further be shown that the closed-loop control system (equation [3.58]) is locally exponentially stable if

$$k_\rho > 0 \ ; \quad k_\beta < 0 \ ; \quad k_\alpha - k_\rho > 0 . \tag{3.60}$$

Proof:

Linearized around the equilibrium ($\cos x = 1$, $\sin x = x$) position, equation (3.58) can be written as

$$
\begin{bmatrix} \dot{\rho} \\ \dot{\alpha} \\ \dot{\beta} \end{bmatrix}
=
\begin{bmatrix}
-k_\rho & 0 & 0 \\
0 & -(k_\alpha - k_\rho) & -k_\beta \\
0 & -k_\rho & 0
\end{bmatrix}
\begin{bmatrix} \rho \\ \alpha \\ \beta \end{bmatrix} ,
\tag{3.61}
$$

hence it is locally exponentially stable if the eigenvalues of the matrix

$$
A =
\begin{bmatrix}
-k_\rho & 0 & 0 \\
0 & -(k_\alpha - k_\rho) & -k_\beta \\
0 & -k_\rho & 0
\end{bmatrix}
\tag{3.62}
$$

all have a negative real part. The characteristic polynomial of the matrix A is

$$(\lambda + k_\rho)(\lambda^2 + \lambda(k_\alpha - k_\rho) - k_\rho k_\beta) \tag{3.63}$$

and all roots have negative real part if

$$k_\rho > 0 \ ; \quad -k_\beta > 0 \ ; \quad k_\alpha - k_\rho > 0 , \tag{3.64}$$

which proves the claim.

For robust position control, it might be advisable to apply the strong stability condition, which ensures that the robot does not change direction during its approach to the goal:

$$k_\rho > 0 \ ; \quad k_\beta < 0 \ ; \quad k_\alpha + \frac{5}{3}k_\beta - \frac{2}{\pi}k_\rho > 0 . \tag{3.65}$$

This implies that $\alpha \in I_1$ for all t, whenever $\alpha(0) \in I_1$ and $\alpha \in I_2$ for all t, whenever $\alpha(0) \in I_2$ respectively. This strong stability condition has also been verified in applications.

3.7 Problems

1. Suppose a differential drive robot has wheels of differing diameters. The left wheel has diameter 2 and the right wheel has diameter 3. $l = 5$ for both wheels. The robot is positioned at $\theta = \pi/4$. The robot spins both wheels at a speed of 6. Compute the robot's instantaneous velocity in the global reference frame. Specify $\dot{x}$, $\dot{y}$, and $\dot{\theta}$.

2. Consider a robot with two powered spherical wheels and a passive castor wheel (no power). Derive the kinematics equations of the form of 3.30 and 3.33 for this robot.

3. Determine the degrees of mobility, steerability, and maneuverability for each of the following: (a) bicycle; (b) dynamically balanced robot with a single spherical wheel; (c) automobile.

4. **Challenge Question**
   Consider the robot of figure 3.10 with dimensions $l = 10$ and $r = 1$. Suppose you wish to command this robot in its local reference frame with speed $\dot{x}$, $\dot{y}$, and $\dot{\theta}$. Write a formula translating the desired $\dot{x}$, $\dot{y}$, $\dot{\theta}$ into speeds for all three wheels. Specifically, identify f_1, f_2, and f_3 such that:

$$\varphi_1 = f_1(\dot{x}, \dot{y}, \dot{\theta})$$

$$\varphi_2 = f_2(\dot{x}, \dot{y}, \dot{\theta})$$

$$\varphi_3 = f_3(\dot{x}, \dot{y}, \dot{\theta})$$

4 Perception

One of the most important tasks of an autonomous system of any kind is to acquire knowledge about its environment. This is done by taking measurements using various sensors and then extracting meaningful information from those measurements.

In this chapter we present the most common sensors used in mobile robots and then discuss strategies for extracting information from the sensors. For more detailed information about many of the sensors used on mobile robots, refer to H.R. Everett's comprehensive book *Sensors for Mobile Robots* [20].

4.1   Sensors for Mobile Robots

A wide variety of sensors is used in mobile robots (figure 4.1). Some sensors are used to measure simple values such as the internal temperature of a robot's electronics or the rotational speed of the motors. Other more sophisticated sensors can be used to acquire information about the robot's environment or even to measure directly a robot's global position. In this chapter we focus primarily on sensors used to extract information about the robot's environment. Because a mobile robot moves around, it will frequently encounter unforeseen environmental characteristics, and therefore such sensing is particularly critical. We begin with a functional classification of sensors. Then, after presenting basic tools for describing a sensor's performance, we proceed to describe selected sensors in detail.

4.1.1   Sensor classification
We classify sensors using two important functional axes: *proprioceptive/exteroceptive* and *passive/active*.

Proprioceptive sensors measure values internal to the system (robot), for example, motor speed, wheel load, robot arm joint angles, and battery voltage.

Exteroceptive sensors acquire information from the robot's environment, for example, distance measurements, light intensity, and sound amplitude. Hence exteroceptive sensor measurements are interpreted by the robot in order to extract meaningful environmental features.

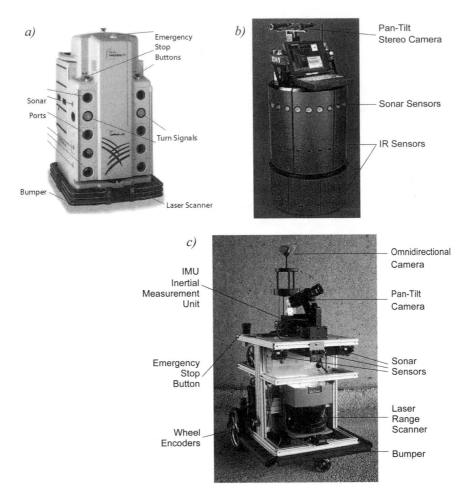

Figure 4.1
Examples of robots with multisensor systems: (a) HelpMate from Transition Research Corporation; (b) B21 from Real World Interface; (c) BIBA Robot, BlueBotics SA.

Passive sensors measure ambient environmental energy entering the sensor. Examples of passive sensors include temperature probes, microphones, and CCD or CMOS cameras.

Active sensors emit energy into the environment, then measure the environmental reaction. Because active sensors can manage more controlled interactions with the environment, they often achieve superior performance. However, active sensing introduces several risks: the outbound energy may affect the very characteristics that the sensor is attempting to measure. Furthermore, an active sensor may suffer from interference between its signal and those beyond its control. For example, signals emitted by other nearby robots, or similar sensors on the same robot, may influence the resulting measurements. Examples of active sensors include wheel quadrature encoders, ultrasonic sensors, and laser rangefinders.

Table 4.1 provides a classification of the most useful sensors for mobile robot applications. The most interesting sensors are discussed in this chapter. The sensor classes in table 4.1 are arranged in ascending order of complexity and descending order of technological maturity. Tactile sensors and proprioceptive sensors are critical to virtually all mobile robots and are well understood and easily implemented. Commercial quadrature encoders, for example, may be purchased as part of a gear-motor assembly used in a mobile robot. At the other extreme, visual interpretation by means of one or more CCD/CMOS cameras provides a broad array of potential functionalities, from obstacle avoidance and localization to human face recognition. However, commercially available sensor units that provide visual functionalities are only now beginning to emerge [172, 346].

4.1.2   Characterizing sensor performance

The sensors we describe in this chapter vary greatly in their performance characteristics. Some sensors provide extreme accuracy in well-controlled laboratory settings but are overcome with error when subjected to real-world environmental variations. Other sensors provide narrow, high-precision data in a wide variety of settings. In order to quantify such performance characteristics, first we formally define the sensor performance terminology that will be valuable throughout the rest of this chapter.

4.1.2.1   Basic sensor response ratings

A number of sensor characteristics can be rated quantitatively in a laboratory setting. Such performance ratings will necessarily be best-case scenarios when the sensor is placed on a real-world robot, but are nevertheless useful.

Dynamic range is used to measure the spread between the lower and upper limits of input values to the sensor while maintaining normal sensor operation. Formally, the dynamic range is the ratio of the maximum input value to the minimum measurable input value. Because this raw ratio can be unwieldy, it is usually measured in *decibels*, which are computed as ten times the common logarithm of the dynamic range. However, there is potential confusion in the calculation of decibels, which are meant to measure the ratio between *powers*, such as watts or horsepower. Suppose your sensor measures motor current and can register values from a minimum of 1 mA to 20 A. The dynamic range of this current sensor is defined as

Table 4.1
Classification of sensors used in mobile robotics applications

General classification (typical use)	Sensor Sensor System	PC or EC	A or P
Tactile sensors (detection of physical contact or closeness; security switches)	Contact switches, bumpers	EC	P
	Optical barriers	EC	A
	Noncontact proximity sensors	EC	A
Wheel/motor sensors (wheel/motor speed and position)	Brush encoders	PC	P
	Potentiometers	PC	P
	Synchros, resolvers	PC	A
	Optical encoders	PC	A
	Magnetic encoders	PC	A
	Inductive encoders	PC	A
	Capacitive encoders	PC	A
Heading sensors (orientation of the robot in relation to a fixed reference frame)	Compass	EC	P
	Gyroscopes	PC	P
	Inclinometers	EC	A/P
Acceleration sensor	Accelerometer	PC	P
Ground beacons (localization in a fixed reference frame)	GPS	EC	A
	Active optical or RF beacons	EC	A
	Active ultrasonic beacons	EC	A
	Reflective beacons	EC	A
Active ranging (reflectivity, time-of-flight, and geometric triangulation)	Reflectivity sensors	EC	A
	Ultrasonic sensor	EC	A
	Laser rangefinder	EC	A
	Optical triangulation (1D)	EC	A
	Structured light (2D)	EC	A
Motion/speed sensors (speed relative to fixed or moving objects)	Doppler radar	EC	A
	Doppler sound	EC	A
Vision sensors (visual ranging, whole-image analysis, segmentation, object recognition)	CCD/CMOS camera(s) Visual ranging packages Object tracking packages	EC	P

A, active; P, passive; P/A, passive/active; PC, proprioceptive; EC, exteroceptive.

$$10 \cdot \log\left[\frac{20}{0.001}\right] = 43 \text{ dB} . \tag{4.1}$$

Now suppose you have a voltage sensor that measures the voltage of your robot's battery, measuring any value from 1 mV to 20 V. Voltage is not a unit of power, but the square of voltage is proportional to power. Therefore, we use 20 instead of 10:

$$20 \cdot \log\left[\frac{20}{0.001}\right] = 86 \text{ dB} . \tag{4.2}$$

Range is also an important rating in mobile robot applications because often robot sensors operate in environments where they are frequently exposed to input values beyond their working range. In such cases, it is critical to understand how the sensor will respond. For example, an optical rangefinder will have a minimum operating range and can thus provide spurious data when measurements are taken with the object closer than that minimum.

Resolution is the minimum difference between two values that can be detected by a sensor. Usually, the lower limit of the dynamic range of a sensor is equal to its resolution. However, in the case of digital sensors, this is not necessarily so. For example, suppose that you have a sensor that measures voltage, performs an analog-to-digital (A/D) conversion, and outputs the converted value as an 8-bit number linearly corresponding to between 0 and 5 V. If this sensor is truly linear, then it has $2^8 - 1$ total output values, or a resolution of $5 \text{ V}(255) = 20 \text{ mV}$.

Linearity is an important measure governing the behavior of the sensor's output signal as the input signal varies. A linear response indicates that if two inputs x and y result in the two outputs $f(x)$ and $f(y)$, then for any values a and b, $f(ax + by) = af(x) + bf(y)$. This means that a plot of the sensor's input/output response is simply a straight line.

Bandwidth or *frequency* is used to measure the speed with which a sensor can provide a stream of readings. Formally, the number of measurements per second is defined as the sensor's frequency in *hertz*. Because of the dynamics of moving through their environment, mobile robots often are limited in maximum speed by the bandwidth of their obstacle detection sensors. Thus, increasing the bandwidth of ranging and vision sensors has been a high-priority goal in the robotics community.

4.1.2.2 In situ sensor performance
These sensor characteristics can be reasonably measured in a laboratory environment with confident extrapolation to performance in real-world deployment. However, a number of important measures cannot be reliably acquired without deep understanding of the complex interaction between all environmental characteristics and the sensors in question. This is

most relevant to the most sophisticated sensors, including active ranging sensors and visual interpretation sensors.

Sensitivity itself is a desirable trait. This is a measure of the degree to which an incremental change in the target input signal changes the output signal. Formally, sensitivity is the ratio of output change to input change. Unfortunately, however, the sensitivity of exteroceptive sensors is often confounded by undesirable sensitivity and performance coupling to other environmental parameters.

Cross-sensitivity is the technical term for sensitivity to environmental parameters that are orthogonal to the target parameters for the sensor. For example, a flux-gate compass can demonstrate high sensitivity to magnetic north and is therefore of use for mobile robot navigation. However, the compass will also demonstrate high sensitivity to ferrous building materials, so much so that its cross-sensitivity often makes the sensor useless in some indoor environments. High cross-sensitivity of a sensor is generally undesirable, especially when it cannot be modeled.

Error of a sensor is defined as the difference between the sensor's output measurements and the true values being measured, within some specific operating context. Given a true value v and a measured value m, we can define *error* as $error = m - v$.

Accuracy is defined as the degree of conformity between the sensor's measurement and the true value, and is often expressed as a proportion of the true value (e.g., 97.5% accuracy). Thus small error corresponds to high accuracy and vice versa:

$$\left(accuracy = 1 - \frac{|error|}{v} \right). \tag{4.3}$$

Of course, obtaining the ground truth, v, can be difficult or impossible, and so establishing a confident characterization of sensor accuracy can be problematic. Furthermore, it is important to distinguish between two different sources of error:

Systematic errors are caused by factors or processes that can in theory be modeled. These errors are, therefore, deterministic (i.e., predictable). Poor calibration of a laser rangefinder, an unmodeled slope of a hallway floor, and a bent stereo camera head due to an earlier collision are all possible causes of systematic sensor errors.

Random errors cannot be predicted using a sophisticated model; neither can they be mitigated by more precise sensor machinery. These errors can only be described in probabilistic terms (i.e., stochastically). Hue instability in a color camera, spurious rangefinding errors, and black level noise in a camera are all examples of random errors.

Precision is often confused with accuracy, and now we have the tools to clearly distinguish these two terms. Intuitively, high precision relates to reproducibility of the sensor results. For example, one sensor taking multiple readings of the same environmental state has high precision if it produces the same output. In another example, multiple copies of this sensor taking readings of the same environmental state have high precision if their outputs agree. Precision does not, however, have any bearing on the accuracy of the sensor's output with respect to the true value being measured. Suppose that the *random error* of a sensor is characterized by some mean value μ and a standard deviation σ. The formal definition of precision is the ratio of the sensor's output range to the standard deviation:

$$precision = \frac{range}{\sigma}.$$ (4.4)

Note that only σ and not μ has impact on precision. In contrast, mean error μ is directly proportional to overall sensor error and inversely proportional to sensor accuracy.

4.1.2.3 Characterizing error: The challenges in mobile robotics

Mobile robots depend heavily on exteroceptive sensors. Many of these sensors concentrate on a central task for the robot: acquiring information on objects in the robot's immediate vicinity so that it may interpret the state of its surroundings. Of course, these "objects" surrounding the robot are all detected from the viewpoint of its local reference frame. Since the systems we study are mobile, their ever-changing position and their motion have a significant impact on overall sensor behavior. In this section, empowered with the terminology of the earlier discussions, we describe how dramatically the sensor error of a mobile robot disagrees with the ideal picture drawn in the previous section.

Blurring of systematic and random errors. Active ranging sensors tend to have failure modes that are triggered largely by specific relative positions of the sensor and environment targets. For example, a sonar sensor will produce specular reflections, producing grossly inaccurate measurements of range, at specific angles to a smooth sheetrock wall. During motion of the robot, such relative angles occur at stochastic intervals. This is especially true in a mobile robot outfitted with a ring of multiple sonars. The chances of one sonar entering this error mode during robot motion is high. From the perspective of the moving robot, the sonar measurement error is a random error in this case. Yet, if the robot were to stop, becoming motionless, then a very different error modality is possible. If the robot's static position causes a particular sonar to fail in this manner, the sonar will fail consistently and will tend to return precisely the same (and incorrect!) reading time after time. Once the robot is motionless, the error appears to be systematic and of high precision.

The fundamental mechanism at work here is the cross-sensitivity of mobile robot sensors to robot pose and robot-environment dynamics. The models for such cross-sensitivity are not, in an underlying sense, truly random. However, these physical interrelationships are rarely modeled, and therefore, from the point of view of an incomplete model, the errors appear random during motion and systematic when the robot is at rest.

Sonar is not the only sensor subject to this blurring of systematic and random error modality. Visual interpretation through the use of a CCD camera is also highly susceptible to robot motion and position because of camera dependence on lighting changes, lighting specularity (e.g., glare), and reflections. The important point is to realize that, while systematic error and random error are well defined in a controlled setting, the mobile robot can exhibit error characteristics that bridge the gap between deterministic and stochastic error mechanisms.

Multimodal error distributions. It is common to characterize the behavior of a sensor's random error in terms of a probability distribution over various output values. In general, one knows very little about the causes of random error, and therefore several simplifying assumptions are commonly used. For example, we can assume that the error is *zero-mean* in that it symmetrically generates both positive and negative measurement error. We can go even further and assume that the probability density curve is Gaussian. Although we discuss the mathematics of this in detail in section 4.1.3, it is important for now to recognize the fact that one frequently assumes *symmetry* as well as *unimodal distribution*. This means that measuring the correct value is most probable, and any measurement that is farther away from the correct value is less likely than any measurement that is closer to the correct value. These are strong assumptions that enable powerful mathematical principles to be applied to mobile robot problems, but it is important to realize how wrong these assumptions usually are.

Consider, for example, the sonar sensor once again. When ranging an object that reflects the sound signal well, the sonar will exhibit high accuracy and will induce random error based on noise, for example, in the timing circuitry. This portion of its sensor behavior will exhibit error characteristics that are fairly symmetric and unimodal. However, when the sonar sensor is moving through an environment and is sometimes faced with materials that cause coherent reflection rather than return the sound signal to the sonar sensor, then the sonar will grossly overestimate the distance to the object. In such cases, the error will be biased toward positive measurement error and will be far from the correct value. The error is not strictly systematic, and so we are left modeling it as a probability distribution of random error. So the sonar sensor has two separate types of operational modes, one in which the signal does return and some random error is possible, and the second in which the signal returns after a multipath reflection and gross overestimation error occurs. The probability distribution could easily be at least bimodal in this case, and since overestimation is more common than underestimation, it will also be asymmetric.

As a second example, consider ranging via stereo vision. Once again, we can identify two modes of operation. If the stereo vision system correctly correlates two images, then the resulting random error will be caused by camera noise and will limit the measurement accuracy. But the stereo vision system can also correlate two images *incorrectly*, matching two fenceposts, for example, that are not the same post in the real world. In such a case stereo vision will exhibit gross measurement error, and one can easily imagine such behavior violating both the unimodal and the symmetric assumptions.

The thesis of this section is that sensors in a mobile robot *may* be subject to multiple modes of operation and, when the sensor error is characterized, unimodality and symmetry may be grossly violated. Nonetheless, as we shall see, many successful mobile robot systems make use of these simplifying assumptions and the resulting mathematical techniques with great empirical success.

4.1.3 Representing uncertainty

In section 4.1.2 we presented a terminology for describing the performance characteristics of a sensor. As mentioned there, sensors are imperfect devices with errors of both systematic and random nature. Random errors, in particular, cannot be corrected, and so they represent atomic levels of sensor uncertainty.

But when you build a mobile robot, you combine information from many sensors, even using the same sensors repeatedly, over time, to build, possibly, a model of the environment. How can we scale up, from characterizing the uncertainty of a single sensor to the uncertainty of the resulting robot system?

We begin by presenting a statistical representation for the random error associated with an individual sensor [14]. With a quantitative tool in hand, the standard Gaussian uncertainty model can be presented and evaluated. Finally, we present a framework for computing the uncertainty of conclusions drawn from a set of quantifiably uncertain measurements, known as the *error propagation law*.

4.1.3.1 Statistical representation

We have already defined *error* as the difference between a sensor measurement and the true value. From a statistical point of view, we wish to characterize the error of a sensor, not for one specific measurement but for any measurement. Let us formulate the problem of sensing as an estimation problem. The sensor has taken a set of n measurements with values ρ_i. The goal is to characterize the estimate of the true value $E[X]$ given these measurements:

$$E[X] = g(\rho_1, \rho_2, ..., \rho_n).$$ (4.5)

From this perspective, the true value is represented by a random (and therefore unknown) variable X. We use a *probability density function* to characterize the statistical properties of the value of X.

In figure 4.2, the density function identifies for each possible value x of X a probability density $f(x)$ along the y-axis. The area under the curve is 1, indicating the complete chance of X having *some* value:

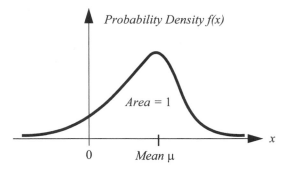

Figure 4.2
A sample probability density function, showing a single probability peak (i.e., unimodal) with asymptotic drops in both directions.

$$\int_{-\infty}^{\infty} f(x)dx = 1 .$$

(4.6)

The probability of the value of X falling between two limits a and b is computed as the bounded integral:

$$p[a < X \leq b] = \int_{a}^{b} f(x)dx$$

(4.7)

The probability density function is a useful way to characterize the possible values of X because it captures not only the range of X but also the comparative probability of different values for X. Using $f(x)$ we can quantitatively define the mean, variance, and standard deviation as follows.

The *mean value* μ is equivalent to the expected value $E[X]$ if we were to measure X an infinite number of times and average all of the resulting values. We can easily define $E[X]$:

$$\mu = E[X] = \int_{-\infty}^{\infty} xf(x)dx .$$

(4.8)

Note in this equation that calculation of $E[X]$ is identical to the weighted average of all possible values of x. In contrast, the *mean square value* is simply the weighted average of the squares of all values of x :

$$E[X^2] = \int_{-\infty}^{\infty} x^2 f(x)dx. \tag{4.9}$$

Characterization of the "width" of the possible values of X is a key statistical measure, and this requires first defining the *variance* σ^2:

$$Var(X) = \sigma^2 = \int_{-\infty}^{\infty} (x - \mu)^2 f(x)dx. \tag{4.10}$$

Finally, the *standard deviation* σ is simply the square root of variance σ^2, and σ^2 will play important roles in our characterization of the error of a single sensor as well as the error of a model generated by combining multiple sensor readings.

Independence of random variables. With the tools presented here, we often evaluate systems with multiple random variables. For instance, a mobile robot's laser rangefinder may be used to measure the position of a feature on the robot's right and, later, another feature on the robot's left. The position of each feature in the real world may be treated as random variables, X_1 and X_2.

Two random variables X_1 and X_2 are *independent* if the particular value of one has no bearing on the particular value of the other. In this case we can draw several important conclusions about the statistical behavior of X_1 and X_2. First, the expected value (or mean value) of the product of random variables is equal to the product of their mean values:

$$E[X_1 X_2] = E[X_1]E[X_2]. \tag{4.11}$$

Second, the variance of their sums is equal to the sum of their variances:

$$Var(X_1 + X_2) = Var(X_1) + Var(X_2). \tag{4.12}$$

In mobile robotics, we often assume the independence of random variables even when this assumption is not strictly true. The simplification that results makes a number of the existing mobile robot-mapping and navigation algorithms tenable, as described in chapter 5. A further simplification, described in the following section, revolves around one specific probability density function used more often than any other when modeling error: the Gaussian distribution.

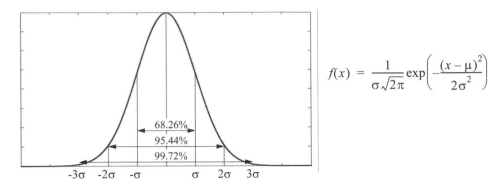

Figure 4.3
The Gaussian function with $\mu = 0$ and $\sigma = 1$. We shall refer to this as the reference Gaussian. The value 2σ is often refereed to as the signal quality; 95.44% of the values fall within $\pm 2\sigma$.

Gaussian distribution. The Gaussian distribution, also called the *normal distribution,* is used across engineering disciplines when a well-behaved error model is required for a random variable for which no error model of greater felicity has been discovered. The Gaussian has many characteristics that make it mathematically advantageous to other ad hoc probability density functions. It is symmetric around the mean μ. There is no particular bias for being larger than or smaller than μ, and this makes sense when there is no information to the contrary. The Gaussian distribution is also unimodal, with a single peak that reaches a maximum at μ (necessary for any symmetric, unimodal distribution). This distribution also has tails (the value of $f(x)$ as x approaches $-\infty$ and ∞) that approach zero only asymptotically. This means that all amounts of error are possible, although very large errors may be highly improbable. In this sense, the Gaussian is conservative. Finally, as seen in the formula for the Gaussian probability density function, the distribution depends on only two parameters:

$$f(x) = \frac{1}{\sigma\sqrt{2\pi}}\exp\left(-\frac{(x-\mu)^2}{2\sigma^2}\right). \tag{4.13}$$

The Gaussian's basic shape is determined by the structure of this formula, and so the only two parameters required to fully specify a particular Gaussian are its mean, μ, and its standard deviation, σ. Figure 4.3 shows the Gaussian function with $\mu = 0$ and $\sigma = 1$.

Figure 4.4
Error propagation in a multiple-input multi-output system with n inputs and m outputs.

Suppose that a random variable X is modeled as a Gaussian. How does one identify the chance that the value of X is within one standard deviation of μ? In practice, this requires integration of $f(x)$, the Gaussian function to compute the area under a portion of the curve:

$$Area = \int_{-\sigma}^{\sigma} f(x)dx. \tag{4.14}$$

Unfortunately, there is no closed-form solution for the integral in equation (4.14), and so the common technique is to use a Gaussian *cumulative probability table*. Using such a table, one can compute the probability for various value ranges of X:

$$p[\mu - \sigma < X \leq \mu + \sigma] = 0.68;$$

$$p[\mu - 2\sigma < X \leq \mu + 2\sigma] = 0.95;$$

$$p[\mu - 3\sigma < X \leq \mu + 3\sigma] = 0.997.$$

For example, 95% of the values for X fall within two standard deviations of its mean. This applies to *any* Gaussian distribution. As is clear from the above progression, under the Gaussian assumption, once bounds are relaxed to 3σ, the overwhelming proportion of values (and, therefore, probability) is subsumed.

4.1.3.2 Error propagation: Combining uncertain measurements

These probability mechanisms may be used to describe the errors associated with a single sensor's attempts to measure a real-world value. But in mobile robotics, one often uses a series of measurements, all of them uncertain, to extract a single environmental measure. For example, a series of uncertain measurements of single points can be fused to extract the position of a line (e.g., a hallway wall) in the environment (figure 4.88).

Consider the system in figure 4.4, where X_i are n input signals with a known probability distribution and Y_i are m outputs. The question of interest is this: What can we say about the probability distribution of the output signals Y_i if they depend with known functions

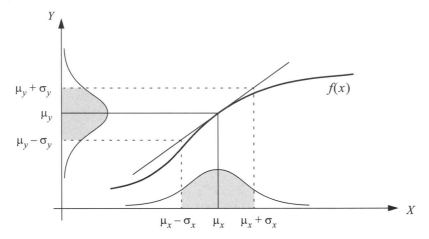

Figure 4.5
One-dimensional case of a nonlinear error propagation problem.

f_i upon the input signals? Figure 4.5 depicts the 1D version of this error propagation problem as an example.

The general solution can be generated using the first-order Taylor expansion of f_i. The output covariance matrix C_Y is given by the error propagation law:

$$C_Y = F_X C_X F_X^T,$$ (4.15)

where

C_X = covariance matrix representing the input uncertainties;

C_Y = covariance matrix representing the propagated uncertainties for the outputs;

F_x is the *Jacobian* matrix defined as

$$F_X = \nabla f = \begin{bmatrix} \dfrac{\partial f_1}{\partial X_1} & \cdots & \dfrac{\partial f_1}{\partial X_n} \\ \vdots & \cdots & \vdots \\ \dfrac{\partial f_m}{\partial X_1} & \cdots & \dfrac{\partial f_m}{\partial X_n} \end{bmatrix}.$$ (4.16)

This is also the transpose of the gradient of $f(X)$.

We will not present a detailed derivation here but will use equation (4.15) to solve an example problem in section 4.7.1.

The preceding sections have presented a terminology with which we can characterize the advantages and disadvantages of various mobile robot sensors. In the following sections, we do the same for a sampling of the most commonly used mobile robot sensors today.

4.1.4 Wheel/motor sensors

Wheel/motor sensors are devices used to measure the internal state and dynamics of a mobile robot. These sensors have vast applications outside of mobile robotics and, as a result, mobile robotics has enjoyed the benefits of high-quality, low-cost wheel and motor sensors that offer excellent resolution. In the next section, we sample just one such sensor, the optical incremental encoder.

4.1.4.1 Optical encoders

Optical incremental encoders have become the most popular device for measuring angular speed and position within a motor drive or at the shaft of a wheel or steering mechanism. In mobile robotics, encoders are used to control the position or speed of wheels and other motor-driven joints. Because these sensors are *proprioceptive*, their estimate of position is best in the reference frame of the robot and, when applied to the problem of robot *localization*, significant corrections are required, as discussed in chapter 5.

An optical encoder is basically a mechanical light chopper that produces a certain number of sine or square wave pulses for each shaft revolution. It consists of an illumination source, a fixed grating that masks the light, a rotor disc with a fine optical grid that rotates with the shaft, and fixed optical detectors. As the rotor moves, the amount of light striking the optical detectors varies based on the alignment of the fixed and moving gratings. In robotics, the resulting sine wave is transformed into a discrete square wave using a threshold to choose between *light* and *dark* states. Resolution is measured in *cycles per revolution* (CPR). The minimum angular resolution can be readily computed from an encoder's CPR rating. A typical encoder in mobile robotics may have 2000 CPR, while the optical encoder industry can readily manufacture encoders with 10,000 CPR. In terms of required bandwidth, it is of course critical that the encoder be sufficiently fast to count at the shaft spin speeds that are expected. Industrial optical encoders present no bandwidth limitation to mobile robot applications.

Usually in mobile robotics the *quadrature encoder* is used. In this case, a second illumination and detector pair is placed 90 degrees shifted with respect to the original in terms of the rotor disc. The resulting twin square waves, shown in figure 4.6, provide significantly more information. The ordering of which square wave produces a rising edge first identifies the direction of rotation. Furthermore, the four detectably different states improve the res-

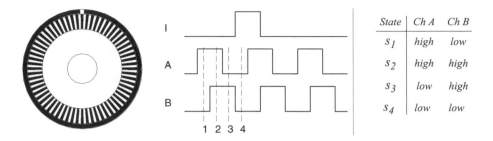

Figure 4.6
Quadrature optical wheel encoder: The observed phase relationship between channel A and B pulse
trains are used to determine the direction of the rotation. A single slot in the outer track generates a
reference (index) pulse per revolution.

olution by a factor of four with no change to the rotor disc. Thus, a 2000 CPR encoder in
quadrature yields 8000 counts. Further improvement is possible by retaining the sinusoidal
wave measured by the optical detectors and performing sophisticated interpolation. Such
methods, although rare in mobile robotics, can yield 1000-fold improvements in resolution.

As with most proprioceptive sensors, encoders are generally in the controlled environment of a mobile robot's internal structure, and so systematic error and cross-sensitivity can
be engineered away. The accuracy of optical encoders is often assumed to be 100% and,
although this may not be entirely correct, any errors at the level of an optical encoder are
dwarfed by errors downstream of the motor shaft.

4.1.5 Heading sensors
Heading sensors can be *proprioceptive* (gyroscope, inclinometer) or *exteroceptive* (compass). They are used to determine the robot's orientation and inclination. They allow us,
together with appropriate velocity information, to integrate the movement to a position estimate. This procedure, which has its roots in vessel and ship navigation, is called *dead reckoning*.

4.1.5.1 Compasses
The two most common modern sensors for measuring the direction of a magnetic field are
the Hall effect and flux gate compasses. Each has advantages and disadvantages, as
described below.

The Hall effect describes the behavior of electric potential in a semiconductor when in
the presence of a magnetic field. When a constant current is applied across the length of a
semiconductor, there will be a voltage difference in the perpendicular direction, across the
semiconductor's width, based on the relative orientation of the semiconductor to magnetic
flux lines. In addition, the sign of the voltage potential identifies the direction of the mag-

Figure 4.7
Digital compass: Sensors such as the digital/analog Hall effect sensor shown, available from Dinsmore, enable inexpensive (< \$ 15) sensing of magnetic fields.

netic field. Thus, a single semiconductor provides a measurement of flux and direction along one dimension. Hall effect digital compasses are popular in mobile robotics, and they contain two such semiconductors at right angles, providing two axes of magnetic field (thresholded) direction, thereby yielding one of eight possible compass directions. The instruments are inexpensive but also suffer from a range of disadvantages. Resolution of a digital Hall effect compass is poor. Internal sources of error include the nonlinearity of the basic sensor and systematic bias errors at the semiconductor level. The resulting circuitry must perform significant filtering, and this lowers the bandwidth of Hall effect compasses to values that are slow in mobile robot terms. For example, the Hall effect compass pictured in figure 4.7 needs 2.5 seconds to settle after a 90-degree spin.

The flux gate compass operates on a different principle. Two small coils are wound on ferrite cores and are fixed perpendicular to one another. When alternating current is activated in both coils, the magnetic field causes shifts in the phase depending on its relative alignment with each coil. By measuring both phase shifts, the direction of the magnetic field in two dimensions can be computed. The flux gate compass can accurately measure the strength of a magnetic field and has improved resolution and accuracy; however, it is both larger and more expensive than a Hall effect compass.

Regardless of the type of compass used, a major drawback concerning the use of Earth's magnetic field for mobile robot applications involves disturbance of that magnetic field by other magnetic objects and man-made structures, as well as the bandwidth limitations of electronic compasses and their susceptibility to vibration. Particularly in indoor environments, mobile robotics applications have often avoided the use of compasses, although a compass can conceivably provide useful *local* orientation information indoors, even in the presence of steel structures.

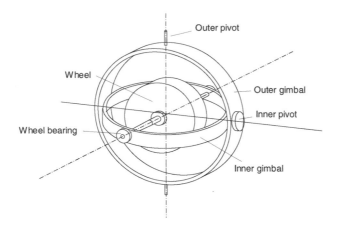

Figure 4.8
Two-axis mechanical gyroscope.

4.1.5.2 Gyroscopes

Gyroscopes are heading sensors that preserve their orientation in relation to a fixed reference frame. Thus, they provide an absolute measure for the heading of a mobile system. Gyroscopes can be classified in two categories, mechanical gyroscopes and optical gyroscopes.

Mechanical gyroscopes. The concept of a mechanical gyroscope relies on the inertial properties of a fast-spinning rotor. The property of interest is known as the gyroscopic precession. If you try to rotate a fast-spinning wheel around its vertical axis, you will feel a harsh reaction in the horizontal axis. This is due to the angular momentum associated with a spinning wheel and will keep the axis of the gyroscope inertially stable. The reactive torque τ and thus the tracking stability with the inertial frame are proportional to the spinning speed ω, the precession speed Ω, and the wheel's inertia I.

$$\tau = I\omega\Omega.\tag{4.17}$$

By arranging a spinning wheel, as seen in figure 4.8, no torque can be transmitted from the outer pivot to the wheel axis. The spinning axis will therefore be space-stable (i.e., fixed in an inertial reference frame). Nevertheless, the remaining friction in the bearings of the gyro axis introduce small torques, thus limiting the long-term space stability and introducing small errors over time. A high quality mechanical gyroscope can cost up to $100,000 and has an angular drift of about 0.1 degrees in 6 hours.

For navigation, the spinning axis has to be initially selected. If the spinning axis is aligned with the north-south meridian, the earth's rotation has no effect on the gyro's horizontal axis. If it points east-west, the horizontal axis reads the earth rotation.

Rate gyros have the same basic arrangement as shown in figure 4.8, but with a slight modification. The gimbals are restrained by a torsional spring with additional viscous damping. This enables the sensor to measure angular speeds instead of absolute orientation.

Optical gyroscopes. Optical gyroscopes are a relatively new innovation. Commercial use began in the early 1980s when they were first installed in aircraft. Optical gyroscopes are angular speed sensors that use two monochromatic light beams, or lasers, emitted from the same source, instead of moving, mechanical parts. They work on the principle that the speed of light remains unchanged and, therefore, geometric change can cause light to take a varying amount of time to reach its destination. One laser beam is sent traveling clockwise through an optical fiber while the other travels counterclockwise. Because the laser traveling in the direction of rotation has a slightly shorter path, it will have a higher frequency. This principle is known as the *Sagnac effect*. The difference in frequency Δf of the two beams is proportional to the angular velocity Ω of the cylinder. To make the difference measurable, the sensor is a coil consisting of as much as 5 km of optical fiber. New solid-state optical gyroscopes based on the same principle are build using microfabrication technology, thereby providing heading information with resolution and bandwidth far beyond the needs of mobile robotic applications. Bandwidth, for instance, can easily exceed 100 kHz, while resolution can be smaller than 0.0001 degrees/hr.

4.1.6 Accelerometers

An accelerometers is a device used to measure all external forces acting upon it, including gravity. Accelerometers belong to the proprioceptive sensors class.

Conceptually, an accelerometer is a spring–mass–damper system (figure 4.9a) in which the three-dimensional position of the proof mass relative to the accelerometer casing can be measured with some mechanism. Assume that an external force is applied on the sensor casing (e.g., gravity) and that we have an ideal spring with a force proportional to its displacement. Then, we can write [118]

$$F_{applied} = F_{inertial} + F_{damping} + F_{spring} = m\ddot{x} + c\dot{x} + kx, \tag{4.18}$$

where m is the proof mass, c is the damping coefficient, k is the spring constant, and x is the equilibrium case relative position. By choosing appropriately the damping material and the mass, the system can be made to converge very quickly to a stable value under the effect

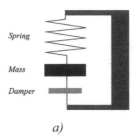
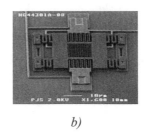

a)   *b)*   *c)*

Figure 4.9 Accelerometers: (a) Working principle of the mechanical accelerometer; (b) An example MEMS accelerometer produced by Sandia National Laboratories; (c) An example commercial MEMS accelerometer.

of a static force. When the stable value is reached, then $\ddot{x} = 0$ and the applied acceleration can be obtained as

$$a_{applied} = \frac{kx}{m}.$$

(4.19)

This is the working principle of a mechanical accelerometer. Modern accelerometers are often small Micro Electro-Mechanical Systems (MEMS) consisting of a springlike structure (*cantilevered beam*) with a proof mass (also known as *seismic mass*). Damping results from the residual gas sealed in the device. When an external force is applied, the proof mass deflects from its neutral position. Depending on the physical principle used to measure this deflection, we can have different types of accelerometers. Capacitive accelerometers measure the deflection by measuring the capacitance between a fixed structure and the proof mass. These accelerometers are reliable and inexpensive (figure 4.9b–c). Another alternative are the piezoelectric accelerometers. They are based on the property exhibited by certain crystals to generate a voltage when a mechanical stress is applied to them. A small mass is positioned on the crystal, and, when an external force is applied, the mass moves, and this induces a voltage that can be measured.

Notice that each accelerometer measures acceleration along a single axis. By mounting three accelerometers orthogonally to one another, an omnidirectional (i.e., three-axis) accelerometer can be obtained.

Also observe that an accelerometer at rest on the Earth's surface will always indicate 1 g along the vertical axis. To obtain the inertial acceleration (due to motion alone), the gravity vector must be subtracted. Conversely, the accelerometer's output will be zero during free fall.

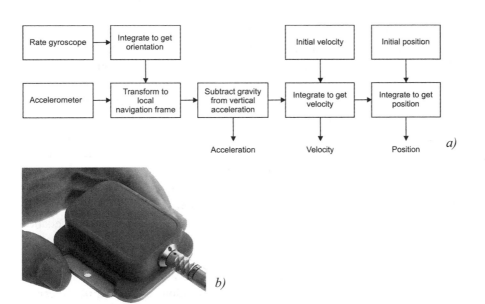

Figure 4.10 (a) IMU block diagram (redrawn from [118]). (b) A commercial IMU produced by Xsens. Image courtesy of Xsens—http://www.xsens.com.

Finally, accelerometers are classified into two categories according to their passband bandwidth: accelerometers for static and dynamic measurements. In the first category are low-pass accelerometers which can measure accelerations from 0 Hz up to usually 500 Hz. This is typical for mechanical and capacitive accelerometers. Typical uses are measurements of the gravitational acceleration or that of a moving vehicle. The second category of accelerometers is used for measuring accelerations of vibrating objects or accelerations during crashes. In this case, the bandwidth ranges between a few Hz up to 50 KHz. Typical accelerometers in this category are those realized with piezoelectric technology.

4.1.7 Inertial measurement unit (IMU)

An inertial measurement unit (IMU) is a device that uses gyroscopes and accelerometers to estimate the relative position, velocity, and acceleration of a moving vehicle. An IMU is also known as an Inertial Navigation System (INS), and it has become a common navigational component of aircraft and ships. An IMU estimates the six-degree-of-freedom (DOF) pose of the vehicle: position (x, y, z) and orientation (roll, pitch, yaw). Nevertheless, heading sensors like compasses and gyroscopes, which conversely only estimate orientation, are often improperly called IMUs.

Besides the 6-DOF pose of the vehicle, commercial IMUs also usually estimate velocity and acceleration. To estimate the velocity, the initial speed of the vehicle needs to be known. The working principle of an IMU is shown in figure 4.10. Let us suppose that our

IMU has three orthogonal accelerometers and three orthogonal gyroscopes. The gyroscope data is integrated to estimate the vehicle orientation while the three accelerometers are used to estimate the instantaneous acceleration of the vehicle. The acceleration is then transformed to the local navigation frame by means of the current estimate of the vehicle orientation relative to gravity. At this point the gravity vector can be subtracted from the measurement. The resulting acceleration is then integrated to obtain the velocity and then integrated again to obtain the position, provided that both the initial velocity and position are a priori known. To overcome the need of knowing of the initial velocity, the integration is typically started at rest (i.e., velocity equal to zero).

Observe that IMUs are extremely sensitive to measurement errors in both gyroscopes and accelerometers. For example, drift in the gyroscope unavoidably undermines the estimation of the vehicle orientation relative to gravity, which results in incorrect cancellation of the gravity vector. Additionally observe that, because the accelerometer data is integrated twice to obtain the position, any residual gravity vector results in a quadratic error in position. Because of this and the fact that any other error is integrated over time, drift is a fundamental problem in IMUs. After long period of operation, all IMUs drift. To cancel this drift, some reference to some external measurement is required. In many robot applications, this has been done using cameras or GPS. In particular, cameras allow the user to annihilate the drift every time a given feature of the environment—whose 3D position in the camera reference frame is known—is reobserved (see sections 4.2.6 or 5.8.5). Similarly, as described in the next section, GPS allows the user to correct the pose estimate every time the GPS signal is received.

4.1.8 Ground beacons

One elegant approach to solving the localization problem in mobile robotics is to use active or passive beacons. Using the interaction of on-board sensors and the environmental beacons, the robot can identify its position precisely. Although the general intuition is identical to that of early human navigation beacons, such as stars, mountains, and lighthouses, modern technology has enabled sensors to localize an outdoor robot with accuracies of better than 5 cm within areas that are kilometers in size.

In the following section, we describe one such beacon system, the global positioning system (GPS), which is extremely effective for outdoor ground and flying robots. Indoor beacon systems have been generally less successful for a number of reasons. The expense of environmental modification in an indoor setting is not amortized over an extremely large useful area, as it is, for example, in the case of the GPS. Furthermore, indoor environments offer significant challenges not seen outdoors, including multipath and environmental dynamics. A laser indoor beacon system, for example, must disambiguate the one true laser signal from possibly tens of other powerful signals reflected off walls, smooth floors, and doors. Confounding this, humans and other obstacles may be constantly changing the envi-

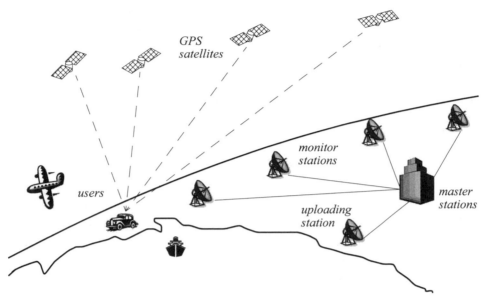

Figure 4.11
Calculation of position and heading based on GPS.

ronment, for example, occluding the one true path from the beacon to the robot. In commercial applications, such as manufacturing plants, the environment can be carefully controlled to ensure success. In less structured indoor settings, beacons have nonetheless been used, and the problems are mitigated by careful beacon placement and the use of passive sensing modalities.

4.1.8.1 The global positioning system

The *global positioning system* (GPS) was initially developed for military use but is now freely available for civilian navigation. There are at least twenty-four operational GPS satellites at all times. The satellites orbit every twelve hours at a height of 20.190 km. Four satellites are located in each of six planes inclined 55 degrees with respect to the plane of the earth's equator (figure 4.11).

Each satellite continuously transmits data that indicate its location and the current time. Therefore, GPS receivers are completely passive but exteroceptive sensors. The GPS satellites synchronize their transmissions so that their signals are sent at the same time. When a GPS receiver reads the transmission of two or more satellites, the arrival time differences inform the receiver as to its relative distance to each satellite. By combining information regarding the arrival time and instantaneous location of four satellites, the receiver can infer its own position. In theory, such triangulation requires only three data points. However, timing is extremely critical in the GPS application because the time intervals being measured are in nanoseconds. It is, of course, mandatory that the satellites be well synchronized. To this end, they are updated by ground stations regularly and each satellite carries on-board atomic clocks for timing.

The GPS receiver clock is also important so that the travel time of each satellite's transmission can be accurately measured. But GPS receivers have a simple quartz clock. So, although three satellites would ideally provide position in three axes, the GPS receiver requires four satellites, using the additional information to solve for four variables: three position axes plus a time correction.

The fact that the GPS receiver must read the transmission of four satellites simultaneously is a significant limitation. GPS satellite transmissions are extremely low-power, and reading them successfully requires direct line-of-sight communication with the satellite. Thus, in confined spaces such as city blocks with tall buildings or in dense forests, one is unlikely to receive four satellites reliably. Of course, most indoor spaces will also fail to provide sufficient visibility of the sky for a GPS receiver to function. For these reasons, the GPS has been a popular sensor in mobile robotics, but it has been relegated to projects involving mobile robot traversal of wide-open spaces and autonomous flying machines.

A number of factors affect the performance of a localization sensor that makes use of the GPS. First, it is important to understand that, because of the specific orbital paths of the GPS satellites, coverage is not geometrically identical in different portions of Earth and therefore resolution is not uniform. Specifically, at the North and South Poles, the satellites are very close to the horizon, and thus, while resolution in the latitude and longitude directions is good, resolution of altitude is relatively poor as compared to more equatorial locations.

The second point is that GPS satellites are merely an information source. They can be employed with various strategies in order to achieve dramatically different levels of localization resolution. The basic strategy for GPS use, called *pseudorange* and described earlier, generally performs at a resolution of 15 m. An extension of this method is *differential GPS (DGPS)*, which makes use of a second receiver that is static and at a known exact position. A number of errors can be corrected using this reference, and so resolution improves to the order of 1 m or less. A disadvantage of this technique is that the stationary receiver must be installed, its location must be measured very carefully, and of course the moving robot must be within kilometers of this static unit in order to benefit from the DGPS technique.

A further improved strategy is to take into account the phase of the carrier signals of each received satellite transmission. There are two carriers, at 19 cm and 24 cm, and therefore significant improvements in precision are possible when the phase difference between multiple satellites is measured successfully. Such receivers can achieve 1 cm resolution for point positions and, with the use of multiple receivers, as in DGPS, sub-1 cm resolution.

A final consideration for mobile robot applications is bandwidth. The GPS will generally offer no better than 200 to 300 ms latency, and so one can expect no better than 5 Hz GPS updates. On a fast-moving mobile robot or flying robot, this can mean that local motion integration will be required for proper control due to GPS latency limitations.

4.1.9 Active ranging

Active ranging sensors continue to be the most popular sensors in mobile robotics. Many ranging sensors have a low price point, and, most important, all ranging sensors provide easily interpreted outputs: direct measurements of distance from the robot to objects in its vicinity. For obstacle detection and avoidance, most mobile robots rely heavily on active ranging sensors. But the local free space information provided by ranging sensors can also be accumulated into representations beyond the robot's current local reference frame. Thus active ranging sensors are also commonly found as part of the localization and environmental modeling processes of mobile robots. It is only with the slow advent of successful visual interpretation competence that we can expect the class of active ranging sensors to gradually lose their primacy as the sensor class of choice among mobile roboticists.

We next present three *time-of-flight* active ranging sensors: the ultrasonic sensor, the laser rangefinder, and the time-of-flight camera. Then, we present two geometric active ranging sensors: the optical triangulation sensor and the structured light sensor.

4.1.9.1 Time-of-flight active ranging

Time-of-flight ranging makes use of the propagation speed of sound or an electromagnetic wave. In general, the travel distance of a sound of electromagnetic wave is given by

$$d = c \cdot t, \tag{4.20}$$

where

d = distance traveled (usually round-trip);

c = speed of wave propagation;

t = time of flight.

It is important to point out that the propagation speed v of sound is approximately 0.3 m/ms whereas the speed of electromagnetic signals is 0.3 m/ns, which is 1 million times faster. The time of flight for a typical distance, say 3 m, is 10 ms for an ultrasonic system but only 10 ns for a laser rangefinder. It is thus evident that measuring the time of flight t with electromagnetic signals is more technologically challenging. This explains why laser range sensors have only recently become affordable and robust for use on mobile robots.

The quality of time-of-flight range sensors depends mainly on

- uncertainties in determining the exact time of arrival of the reflected signal;

- inaccuracies in the time-of-flight measurement (particularly with laser range sensors);

- the dispersal cone of the transmitted beam (mainly with ultrasonic range sensors);

- interaction with the target (e.g., surface absorption, specular reflections);

- variation of propagation speed;

- the speed of the mobile robot and target (in the case of a dynamic target);

As discussed in the following, each type of time-of-flight sensor is sensitive to a particular subset of this list of factors.

The ultrasonic sensor (time-of-flight, sound). The basic principle of an ultrasonic sensor is to transmit a packet of (ultrasonic) pressure waves and to measure the time it takes for this wave packet to reflect and return to the receiver. The distance d of the object causing the reflection can be calculated based on the propagation speed of sound c and the time of flight t.

$$d = \frac{c \cdot t}{2}. \tag{4.21}$$

The speed of sound c in air is given by

$$c = \sqrt{\gamma R T}, \tag{4.22}$$

where

γ = ratio of specific heats;

R = gas constant;

T = temperature in degrees Kelvin.

In air at standard pressure and 20° C the speed of sound is approximately c = 343 m/s.

Figure 4.12 shows the different signal output and input of an ultrasonic sensor. First, a series of sound pulses are emitted, comprising the *wave packet*. An integrator also begins to linearly climb in value, measuring the time from the transmission of these sound waves to detection of an echo. A threshold value is set for triggering an incoming sound wave as a valid echo. This threshold is often decreasing in time, because the amplitude of the expected echo decreases over time based on dispersal as it travels longer. But during transmission of the initial sound pulses and just afterward, the threshold is set very high to suppress triggering the echo detector with the outgoing sound pulses. A transducer will

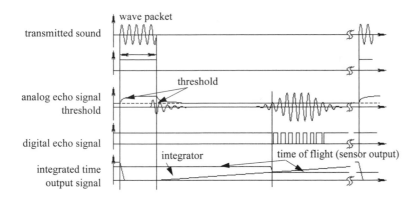

Figure 4.12
Signals of an ultrasonic sensor.

continue to ring for up to several milliseconds after the initial transmission, and this governs the *blanking time* of the sensor. Note that if, during the blanking time, the transmitted sound were to reflect off of an extremely close object and return to the ultrasonic sensor, it may fail to be detected.

However, once the blanking interval has passed, the system will detect any above-threshold reflected sound, triggering a digital signal and producing the distance measurement using the integrator value.

The ultrasonic wave typically has a frequency between 40 and 180 kHz and is usually generated by a piezo or electrostatic transducer. Often the same unit is used to measure the reflected signal, although the required blanking interval can be reduced through the use of separate output and input devices. Frequency can be used to select a useful range when choosing the appropriate ultrasonic sensor for a mobile robot. Lower frequencies correspond to a longer range, but with the disadvantage of longer post-transmission ringing and, therefore, the need for longer blanking intervals. Most ultrasonic sensors used by mobile robots have an effective range of roughly 12 cm to 5 m. The published accuracy of commercial ultrasonic sensors varies between 98% and 99.1%. In mobile robot applications, specific implementations generally achieve a resolution of approximately 2 cm.

In most cases one may want a narrow opening angle for the sound beam in order to also obtain precise directional information about objects that are encountered. This is a major limitation, since sound propagates in a conelike manner (figure 4.13) with opening angles around 20 to 40 degrees. Consequently, when using ultrasonic ranging one does not acquire depth data points but, rather, entire regions of constant depth. This means that the sensor tells us only that there is an object at a certain distance within the area of the measurement cone. The sensor readings must be plotted as segments of an arc (sphere for 3D) and not as

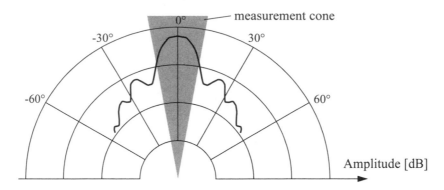

Figure 4.13
Typical intensity distribution of an ultrasonic sensor.

point measurements (figure 4.14). However, recent research developments show significant improvement of the measurement quality in using sophisticated echo processing [149].

Ultrasonic sensors suffer from several additional drawbacks, namely in the areas of error, bandwidth, and cross-sensitivity. The published accuracy values for ultrasonics are nominal values based on successful, perpendicular reflections of the sound wave off an acoustically reflective material. This does not capture the effective error modality seen on a mobile robot moving through its environment. As the ultrasonic transducer's angle to the object being ranged varies away from perpendicular, the chances become good that the sound waves will coherently reflect away from the sensor, just as light at a shallow angle reflects off of a smooth surface. Therefore, the true error behavior of ultrasonic sensors is compound, with a well-understood error distribution near the true value in the case of a successful retroreflection, and a more poorly understood set of range values that are grossly larger than the true value in the case of coherent reflection. Of course, the acoustic properties of the material being ranged have direct impact on the sensor's performance. Again, the impact is discrete, with one material possibly failing to produce a reflection that is sufficiently strong to be sensed by the unit. For example, foam, fur, and cloth can, in various circumstances, acoustically absorb the sound waves.

A final limitation of ultrasonic ranging relates to bandwidth. Particularly in moderately open spaces, a single ultrasonic sensor has a relatively slow cycle time. For example, measuring the distance to an object that is 3 m away will take such a sensor 20 ms, limiting its operating speed to 50 Hz. But if the robot has a ring of twenty ultrasonic sensors, each firing sequentially and measuring to minimize interference between the sensors, then the ring's cycle time becomes 0.4 seconds and the overall update frequency of any one sensor is just 2.5 Hz. For a robot conducting moderate speed motion while avoiding obstacles

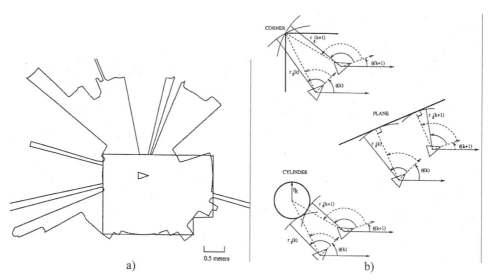

Figure 4.14
Typical readings of an ultrasonic system: (a) 360 degree scan; (b) results from different geometric primitives [35]. Courtesy of John Leonard, MIT.

using ultrasonics, this update rate can have a measurable impact on the maximum speed possible while still sensing and avoiding obstacles safely.

Laser rangefinder (time-of-flight, electromagnetic). The laser rangefinder is a time-of-flight sensor that achieves significant improvements over the ultrasonic range sensor owing to the use of laser light instead of sound. This type of sensor consists of a transmitter that illuminates a target with a collimated beam (e.g., laser), and a receiver capable of detecting the component of light, which is essentially coaxial with the transmitted beam. Often referred to as optical radar or *lidar* (light detection and ranging), these devices produce a range estimate based on the time needed for the light to reach the target and return. A mechanical mechanism with a mirror sweeps the light beam to cover the required scene in a plane or even in three dimensions, using a rotating, nodding mirror.

One way to measure the time of flight for the light beam is to use a pulsed laser and then measure the elapsed time directly, just as in the ultrasonic solution described earlier. Electronics capable of resolving picoseconds are required in such devices and they are therefore

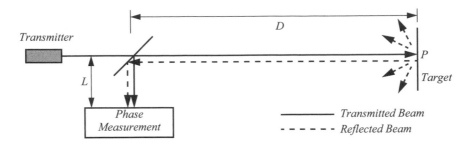

Figure 4.15
Schematic of laser rangefinding by phase-shift measurement.

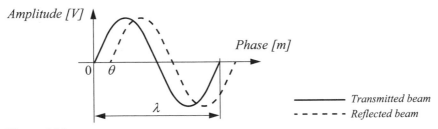

Figure 4.16
Range estimation by measuring the phase shift between transmitted and received signals.

very expensive. A second method is to measure the beat frequency between a frequency-modulated continuous wave (FMCW) and its received reflection. Another, even easier method is to measure the phase shift of the reflected light. We describe this third approach in detail.

Phase-shift measurement. Near-infrared light (from a light-emitting diode [LED] or laser) is collimated and transmitted from the transmitter in figure 4.15 and hits a point P in the environment. For surfaces having a roughness greater than the wavelength of the incident light, diffuse reflection will occur, meaning that the light is reflected almost isotropically. The wavelength of the infrared light emitted is 824 nm, and so most surfaces, with the exception of only highly polished reflecting objects, will be diffuse reflectors. The component of the infrared light that falls within the receiving aperture of the sensor will return almost parallel to the transmitted beam for distant objects.

The sensor transmits 100% amplitude-modulated light at a known frequency and measures the phase shift between the transmitted and reflected signals. Figure 4.16 shows how this technique can be used to measure range. The wavelength of the modulating signal obeys the equation $c = f \cdot \lambda$ where c is the speed of light and f the modulating frequency.

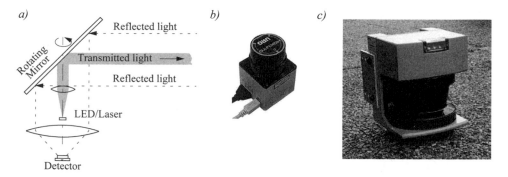

Figure 4.17
(a) Schematic drawing of laser range sensor with rotating mirror; (b) 240-degree laser rangefinder
from Hokuyo Ltd.; (c) Industrial 180 degree laser range sensor from Sick Inc., Germany.

For f = 5 MHz (as in the AT&T sensor), λ = 60 m. The total distance D' covered by the
emitted light is

$$D' = L + 2D = L + \frac{\theta}{2\pi}\lambda, \tag{4.23}$$

where D and L are the distances defined in figure 4.15. The required distance D between
the beam splitter and the target is therefore given by

$$D = \frac{\lambda}{4\pi}\theta, \tag{4.24}$$

where θ is the electronically measured phase difference between the transmitted and
reflected light beams, and λ the known modulating wavelength. It can be seen that the
transmission of a single frequency-modulated wave can theoretically result in ambiguous
range estimates since, for example, if λ = 60 m, a target at a range of 5 m would give an
indistinguishable phase measurement from a target at 35 m, since each phase angle would
be 360 degrees apart. We therefore define an "ambiguity interval" of λ, but in practice we
note that the range of the sensor is much lower than λ due to the attenuation of the signal
in air.

It can be shown that the confidence in the range (phase estimate) is inversely proportional to the square of the received signal amplitude, directly affecting the sensor's accuracy. Hence dark, distant objects will not produce as good range estimates as close, bright
objects.

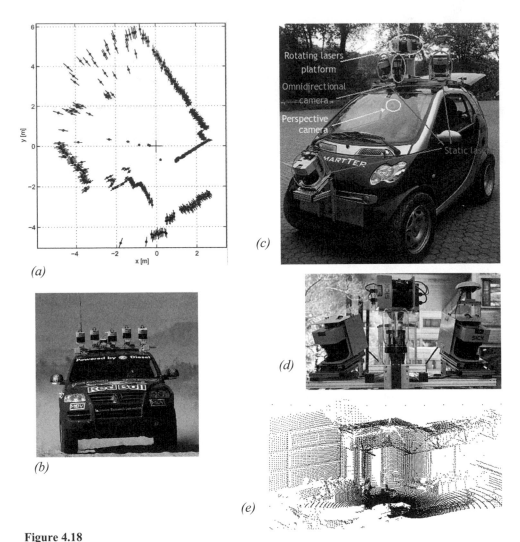

Figure 4.18
(a) Typical range image of a 2D laser range sensor with a rotating mirror. The length of the lines through the measurement points indicate the uncertainties. (b) Stanley, the autonomous car from Stanford winning the 2005 Darpa Grand Challenge. (c) The Smarter, the autonomous car developed at the ASL (ETH Zurich). (d) A close view at the Sicks used on the Smarter. (e) A 3D laser-point-cloud built from the rotating Sicks.

In figure 4.17, the schematic of a typical 360-degree laser range sensor and two examples are shown. Figure 4.18a shows a typical range image of a 360-degree scan taken with a laser range sensor.

As expected, the angular resolution of laser rangefinders far exceeds that of ultrasonic sensors. The Sick LMS 200 laser scanner shown in figure 4.17c achieves an angular resolution of 0.25 degree. Depth resolution ranges between 10 and 15 mm and the typical accuracy is 35 mm, over a range from 5 cm up to 20 m or more (up to 80 m), depending on the reflectivity of the object being ranged. This device performs seventy five 180-degrees scans per second but has no mirror nodding capability for the vertical dimension.

As an example of use in mobile robotics, five Sick lasers were used for short range detection on Stanley (figure 4.18b), the autonomous car that won the 2005 DARPA Grand Challenge. In a different configuration, five Sick lasers were also used on the Smarter (figure 4.18c), the autonomous car developed at the ASL (ETH Zurich), which participated in ELROB 2006, the European Land Robot Trial. On the Smarter, one Sick laser at the lower front was used for close obstacle avoidance, while two lasers on the roof (figure 4.18d), slightly canted to the sides, were used for local navigation. Finally, another two lasers, mounted vertically on a turntable (figure 4.18d), were used as a 3D range scanner for 3D mapping.

As with ultrasonic ranging sensors, an important error mode involves coherent reflection of the energy. With light, this will occur only when striking a highly polished surface. Practically, a mobile robot may encounter such surfaces in the form of a polished desktop, file cabinet or, of course, a mirror. Unlike ultrasonic sensors, laser rangefinders cannot detect the presence of optically transparent materials such as glass, and this can be a significant obstacle in environments like, for example, museums, where glass is commonly used

3D laser rangefinders. A 3D laser rangefinder is a laser scanner that acquires scan data in more than a single plane. Custom-made 3D scanners are typically built by nodding or rotating a 2D scanner in a stepwise or continuous manner around an axis parallel to the scanning plane. An example custom-built 3D scanner was developed at the ASL for the Smarter (figure 4.18d). In this case, two Sick lasers were positioned to look into opposite directions. This way, after half rotation of the turntable, a full 3D scan of the environment around the vehicle could be acquired. This data was mainly used to compute a consistent 3D digital terrain model of the environment (figure 4.18e). By lowering the rotational speed of the turntable, the angular resolution in the horizontal direction can be made as small as desired. The advantage of this setting is that the full spherical field of view can be covered (360° in azimuth and 90° in elevation). The drawback is that the acquisition time for a full 3D scan can take up to several seconds depending on the desired resolution. For instance, consider that our Sick scanner acquires 75 vertical plane scans per second and that we need an azimuthal angular resolution of 0.25 degrees. The period for a half-rotation of the turntable necessary to capture a spherical 3D scan with two Sicks is then 360 / 0.25 / 75 / 2 = 9.6 seconds. If one is satisfied with an azimuthal angular resolution of 1 degree, then the acquisition time drops down to 2.4 seconds. This, of course, limits the use of this configuration

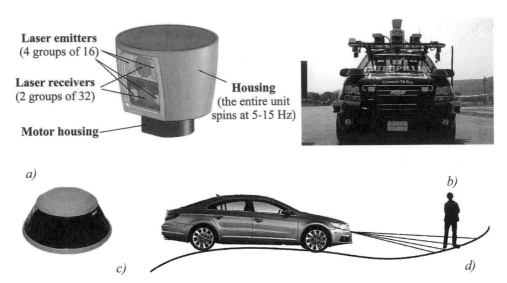

Figure 4.19 (a) The Velodyne HDL-64E unit features 64 laser beams and spins up to 15 Hz to gather data (image courtesy of Velodyne—http://www.velodyne.com/lidar). It delivers over 1.3 million data points per second. (d) The working principle of the Ibeo Alasca XT (c), which uses a four-layer laser beam. The Alasca and Velodyne were used by the CMU Tartan Racing team (b) (image courtesy of the Tartan Racing team).

to static environments. As a matter of fact, the rotating Sicks developed at the ASL (ETH Zurich) were used on an autonomous car running at 10 km/h. In this case, very accurate (up to centimeter) vehicle motion estimation was necessary to correct the errors in the 3D data caused by the movement of the car.

The Velodyne HDL-64E (figure 4.19a) overcomes the drawbacks of custom-made 3D laser range finders. This sensor is a 3D lidar that uses 64 laser emitters instead of the single one used in the Sick. This device spins at rates of 5–15 Hz and delivers more than 1.3 million data points per second. The field of view is 360° in azimuth and 26.8° in elevation and the angular resolution is 0.09° and 0.4° respectively. The distance accuracy is better than 2 cm and can measure depth up to 50 m, with 10% reflectivity, or up to 120 m, with 80% reflectivity. This sensor was the primary means of terrain map construction and obstacle detection for all the top DARPA 2007 Urban Challenge teams. However, the Velodyne is currently still much more expensive than Sick laser rangefinders.

The laser scanner Alasca XT, produced by Ibeo (figure 4.19c–d), on the other hand, splits the laser beam into four vertical layers. Distance measurements are taken independently for each of these layers with an aperture angle of 3.2°. This sensor is typically used for obstacle and pedestrian detection on cars. Because of its multilayer scanning principle, it allows us any pitching of the vehicle (caused by an uneven surface or driving manoeuvres

(a)   *(b)*   *(c)*

Figure 4.20 (a) The ZCAM produced by the Israeli developer 3DV Systems; (b) the Swiss Ranger SR3000 produced by the Swiss company MESA; (c) Range image of a chair captured with a time-of-flight camera (image courtesy of S. Gächter).

such as braking and accelerating) to be fully compensated. This sensor was used by the Tartan Racing team in the autonomous car from CMU that won the 2007 Urban Grand Challenge. Additionally, it also used the Velodyne sensor (figure 4.19b).

Time-of-flight camera. A Time-of-Flight camera (TOF camera, figure 4.20) works similarly to a lidar with the advantage that the whole 3D scene is captured at the same time and that there are no moving parts. This device uses a modulated infrared lighting source to determine the distance for each pixel of a Photonic Mixer Device (PMD) sensor. As the illumination source is placed just next to the lens (figure 4.20), the whole system is very compact compared to lidars, stereo vision, or triangulation sensors (see below). In the presence of background light, the image sensor receives an additional illumination signal which disturbs the distance measurement. To eliminate the background part of the signal, the acquisition is done a second time with the illumination switched off. As the scene is captured in one shot, the camera reaches up to 100 frames per second and is therefore ideally suited for real-time applications.

The PMD sensor appeared the first time in 1997, but TOF cameras became popular only a few years later, when the semiconductor processes became fast enough for such devices. This sensor typically covers ranges from 0.5 m up to 8 m, but even larger ranges are possible. The distance resolution is about 1 cm. Typical images sizes are quite small: in the two examples shown in figure 4.20, the Swiss Ranger SR3000 by MESA has 174×144 pixels, while the ZCAM by 3DV Systems has 320×240 pixels with 256 depth levels per pixel. The Swiss Ranger has found many robotic applications including map building, obstacle avoidance, and recognition [134, 330]; the ZCAM has been used in video game consoles for player motion detection and activity recognition.

Because it is very easy to extract distance information from the TOF sensor, TOF cameras use less processing power than stereo vision, where complex correlation algorithms are

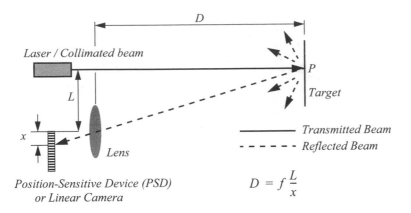

Figure 4.21
Principle of 1D laser triangulation.

used. Additionally, the extracted range information is not disturbed by the patterns on the object as it happens in stereovision.

4.1.9.2 Triangulation active ranging

Triangulation ranging sensors use geometric properties manifest in their measuring strategy to establish distance readings to objects. The simplest class of triangulation rangers are *active* because they project a known light pattern (e.g., a point, a line, or a texture) onto the environment. The reflection of the known pattern is captured by a receiver and, together with known geometric values, the system can use simple triangulation to establish range measurements. If the receiver measures the position of the reflection along a single axis, we call the sensor an optical triangulation sensor in 1D. If the receiver measures the position of the reflection along two orthogonal axes, we call the sensor a structured light sensor. These two sensor types are described in the two sections below.

Optical triangulation (1D sensor). The principle of optical triangulation in 1D is straightforward, as depicted in figure 4.21. A collimated beam (e.g., focused infrared LED, laser beam) is transmitted toward the target. The reflected light is collected by a lens and projected onto a position-sensitive device (PSD) or linear camera. Given the geometry of figure 4.21, the distance D is given by

$$D = f\frac{L}{x}.$$

(4.25)

Figure 4.22
A commercially available, low-cost optical triangulation sensor: the Sharp GP series infrared range-finders provide either analog or digital distance measures and cost only about $15.

The distance is proportional to $1/x$; therefore the sensor resolution is best for close objects and becomes poor at a distance. Sensors based on this principle are used in range sensing up to 1 or 2 m, but also in high-precision industrial measurements with resolutions far below 1 μm.

Optical triangulation devices can provide relatively high accuracy with very good resolution (for close objects). However, the operating range of such a device is normally fairly limited by geometry. For example, the optical triangulation sensor pictured in figure 4.22 operates over a distance range of between 8 and 80 cm. It is inexpensive compared to ultrasonic and laser rangefinder sensors. Although more limited in range than sonar, the optical triangulation sensor has high bandwidth and does not suffer from cross-sensitivities that are more common in the sound domain.

Structured light (2D sensor). If one replaces the linear camera or PSD of an optical triangulation sensor with a 2D receiver such as a CCD or CMOS camera, then one can recover distance to a large set of points instead of to only one point. The emitter must project a known pattern, or *structured light*, onto the environment. Many systems exist which either project light textures (figure 4.23b) or emit collimated light (possibly laser) by means of a rotating mirror. Yet another popular alternative is to project a laser stripe (figure 4.23a) by turning a laser beam into a plane using a prism. Regardless of how it is created, the projected light has a known structure, and therefore the image taken by the CCD or CMOS receiver can be filtered to identify the pattern's reflection.

Note that the problem of recovering depth is in this case far simpler than the problem of passive image analysis. In passive image analysis, as we discuss later, existing features in the environment must be used to perform *correlation*, while the present method projects a

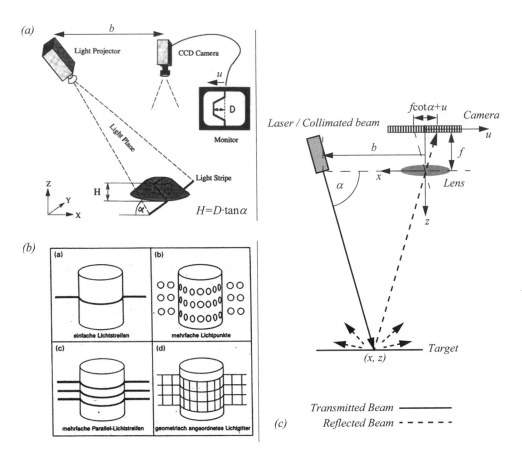

Figure 4.23
(a) Principle of active two dimensional triangulation. (b) Other possible light structures. (c) 1D schematic of the principle. Images (a) and (b) courtesy of Albert-Jan Baerveldt, Halmstad University.

known pattern upon the environment and thereby avoids the standard correlation problem altogether. Furthermore, the structured light sensor is an active device so it will continue to work in dark environments as well as environments in which the objects are featureless (e.g., uniformly colored and edgeless). In contrast, stereo vision would fail in such texture-free circumstances.

Figure 4.23c shows a 1D active triangulation geometry. We can examine the trade-off in the design of triangulation systems by examining the geometry in figure 4.23c. The measured values in the system are α and u, the distance of the illuminated point from the origin in the imaging sensor. Note the imaging sensor here can be a camera or an array of photo diodes of a position-sensitive device (e.g., a PSD).

From figure 4.23c, simple geometry shows that

$$x = \frac{b \cdot u}{f \cot \alpha + u} \; ; \quad z = \frac{b \cdot f}{f \cot \alpha + u}, \tag{4.26}$$

where f is the distance of the lens to the imaging plane. In the limit, the ratio of image resolution to range resolution is defined as the triangulation gain G_p and from equation (4.26) is given by

$$\frac{\partial u}{\partial z} = G_p = \frac{b \cdot f}{z^2}. \tag{4.27}$$

This shows that the ranging accuracy, for a given image resolution, is proportional to source/detector separation b and focal length f, and decreases with the square of the range z. In a scanning ranging system, there is an additional effect on the ranging accuracy, caused by the measurement of the projection angle α. From equation 4.26 we see that

$$\frac{\partial \alpha}{\partial z} = G_\alpha = \frac{b \sin \alpha^2}{z^2}. \tag{4.28}$$

We can summarize the effects of the parameters on the sensor accuracy as follows:

- *Baseline length (b)*: the smaller b is, the more compact the sensor can be. The larger b is, the better the range resolution will be. Note also that although these sensors do not suffer from the correspondence problem, the disparity problem still occurs. As the baseline length b is increased, one introduces the chance that, for close objects, the illuminated point(s) may not be in the receiver's field of view.

- *Detector length and focal length (f)*: A larger detector length can provide either a larger field of view or an improved range resolution or partial benefits for both. Increasing the detector length, however, means a larger sensor head and worse electrical characteristics (increase in random error and reduction of bandwidth). Also, a short focal length gives a large field of view at the expense of accuracy, and vice versa.

At one time, laser stripe structured light sensors were common on several mobile robot bases as an inexpensive alternative to laser rangefinding devices. However, with the increasing quality of laser rangefinding sensors in the 1990s, the structured light system has become relegated largely to vision research rather than applied mobile robotics. However, new possibilities of applications for robotics have recently been opened by Kinect, the sensor released in 2010 within the Microsoft Xbox 360 videogame console, and produced

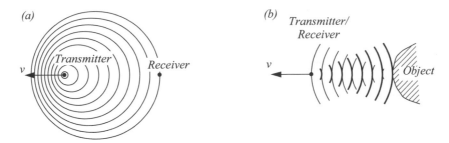

Figure 4.24
Doppler effect between two moving objects (a) or a moving and a stationary object (b).

by the Israeli company PrimeSense. Kinect is a very cheap range camera that uses the structured-light principle explained before. An infrared laser emitter is used to make the projected pattern invisible to the human eye [132].

4.1.10 Motion/speed sensors

Some sensors measure directly the relative motion between the robot and its environment. Since such motion sensors detect relative motion, so long as an object is moving relative to the robot's reference frame, it will be detected and its speed can be estimated. There are a number of sensors that inherently measure some aspect of motion or change. For example, a pyroelectric sensor detects change in heat. When a human walks across the sensor's field of view, his or her motion triggers a change in heat in the sensor's reference frame. In the next section, we describe an important type of motion detector based on the Doppler effect. These sensors represent a well-known technology with decades of general applications behind them. For fast-moving mobile robots such as autonomous highway vehicles and unmanned flying vehicles, Doppler motion detectors are the obstacle detection sensor of choice.

4.1.10.1 Doppler effect sensing (radar or sound)

Anyone who has noticed the change in siren pitch that occurs when an approaching fire engine passes by and recedes is familiar with the Doppler effect.

A transmitter emits an electromagnetic or sound wave with a frequency f_t. It is either received by a receiver (figure 4.24a) or reflected from an object (figure 4.24b). The measured frequency f_r at the receiver is a function of the relative speed v between transmitter and receiver according to

$$f_r = f_t \frac{1}{1 + v/c},$$

$$(4.29)$$

if the transmitter is moving and

$$f_r = f_t(1 + v/c), \tag{4.30}$$

if the receiver is moving.

In the case of a reflected wave (figure 4.24b) there is a factor of 2 introduced, since any change x in relative separation affects the round-trip path length by $2x$. Furthermore, in such situations it is generally more convenient to consider the change in frequency Δf, known as the *Doppler shift*, as opposed to the *Doppler frequency* notation above.

$$\Delta f = f_t - f_r = \frac{2f_t v \cos\theta}{c}, \tag{4.31}$$

$$v = \frac{\Delta f \cdot c}{2f_t \cos\theta}, \tag{4.32}$$

where

Δf = Doppler frequency shift;

θ = relative angle between direction of motion and beam axis.

The Doppler effect applies to sound and electromagnetic waves. It has a wide spectrum of applications:

- *Sound waves*: for example, industrial process control, security, fish finding, measure of ground speed.

- *Electromagnetic waves*: for example, vibration measurement, radar systems, object tracking.

A current application area is both autonomous and manned highway vehicles. Both microwave and laser radar systems have been designed for this environment. Both systems have equivalent range, but laser can suffer when visual signals are deteriorated by environmental conditions such as rain, fog, and so on. Commercial microwave radar systems are already available for installation on highway trucks. These systems are called VORAD (vehicle on-board radar) and have a total range of approximately 150 m. With an accuracy of approximately 97%, these systems report range rates from 0 to 160 km/hr with a resolution of 1 km/hr. The beam is approximately 4 degrees wide and 5 degrees in elevation. One of the key limitations of radar technology is its bandwidth. Existing systems can provide information on multiple targets at approximately 2 Hz.

4.1.11 Vision sensors

Vision is our most powerful sense. It provides us with an enormous amount of information about the environment and enables rich, intelligent interaction in dynamic environments. It is therefore not surprising that a great deal of effort has been devoted to providing machines with sensors that mimic the capabilities of the human vision system. The first step in this process is the creation of sensing devices that capture the light and convert it into a digital image. The second step is the processing of the digital image in order to get salient information like depth computation, motion detection, color tracking, feature detection, scene recognition, and so on. Because vision sensors have become very popular in robotic applications, the remaining sections of this chapter will be dedicated to the fundamentals of computer vision and image processing and their use in robotics.

4.2 Fundamentals of Computer Vision

4.2.1 Introduction

The analysis of images and their processing are two major fields that are known as computer vision and image processing. The years between 1980 and 2010 have seen significant advances and new theoretical findings in these fields and some of the most sophisticated computer vision and image processing techniques have found many industrial applications in consumer cameras, photography, defect inspection, monitoring and surveillance, video games, movies, and the like. For more information on the computer vision industry, see [346].

The remaining parts of this chapter are dedicated to these two fields. First, we will introduce the working principle of the digital camera, the imaging sensors, the optics, and the image formation; then, we will present two ways of estimating the depth, which are depth from focus and stereo vision. Next, we will detail some of the most important tools used in image processing. Finally, we will close this chapter by presenting state-of-the-art algorithms for feature extraction and place recognition from digital images.

For an in-depth study of computer vision, we refer the reader to the following books: [21, 29, 36, 49, 53].

4.2.2 The digital camera

After starting from one or more light sources, reflecting off of one or more surfaces in the world, and passing through the camera's optics (lenses), light finally reaches the imaging sensor. How are the photons arriving at this sensor converted into the digital (R,G,B) values that we observe when we look at a digital image?

Light falling on an imaging sensor is usually picked up by an active sensing area, integrated for the duration of the exposure (usually expressed as the shutter speed, e.g., 1/125, 1/60, 1/30 of a second), and then passed to a set of sense amplifiers. The two main kinds of

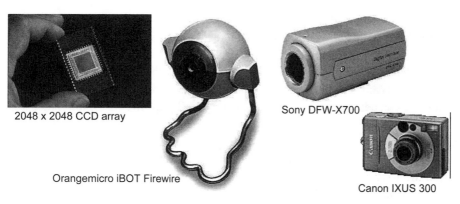

2048 x 2048 CCD array

Sony DFW-X700

Orangemicro iBOT Firewire

Canon IXUS 300

Figure 4.25
Commercially available CCD chips and CCD cameras. Because this technology is relatively mature, cameras are available in widely varying forms and costs.

sensors used in digital still and video cameras today are CCD (charge coupled device) and CMOS (complementary metal oxide on silicon). Below, we review the advantages and drawbacks of these two technologies.

CCD cameras. The CCD chip (see figure 4.25) is an array of light-sensitive picture elements, or pixels, usually with between 20,000 and several million pixels total. Each pixel can be thought of as a light-sensitive, discharging capacitor that is 5 to 25 µm in size. First, the capacitors of all pixels are charged fully, then the integration period begins. As photons of light strike each pixel, they liberate electrons, which are captured by electric fields and retained at the pixel. Over time, each pixel accumulates a varying level of charge based on the total number of photons that have struck it. After the integration period is complete, the relative charges of all pixels need to be frozen and read. In a CCD, the reading process is performed at one corner of the CCD chip. The bottom row of pixel charges is transported to this corner and read, then the rows above shift down and the process is repeated. This means that each charge must be transported across the chip, and it is critical that the value be preserved. This requires specialized control circuitry and custom fabrication techniques to ensure the stability of transported charges.

The photodiodes used in CCD chips (and CMOS chips as well) are not equally sensitive to all frequencies of light. They are sensitive to light between 400 and 1000 nm wavelength. It is important to remember that photodiodes are less sensitive to the ultraviolet end of the spectrum (e.g., blue) and are overly sensitive to the infrared portion (e.g., heat).

The CCD camera has several camera parameters that affect its behavior. In some cameras, these values are fixed. In others, the values are constantly changing based on built-in feedback loops. In higher-end cameras, the user can modify the values of these parameters via software. The *iris position* and *shutter speed* regulate the amount of light being measured by the camera. The iris is simply a mechanical aperture that constricts incoming light, just as in standard 35 mm cameras. Shutter speed regulates the integration period of the chip. In higher-end cameras, the effective shutter speed can be as brief at 1/30,000 seconds and as long as 2 seconds. *Camera gain* controls the overall amplification of the analog signal, prior to A/D conversion. However, it is very important to understand that even though the image may appear brighter after setting high gain, the shutter speed and iris may not have changed at all. Thus gain merely amplifies the signal, and it amplifies along with the signal all of the associated noise and error.

The key disadvantages of CCD cameras are primarily in the areas of inconstancy and dynamic range. As mentioned earlier, a number of parameters can change the brightness and colors with which a camera creates its image. Manipulating these parameters in a way to provide consistency over time and over environments, for example, ensuring that a green shirt always looks green, and something dark gray is always dark gray, remains an open problem in the vision community. For more details on the fields of color constancy and luminosity constancy, consult [65].

The second class of disadvantages relates to the behavior of a CCD chip in environments with extreme illumination. In cases of very low illumination, each pixel will receive only a small number of photons. The longest possible integration period (i.e., shutter speed) and camera optics (i.e., pixel size, chip size, lens focal length and diameter) will determine the minimum level of light for which the signal is stronger than random error noise. In cases of very high illumination, a pixel fills its well with free electrons. As the well reaches its limit, the probability of trapping additional electrons falls, and therefore the linearity between incoming light and electrons in the well degrades. Termed *saturation*, this can indicate the existence of a further problem related to cross-sensitivity. When a well has reached its limit, then additional light within the remainder of the integration period may cause further charge to leak into neighboring pixels, causing them to report incorrect values or even reach secondary saturation. This effect, called *blooming*, means that individual pixel values are not truly independent.

The camera parameters may be adjusted for an environment with a particular light level, but the problem remains that the dynamic range of a camera is limited by the well capacity (also called *well depth*) of the individual pixels. The well depth typically ranges between 20,000 and 350,000 electrons. For example, a high-quality CCD may have pixels that can hold 40,000 electrons. The noise level for reading the well may be 11 electrons, and therefore the dynamic range will be 40,000:11, or 3600:1, which is 35 dB.

CMOS cameras. The complementary metal oxide semiconductor chip is a significant departure from the CCD. It, too, has an array of pixels, but located along the side of each

Figure 4.26
A commercially available, low-cost CMOS camera with lens attached.

pixel are several transistors specific to that pixel. As in CCD chips, all of the pixels accumulate charge during the integration period. During the data collection step, the CMOS takes a new approach: the pixel-specific circuitry next to every pixel measures and amplifies the pixel's signal, all in parallel for every pixel in the array. Using more traditional traces from general semiconductor chips, the resulting pixel values are all carried to their destinations.

CMOS has a number of advantages over CCD technologies. First and foremost, there is no need for the specialized clock drivers and circuitry required in the CCD to transfer each pixel's charge down all of the array columns and across all of its rows. This also means that specialized semiconductor manufacturing processes are not required to create CMOS chips. Therefore, the same production lines that create microchips can create inexpensive CMOS chips as well (see figure 4.26). The CMOS chip is so much simpler that it consumes significantly less power; incredibly, it operates with a power consumption that is one-hundredth the power consumption of a CCD chip. In a mobile robot, especially flying, power is a scarce resource and therefore this is an important advantage.

Traditionally, CCD sensors outperformed CMOS in quality sensitive applications such as digital single-lens-reflex cameras, while CMOS was better for low-power applications, but today, CMOS is used in most digital cameras.

Given this summary of the mechanism behind CCD and CMOS chips, one can appreciate the sensitivity of any vision robot sensor to its environment. As compared to the human eye, these chips all have far poorer adaptation, cross-sensitivity, and dynamic range. As a result, vision sensors today continue to be fragile. Only over time, as the underlying performance of imaging chips improves, will significantly more robust vision sensors for mobile robots be available.

Camera output considerations. Although digital cameras have inherently digital output, throughout the 1980s and early 1990s, most affordable vision modules provided analog output signals, such as NTSC (National Television Standards Committee) and PAL (Phase Alternating Line). These camera systems included a D/A converter which, ironically, would be counteracted on the computer using a *frame grabber*, effectively an A/D converter board situated, for example, on a computer's bus. The D/A and A/D steps are far from noise free, and furthermore the color depth of the analog signal in such cameras was optimized for human vision, not computer vision.

More recently, both CCD and CMOS technology vision systems provide digital signals that can be directly utilized by the roboticist. At the most basic level, an imaging chip provides parallel digital I/O (input/output) pins that communicate discrete pixel level values. Some vision modules make use of these direct digital signals, which must be handled subject to hard-time constraints governed by the imaging chip. To relieve the real-time demands, researchers often place an *image buffer chip* between the imager's digital output and the computer's digital inputs. Such chips, commonly used in webcams, capture a complete image snapshot and enable non-real-time access to the pixels, usually in a single, ordered pass.

At the highest level, a roboticist may choose instead to utilize a higher-level digital transport protocol to communicate with an imager. Most common are the IEEE 1394 (Firewire) standard and the USB (and USB 2.0) standards, although some older imaging modules also support serial (RS-232). To use any such high-level protocol, one must locate or create driver code both for that communication layer and for the particular implementation details of the imaging chip. Take note, however, of the distinction between lossless digital video and the standard digital video stream designed for human visual consumption. Most digital video cameras provide digital output, but often only in compressed form. For vision researchers, such compression must be avoided as it not only discards information but even introduces image detail that does not actually exist, such as MPEG (Moving Picture Experts Group) discretization boundaries.

Color camera. The basic light-measuring process described before is colorless: it is just measuring the total number of photons that strike each pixel in the integration period. There are two common approaches for creating *color* images, which use a single chip or three separate chips.

The single chip technology uses the so-called *Bayer* filter. The pixels on the chip are grouped into 2×2 sets of four, then red, green, and blue color filters are applied so that each individual pixel receives only light of one color. Normally, two pixels of each 2×2 block measure green while the remaining two pixels measure red and blue light intensity (figure 4.27). The reason there are twice as many green filters as red and blue is that the

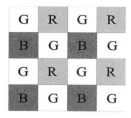

Figure 4.27 Bayer color filter array.

luminance signal is mostly determined by green values, and the visual system is much more sensitive to high frequency detail in luminance than in chrominance. The process of interpolating the missing color values so that we have valid RGB values as all the pixels is known as *demosaicing*. Of course, this one-chip technology has a geometric resolution disadvantage. The number of pixels in the system has been effectively cut by a factor of four, and therefore the image resolution output by the camera will be sacrificed.

The three-chip color camera avoids these problems by splitting the incoming light into three complete (lower intensity) copies. Three separate chips receive the light, with one red, green, or blue filter over each entire chip. Thus, in parallel, each chip measures light intensity for one color, and the camera must combine the chips' outputs to create a joint color image. Resolution is preserved in this solution, although the three-chip color cameras are, as one would expect, significantly more expensive and therefore more rarely used in mobile robotics.

Both three-chip and single-chip color cameras suffer from the fact that photodiodes are much more sensitive to the near-infrared end of the spectrum. This means that the overall system detects blue light much more poorly than red and green. To compensate, the gain must be increased on the blue channel, and this introduces greater absolute noise on blue than on red and green. It is not uncommon to assume at least one to two bits of additional noise on the blue channel. Although there is no satisfactory solution to this problem today, over time the processes for blue detection have been improved, and we expect this positive trend to continue.

In color cameras, an additional control exists for *white balance*. Depending on the source of illumination in a scene (e.g., fluorescent lamps, incandescent lamps, sunlight, underwater filtered light, etc.), the relative measurements of red, green, and blue light that define pure white light will change dramatically. The human eye compensates for all such effects in ways that are not fully understood, but, the camera can demonstrate glaring inconsistencies in which the same table looks blue in one image, taken during the night, and yellow in another image, taken during the day. White balance controls enable the user to change the relative gains for red, green, and blue in order to maintain more consistent color definitions in varying contexts.

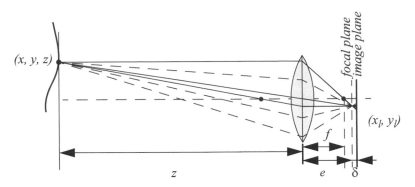

Figure 4.28
Depiction of the camera optics and its impact on the image. In order to get a sharp image, the image plane must coincide with the focal plane. Otherwise the image of the point (x,y,z) will be blurred in the image, as can be seen in the drawing above.

4.2.3 Image formation
Before we can intelligently analyze and manipulate images, we need to understand the image formation process that produced a particular image.

4.2.3.1 Optics
Once the light from the scene reaches the camera, it must still pass through the lens before reaching the sensor. Figure 4.28 shows a diagram of the most basic lens model, which is the thin lens. This lens is composed of a single piece of glass with very low, equal curvature on both sides. According to the lens law (which can be derived using simple geometric arguments on light ray refraction), the relationship between the distance to an object z and the distance behind the lens at which a focused image is formed e can be expressed as

$$\frac{1}{f} = \frac{1}{z} + \frac{1}{e}, \tag{4.33}$$

where f is the focal length. As you can perceive, this formula can also be used to estimate the distance to an object by knowing the focal length and the current distance of the image plane to the lens. This technique is called *depth from focus*.

If the image plane is located at distance e from the lens, then for the specific object voxel depicted, all light will be focused at a single point on the image plane and the object voxel will be *focused*. However, when the image plane is not at e, as is depicted in figure 4.28, then the light from the object voxel will be cast on the image plane as a *blur circle* (or *circle of confusion*). To a first approximation, the light is homogeneously distributed throughout this blur circle, and the radius R of the circle can be characterized according to the equation

$$R = \frac{L\delta}{2e}.$$ (4.34)

L is the diameter of the lens or aperture, and δ is the displacement of the image plane from the focal point.

Given these formulas, several basic optical effects are clear. For example, if the aperture or lens is reduced to a point, as in a pinhole camera, then the radius of the blur circle approaches zero. This is consistent with the fact that decreasing the iris aperture opening causes the *depth of field* to increase until all objects are in focus. Of course, the disadvantage of doing so is that we are allowing less light to form the image on the image plane, and so this is practical only in bright circumstances.

The second property that can be deduced from these optics equations relates to the sensitivity of blurring as a function of the distance from the lens to the object. Suppose the image plane is at a fixed distance 1.2 from a lens with diameter $L = 0.2$ and focal length $f = 0.5$. We can see from equation (4.34) that the size of the blur circle R changes proportionally with the image plane displacement δ. If the object is at distance $z = 1$, then from equation (4.33) we can compute $e = 1$, and therefore $\delta = 0.2$. Increase the object distance to $z = 2$ and as a result $\delta = 0.533$. Using equation (4.34) in each case, we can compute $R = 0.02$ and $R = 0.08$ respectively. This demonstrates high sensitivity for defocusing when the object is close to the lens.

In contrast, suppose the object is at $z = 10$. In this case we compute $e = 0.526$. But if the object is again moved one unit, to $z = 11$, then we compute $e = 0.524$. The resulting blur circles are $R = 0.117$ and $R = 0.129$, far less than the quadrupling in R when the obstacle is one-tenth the distance from the lens. This analysis demonstrates the fundamental limitation of depth from focus techniques: they lose sensitivity as objects move farther away (given a fixed focal length). Interestingly, this limitation will turn out to apply to virtually all visual ranging techniques, including depth from stereo (section 4.2.5) and depth from motion (section 4.2.6).

Nevertheless, camera optics can be customized for the depth range of the intended application. For example, a zoom lens with a very large focal length f will enable range resolution at significant distances, of course at the expense of field of view. Similarly, a large lens diameter, coupled with a very fast shutter speed, will lead to larger, more detectable blur circles.

Given the physical effects summarized by the above equations, one can imagine a visual ranging sensor that makes use of multiple images in which camera optics are varied (e.g., image plane displacement δ) and the same scene is captured (see figure 4.29). In fact, this

Figure 4.29
Two images of the same scene taken with a camera at two different focusing positions. Note the significant change in texture sharpness between the near surface and far surface. The scene is an outdoor concrete step.

approach is not a new invention. The human visual system uses an abundance of cues and techniques, and one system demonstrated in humans is depth from focus. Humans vary the focal length of their lens continuously at a rate of about 2 Hz. Such approaches, in which the lens optics are actively searched in order to maximize focus, are technically called *depth from focus* [241]. In contrast, *depth from defocus* means that depth is recovered using a series of images that have been taken with different camera geometries, and hence different focusing positions.

The depth from focus method is one of the simplest visual ranging techniques. To determine the range to an object, the sensor simply moves the image plane (via focusing) until maximizing the sharpness of the object. When the sharpness is maximized, the corresponding position of the image plane directly reports range. Some autofocus cameras and virtually all autofocus video cameras use this technique. Of course, a method is required for measuring the sharpness of an image or an object within the image.

An example application of depth-from-focus to robotics has been shown in [250], where the authors demonstrated obstacle avoidance in a variety of environments, as well as avoidance of concave obstacles such as steps and ledges.

4.2.3.2 Pinhole camera model

The pinhole camera, or *camera obscura*, has been the first example of camera in the history, which led to the invention of photography [27]. A pinhole camera has no lens, but a single very small aperture. In short, it is a lightproof box with a small hole in one side. Light from the scene passes through this single point and projects an inverted image on the opposite side of the box (figure 4.30). The working principle of this camera was already known as far back as the 4th century BC by the Greek Aristotle and Euclid and the Chinese Mozi.

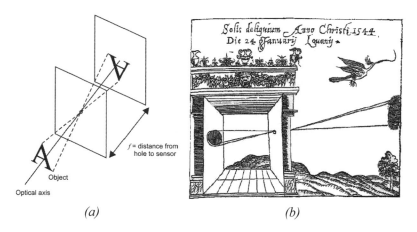

(a)   *(b)*

Figure 4.30 (a) When $d \gg f$ and $d \gg L$ the camera can be modeled as a pinhole camera. (b) The *camera obscura* in a drawing from mathematician Reinerus Gemma-Frisius (1508–1555), who used this illustration in his book *De Radio Astronomica et Geometrica* (1545) to describe an eclipse of the sun at Louvain on January 24, 1544. It is thought to be the first published illustration of a camera obscura [27].

The pinhole projection model was also used as drawing aid by artists such as Leonardo da Vinci (1452–1519).

The importance of the pinhole camera is that its principle has also been adopted as a standard model for perspective cameras. This model can be derived directly from equation (4.33). In fact, notice that if we let $z \to \infty$, i.e. we adjust the lens (move the image plane) so that objects at infinity are in focus (i.e. $z \gg f$ and $z \gg L$), we get $e = f$, which is why we can think of a lens of focal length f as being equivalent (to a first approximation) to a pinhole a distance f from the focal plane (figure 4.31a).

When using the pinhole camera model, it is very important to remember that the pinhole corresponds to the center of the lens. This point is also commonly called *center of projection* of *optical center* (indicated with C in figure 4.31). The axis perpendicular to the *image plane* Π, which passes through the center of projection is called *optical axis*.

For convenience, the pinhole camera is commonly represented with the image plane between the center of projection and the scene (figure 4.31(b)). This is done for the image to preserve the same orientation as the object, that is, the image is not flipped. The intersection O between the optical axis and the image plane is called *principal point*.

As shown in figure 4.31b, observe that a camera does not measure distances but angles and therefore it can be thought as a *bearing* sensor.

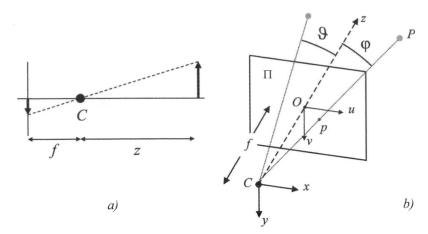

Figure 4.31 (a) Pinhole camera model used for representing standard perspective cameras. (b) The pinhole model is more commonly described with the image plane between the center of projection and the scene for the image to preserve the same orientation as the object.

4.2.3.3 Perspective projection

To describe analytically the *perspective projection* operated by the camera, we have to introduce some opportune reference system wherein we can express the 3D coordinates of the *scene point P* and the coordinates of its projection p on the image plane. We will first consider a simplified model and finally the general model.

Simplified model. Let (x, y, z) be the *camera reference frame* with origin in C and z-axis coincident with the optical axis. Assume also that the camera reference frame coincides with the *world* reference frame. This implies that the coordinates of the scene point P are already expressed in the camera frame.

Let us also introduce a two-dimensional reference frame (u, v) for the image plane Π with origin in O and the u and v axes aligned as x and y respectively as shown in figure 4.31b.

Finally, let $P = (x, y, z)$ and $p = (u, v)$. By means of simple considerations on the similarity of triangles, we can write

$$\frac{f}{z} = \frac{u}{x} = \frac{v}{y},$$

(4.35)

and therefore

$$u = \frac{f}{z} \cdot x \ , \tag{4.36}$$

$$v = \frac{f}{z} \cdot y \ . \tag{4.37}$$

This is the *perspective projection*. The mapping from 3D coordinates to 2D coordinates is clearly nonlinear. However, using *homogeneous coordinates* instead allows us to obtain linear equations. Let

$$\tilde{p} = \begin{bmatrix} u \\ v \\ 1 \end{bmatrix} \text{ and } \tilde{P} = \begin{bmatrix} x \\ y \\ z \\ 1 \end{bmatrix} \ , \tag{4.38}$$

be the homogeneous coordinates of p and P respectively. We will henceforth use the superscript ~ to denote homogeneous coordinates[3]. The projection equation, in this simplified case, can be written as:

$$\begin{bmatrix} \lambda u \\ \lambda v \\ \lambda \end{bmatrix} = \begin{bmatrix} fx \\ fy \\ z \end{bmatrix} = \begin{bmatrix} f & 0 & 0 & 0 \\ 0 & f & 0 & 0 \\ 0 & 0 & 1 & 0 \end{bmatrix} \begin{bmatrix} x \\ y \\ z \\ 1 \end{bmatrix} \ . \tag{4.39}$$

Note that λ is equal to the third coordinate of P, which—in this special reference frame—coincides with the distance of the point to the plane xy. Note that this equation also shows that every image point is the projection of all infinite 3D points lying on the ray passing through the same image point and the center of projection (figure 4.31b). Therefore, using a single pinhole camera it is not possible to estimate the distance to a point, but we need two cameras (i.e., *stereo camera*, section 4.2.5.2).

General model. A realistic camera model that describes the transformation from 3D coordinates to pixel coordinates must also take into account

---

3. In homogeneous coordinates we denote 2D points in the image plane as (x_1, x_2, x_3) with $(x_1/x_3, x_2/x_3)$ being the corresponding Cartesian coordinates. Therefore, there is a one to many correspondence between Cartesian and homogeneous coordinates. Homogeneous coordinates can represent the usual Euclidean points plus the points at infinity, which are points with the last component equal to zero that do not have a Cartesian counterpart.

- the *pixelization*, that is, shape (size) of the CCD and its position with respect to the optical center,

- the rigid body transformation between the camera and the scene (i.e., world).

The pixelization takes into account the fact that:

1. The camera optical center has pixel coordinates (u_0, v_0) with respect to the upper left corner of the image, which is commonly assumed as origin of the image coordinate system. Note, the optical center in general does not correspond to the center of the CCD.

2. The coordinates of a point on the image plane are measured in pixels. Therefore, we must introduce a scale factor.

3. The shape of the pixel is in general assumed not perfectly squared and therefore we must use two different scale factors k_u and k_v along the horizontal and vertical directions respectively.

4. The u and v axes might not be orthogonal but misaligned of an angle θ. This models, for instance, the fact that the lens may not be parallel to the CCD.

The first three points are addressed by means of the translation of the optical center and the individual rescaling of the u and v axes:

$$u = k_u \frac{f}{z} \cdot x + u_0 \qquad (4.40)$$

$$v = k_v \frac{f}{z} \cdot y + v_0 \;, \qquad (4.41)$$

where (u_0, v_0) are the coordinates of the principal point, k_u (k_v) is the inverse of the effective pixel size along the u (v) direction and is measured in $pixel \cdot m^{-1}$.

After this update the perspective projection equations become:

$$\begin{bmatrix} \lambda u \\ \lambda v \\ \lambda \end{bmatrix} = \begin{bmatrix} fk_u & 0 & u_0 & 0 \\ 0 & fk_v & v_0 & 0 \\ 0 & 0 & 1 & 0 \end{bmatrix} \begin{bmatrix} x \\ y \\ z \\ 1 \end{bmatrix} . \qquad (4.42)$$

Observe that we can pose $\alpha_u = fk_u$ and $\alpha_v = fk_v$ which describe the focal lengths expressed in horizontal and vertical pixels respectively.

To take into account the fact that in general the world reference system $(x_w, y_w z_w)$ does not coincide with the camera reference system (x, y, z), we have to introduce the rigid body

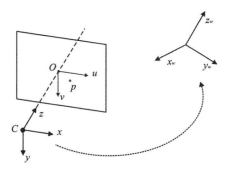

Figure 4.32 Coordinate change between camera and world reference frame.

transformation between the two reference frames (figure 4.32). Let us therefore introduce a coordinate change composed of a rotation R followed by a translation t, therefore

$$\begin{bmatrix} x \\ y \\ z \end{bmatrix} = R \begin{bmatrix} x_w \\ y_w \\ z_w \end{bmatrix} + t. \tag{4.43}$$

Using this transformation, equation (4.42) can be rewritten as

$$\begin{bmatrix} \lambda u \\ \lambda v \\ \lambda \end{bmatrix} = \begin{bmatrix} \alpha_u & 0 & u_0 \\ 0 & \alpha_v & v_0 \\ 0 & 0 & 1 \end{bmatrix} \begin{bmatrix} r_{11} & r_{12} & r_{13} & t_1 \\ r_{21} & r_{22} & r_{23} & t_2 \\ r_{31} & r_{32} & r_{33} & t_3 \end{bmatrix} \begin{bmatrix} x_w \\ y_w \\ z_w \\ 1 \end{bmatrix}, \tag{4.44}$$

or, using the homogeneous coordinates (4.38),

$$\lambda \tilde{p} = A[R|t]\tilde{P}_w, \tag{4.45}$$

where

$$A = \begin{bmatrix} \alpha_u & 0 & u_0 \\ 0 & \alpha_v & v_0 \\ 0 & 0 & 1 \end{bmatrix} \tag{4.46}$$

is the *intrinsic parameter matrix*.

As anticipated, the most general model also takes into consideration the possibility that the u and v axes are not orthogonal but are inclined of an angle θ. Therefore, the most general form for A is

$$
A = \begin{bmatrix} \alpha_u & \alpha_u \cot\theta & u_0 \\ 0 & \alpha_v & v_0 \\ 0 & 0 & 1 \end{bmatrix},
\tag{4.47}
$$

where $\alpha_u \cot\theta$ can be absorbed into a single parameter α_c.

α_c, α_u, α_{uv}, u_0, v_0 are called *camera intrinsic parameters*. The rotation and translation parameters R and t are called *camera extrinsic parameters*. The intrinsic and extrinsic parameters can be estimated using a procedure called *camera calibration* that we will shortly describe in section 4.2.3.4.

Radial distortion. The aforementioned image projection model assumes that the camera obeys a linear projection model where straight lines in the world result in straight lines in the image. Unfortunately, many wide-angle lenses have noticeable radial distortion, which manifests itself as a visible curvature in the projection of straight lines. An accurate model of the camera should therefore also take into account the radial distortion of the lens, especially for lenses with short focal length (i.e., large field of view) (figure 4.33).

The standard model of radial distortion is a transformation from the ideal coordinates (i.e., undistorted) (u, v) to the real observable coordinates (distorted) (u_d, v_d). Depending on the type of radial distortion, the coordinates in the observed images are displaced away (*barrel* distortion) or toward (*pincushion* distortion) the image center. The amount of distortion of the coordinates of the observed image is a nonlinear function of their radial distance r. For most lenses, a simple quadratic model of distortion produces good results:

$$
\begin{bmatrix} u_d \\ v_d \end{bmatrix} = (1 + k_1 r^2) \begin{bmatrix} u - u_0 \\ v - v_0 \end{bmatrix} + \begin{bmatrix} u_0 \\ v_0 \end{bmatrix},
\tag{4.48}
$$

where

$$
r^2 = (u - u_0)^2 + (v - v_0)^2.
\tag{4.49}
$$

and k_1 is the radial distortion parameter, which can be estimated by camera calibration. The radial distortion parameter is also an intrinsic parameter of the camera.

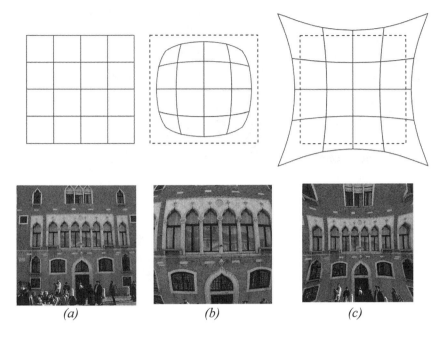

Figure 4.33 Example of radial lens distortion: (a) no distortion, (b) barrel distortion, (c) pincushion.

Sometimes the above simplified model does not model the true distortions produced by complex lenses accurately enough (especially at very wide angles). A more complete analytic model also includes *tangential distortions* and *decentering distortions* [48], but these will not be covered in this book. Fisheye lenses require a different model than traditional polynomial models of radial distortion and will be introduced in section 4.2.4.2.

4.2.3.4 Camera calibration

Calibration consists in measuring accurately the intrinsic and extrinsic parameters of the camera model. As these parameters govern the way the scene points are mapped to their corresponding image points, the idea is that by knowing the pixel coordinates of the image points $\tilde{p}$ and the 3D coordinates of the corresponding scene points $\tilde{P}$, it is possible to compute the unknown parameters A, R, and t by solving the perspective projection equation (4.45).

One of the first and most used camera calibration techniques was proposed in 1987 by Tsai [319]. Its implementation needs corresponding 3D point coordinates and 2D pixels coordinates in the image. It uses a two-stage technique to compute, first, the position and orientation and, second, the internal parameters of the camera.

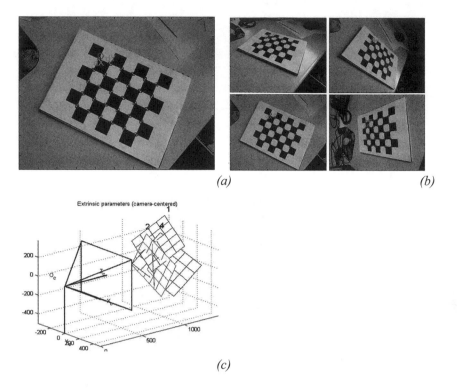

(a)   *(b)*

(c)

Figure 4.34 Pictures of the Camera Calibration Toolbox for Matlab developed by J. Y. Bouguet. (a) An example checkerboard-like pattern used in camera calibration with extracted corners. (b) Several pictures of the pattern with different positions and orientations. (c) The reconstructed position and orientations of the pattern after calibration.

However, in the last decade an alternative camera calibration technique has been proposed by Zhang [337] that, instead of a three-dimensional calibration object, uses a planar grid. The most common planar grid is a chessboard-like pattern due to the ease of extracting its corners, which are then used for calibration (figure 4.34). This method is known as *calibration from planar grid* and is very simple and practical to execute for both expert and non-expert users. The method requires the user to take several pictures of the pattern shown at different positions and orientations[4]. By knowing the 2D position of the corners on the real pattern and the pixel coordinates of their corresponding corners on each image, the intrinsic and extrinsic parameters (including radial and tangential distortion) are determined simultaneously by solving a least-square linear minimization followed by a nonlin-

---

4. Note that in this case the number of extrinsic parameters is different for each position of the grid while the intrinsic parameter are obviously the same.

ear refinement (i.e., Gauss-Newton). As pointed out by Zhang, the accuracy of the calibration results increases with the number of images used. It is also important that the images cover as much of the field of view of the camera as possible and that the range of orientations is wide.

This calibration method has been implemented in a very successful open source toolbox for Matlab (which can be downloaded for free [347]) that is also available in C in the open source Computer Vision Library (OpenCV) [343]. This toolbox has been used by thousands of users all around the world and is considered one of the most practical and easy-to-use camera calibration softwares for standard perspective cameras. In this section, we used the same model as the Matlab toolbox. This should facilitate the understanding and implementation of the interested reader. Alternatively, a complete list of all available camera calibration softwares can be found in [348].

4.2.4 Omnidirectional cameras

4.2.4.1 Introduction

In the previous section, we described the image formation of the pinhole camera, which is modeled as a perspective projection. However, there are projection systems whose geometry cannot be described using the conventional pinhole model because of the very high distortion introduced by the imaging device. Some of these systems are omnidirectional cameras.

An omnidirectional camera is a camera that provides wide field of view, at least more than 180 degrees. There are several ways to build an omnidirectional camera. Dioptric cameras use a combination of shaped lenses (e.g., fisheye lenses; see figure 4.35a) and typically can reach a field of view slightly larger than 180 degrees. Catadioptric cameras combine a standard camera with a shaped mirror, like a parabolic, hyperbolic, or elliptical mirror and are able to provide much more than 180 degrees field-of-view in elevation and 360 in the azimuthal direction. In figure 4.35b you can see an example catadioptric camera using a hyperbolic mirror. Finally, polydioptric cameras use multiple cameras with overlapping field of view (figure 4.35c) and so far are the only cameras able to provide a real omnidirectional (spherical) view (i.e., 4π steradians).

Catadioptric cameras were first introduced in robotics in 1990 by Yagi and Kawato [333], who utilized them for localizing robots. Fisheye cameras started to spread over only in 2000 thanks to new manufacturing techniques and precision tools that led to an increase of their field of view up to 180 degrees. However, it is only since 2005 that these cameras have been miniaturized to the size of 1–2 centimeters, and their field of view has been increased up to 190 degrees (see, for instance, figure 4.36a).

Thanks to the camera miniaturization, to the recent developments in optics manufacturing, and to the decreasing prices in the cameras' market, catadioptric and dioptric omni-

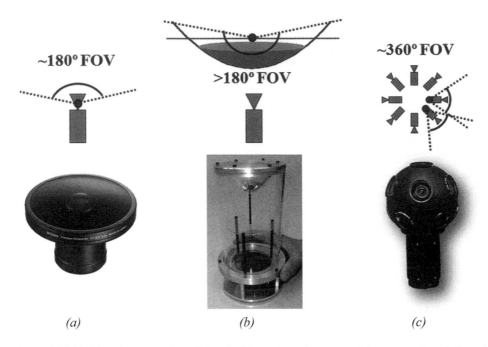

(a)   *(b)*   *(c)*

Figure 4.35 (a) Dioptric camera (e.g. fisheye); (b) catadioptric camera; (c) an example polydioptric camera produced by Immersive Media.

directional cameras are being more and more used in different research fields. Miniature dioptric and catadioptric cameras are now used by the automobile industry in addition to sonars for improving safety, by providing to the driver an omnidirectional view of the surrounding environment. Miniature fisheye cameras are used in endoscopes for surgical operations or on board microaerial vehicles for pipeline inspection as well as rescue operations. Other examples involve meteorology for sky observation.

Roboticists have also been using omnidirectional vision with very successful results on robot localization, mapping, and aerial and ground robot navigation [76, 80, 107, 278, 279, 307]. Omnidirectional vision allows the robot to recognize places more easily than with standard perspective cameras [276]. Furthermore, landmarks can be tracked in all directions and over longer periods of time, making it possible to estimate motion and build maps of the environment with better accuracy than with standard cameras, see figure 4.36 for some of examples of miniature omnidirectional cameras used on state-of-the-art micro aerial vehicles. Several companies, like Google, are using omnidirectional cameras to build photorealistic street views and three-dimensional reconstructions of cities along with texture. Two example omnidirectional images are shown in figure 4.37.

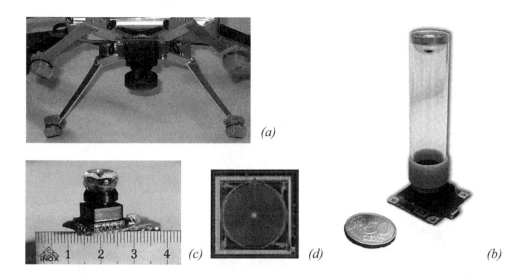

Figure 4.36 (a) The fisheye lens from Omnitech Robotics (www.omnitech.com) provides a field of view of 190 deg. This lens has a diameter of 1.7 cm. This camera has been used on the sFly autonomous helicopter at the ETH Zurich, (section 2.4.3) [76]. (b) A miniature catadioptric camera built at the ETH Zurich, which is also used for autonomous flight. It uses a spherical mirror and a transparent plastic support. The camera measures 2 cm in diameter and 8 cm in height. (c) The muFly camera built by CSEM, which is used on the muFly helicopter at the ETH Zurich (section 2.4.3). This is one of the smallest catadioptric cameras ever built. Additionally, it uses a polar CCD (d) where pixels are arranged radially.

In the next sections we will give an overview of omnidirectional camera models and calibration. For an in-depth study on omnidirectional vision, we refer the reader to [4, 15, 273].

4.2.4.2 Central omnidirectional cameras

A vision system is said to be central when the optical rays to the viewed objects intersect in a single point in 3D called projection center or *single effective viewpoint* (figure 4.38). This property is called *single effective viewpoint property*. The perspective camera is an example of a central projection system because all optical rays intersect in one point, that is, the camera optical center.

All modern fisheye cameras are central, and hence, they satisfy the single effective viewpoint property. Central catadioptric cameras conversely can be built only by opportunely choosing the mirror shape and the distance between the camera and the mirror. As proven by Baker and Nayar [64], the family of mirrors that satisfy the single viewpoint property is the class of rotated (swept) conic sections, that is, hyperbolic, parabolic, and elliptical mirrors. In the case of hyperbolic and elliptical mirrors, the single view point

(a)

(b)

Figure 4.37 (a) A catadioptric omnidirectional camera using a hyperbolic mirror. The image is typically unwrapped into a cylindrical panorama. The field of view is typically 100 degrees in elevation and 360 degrees in azimuth. (b) Nikon fisheye lens FC-E8. This lens provides a hemispherical (180 deg) field of view.

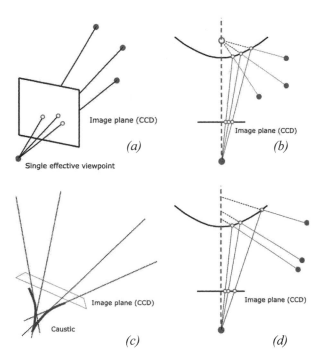

Figure 4.38 (a–b) Example of central cameras: perspective projection and catadioptric projection through a hyperbolic mirror. (c–d) Example of noncentral cameras: the envelope of the optical rays forms a *caustic*.

property is achieved by ensuring that the camera center (i.e., the pinhole or the center of the lens) coincides with one of the foci of the hyperbola (ellipse) (figure 4.39). In the case of parabolic mirrors, an orthographic lens must be interposed between the camera and the mirror, this makes it possible that parallel rays reflected by the parabolic mirror converge to the camera center (figure 4.39).

The reason a single effective viewpoint is so desirable is that it allows us to generate geometrically correct perspective images from the pictures captured by the omnidirectional camera (figure 4.40). This is possible because, under the single view point constraint, every pixel in the sensed image measures the irradiance of the light passing through the viewpoint in one particular direction. When the geometry of the omnidirectional camera is known, that is, when the camera is calibrated, one can precompute this direction for each pixel. Therefore, the irradiance value measured by each pixel can be mapped onto a plane at any distance from the viewpoint to form a planar perspective image. Additionally, the image can be mapped on to a sphere centered on the single viewpoint, that is, spherical projection (figure 4.40, bottom).

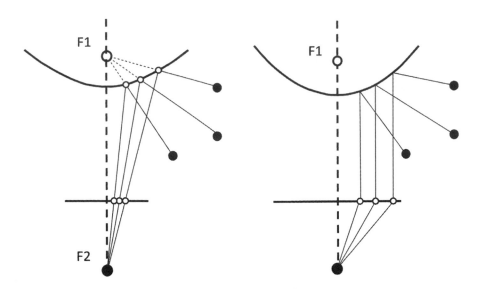

Figure 4.39 Central catadioptric cameras can be built by using hyperbolic and parabolic mirrors. The parabolic mirror requires the use of an orthographic lens.

Another reason why the single view point property is so important is that it allows us to apply the well known theory of *epipolar geometry* (see section 4.2.6.1), which easily allows us to perform *structure from stereo* (section 4.2.5) and *structure from motion* (section 4.2.6). As we will see, epipolar geometry holds for any central camera, both perspective and omnidirectional. Therefore, in those sections we will not make any distinction about the camera.

4.2.4.3 Omnidirectional camera model and calibration

Intuitively, the model of an omnidirectional camera is a little more complicated than a standard perspective camera. The model should indeed take into account the reflection operated by the mirror in the case of a catadioptric camera or the refraction caused by the lens in the case of a fisheye camera. Because the literature in this field is quite large, here we review two different projection models that have become standards in omnidirectional vision and robotics. Additionally, Matlab toolboxes have been developed for these two models, which are used worldwide by both specialists and non-experts.

The first model is known as the *unified projection model for central catadioptric cameras*. It was developed in 2000 by Geyer and Daniilidis [137] (later refined by Barreto and Araujo [66]), who have the merit of having proposed a model that encompasses all three types of central catadioptric cameras, that is, cameras using hyperbolic, parabolic, or elliptical mirror. This model was developed specifically for central catadioptric cameras and is

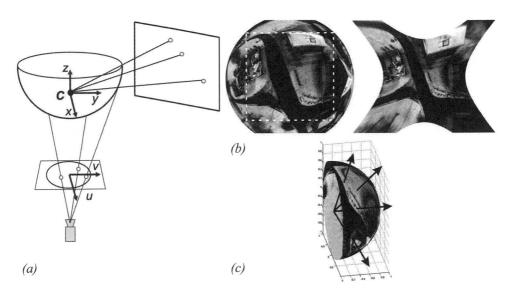

(b)

(a)

(c)

Figure 4.40 Central cameras allows us to remap regions of the omnidirectional image into a perspective image. This can be done straightforwardly by intersecting the optical rays with a plane specified arbitrarily by the user (a). Of course we cannot project the whole image onto a plane but only subregions of it (b). Another possible projection is that onto a sphere (c).

not valid for fisheye cameras. The approximation of a fisheye lens model by a catadioptric one is usually possible, however, with limited accuracy only [335].

Conversely, the second model unifies both central catadioptric cameras and fisheye cameras under a general model also known as Taylor model. It was developed in 2006 by Scaramuzza et al. [274, 275] and has the advantage that both catadioptric and dioptric cameras can be described through the same model, namely a Taylor polynomial.

Unified model for central catadioptric cameras. With their landmark paper from 2000, Geyer and Daniilidis showed that every catadioptric (parabolic, hyperbolic, elliptical) and standard perspective projection is equivalent to a projective mapping from a sphere, centered in the single viewpoint, to a plane with the projection center on the perpendicular to the plane and distant ε from the center of the sphere. This is summarized in figure 4.41.

As we did for the perspective camera, the goal is again to find the relation between the viewing direction to the scene point and the pixel coordinates of the corresponding image point. The projection model of Geyer and Daniilidis follows a four-step process. Let again $P = (x, y, z)$ be a scene point in the mirror reference frame[5] centered in C (figure 4.41).

1. The first step consists in projecting the scene point onto the unit sphere; therefore:

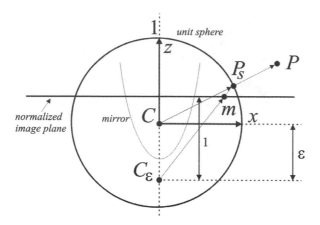

Mirror type	ε
Parabola	1
Hyperbola	$\dfrac{d}{\sqrt{d^2 + 4l^2}}$
Ellipse	$\dfrac{d}{\sqrt{d^2 + 4l^2}}$
Perspective	0

Figure 4.41 Unified projection model for central catadioptric cameras of Geyer and Daniilidis.

$$P_s = \frac{P}{\|P\|} = (x_s, y_s, z_s).$$ (4.50)

2. The point coordinates are then changed to a new reference frame centered in $C_\varepsilon = (0, 0, -\varepsilon)$; therefore:

$$P_\varepsilon = (x_s, y_s, z_s + \varepsilon).$$ (4.51)

Observe that ε ranges between 0 (planar mirror) and 1 (parabolic mirror). The correct value of ε can be obtained knowing the distance d between the foci of the conic and the latus rectum[6] l as summarized in the table of figure 4.41.

3. P_ε is then projected onto the normalized image plane distant 1 from C_ε; therefore,

$$\tilde{m} = (x_m, y_m, 1) = \left(\frac{x_s}{z_s + \varepsilon}, \frac{y_s}{z_s + \varepsilon}, 1 \right) = g^{-1}(P_s).$$ (4.52)

4. Finally, the point $\tilde{m}$ is mapped to the camera image point $\tilde{p} = (u, v, 1)$ through the intrinsic parameter matrix A; therefore,

---

5. For convenience we assume that the mirror axis of symmetry is perfectly aligned with the camera optical axis. We also assume that the x-y axes of the camera and mirror are aligned. Therefore, the camera and mirror reference frames differ only by a translation along z.
6. The latus rectum of a conic section is the chord through a focus parallel to the conic section directrix.

$$\tilde{p} = A \cdot \tilde{m}, \qquad (4.53)$$

where A is given by (4.47), that is,

$$A = \begin{bmatrix} \alpha_u & \alpha_u \cot\theta & u_0 \\ 0 & \alpha_v & v_0 \\ 0 & 0 & 1 \end{bmatrix}. \qquad (4.54)$$

It is easy to show that function g^{-1} is bijective and that its inverse g is given by[7]:

$$P_s = g(m) \sim \begin{bmatrix} x_m \\ y_m \\ 1 - \varepsilon \dfrac{x_m^2 + y_m^2 + 1}{\varepsilon + \sqrt{1 + (1-\varepsilon^2)(x_m^2 + y_m^2)}} \end{bmatrix}, \qquad (4.55)$$

where $\sim$ indicates that g is proportional to the quantity on the right-hand side. To obtain the scale factor, it is sufficient to normalize $g(m)$ onto the unit sphere.

Observe that equation (4.55) is the core of the projection model of central catadioptric cameras. It expresses the relation between the point m on the normalized image plane and the unit vector P_s in the mirror reference frame. Note that in the case of planar mirror, we have $\varepsilon = 0$ and (4.55) becomes the projection equation of perspective cameras $P_s \sim (x_m, y_m, 1)$.

This model has proved to be able to describe accurately all central catadioptric cameras (parabolic, hyperbolic, and elliptical mirror) and standard perspective cameras. An extension of this model for fisheye lenses was proposed in 2004 by Ying and Hu [335]. However, the approximation of a fisheye camera through a catadioptric one works only with limited accuracy. This is mainly due because, while the three types of central catadioptric cameras can be represented through an exact parametric function (parabola, hyperbola, ellipse), the projective models of fisheye vary from camera to camera and depend on the lens field-of-view. To overcome this problem, a new unified model has been proposed, which will be described in the next section.

---

7. Equation (4.55) can be obtained by inverting (4.52) and imposing the constraint that P_s must lie on the unit sphere and, thus, $x_s^2 + y_s^2 + z_s^2 = 1$. From this constraint you will then get an expression of z_s as a function of ε, x_m, and y_m. More details can be found in [66].

Unified model for catadioptric and fisheye cameras. This unified model was proposed by Scaramuzza et al. in 2006 [274, 275]. The main difference with the previous model lies in the choice of the function g. To overcome the lack of knowledge of a parametric model for fisheye cameras, the authors proposed the use of a Taylor polynomial, whose coefficients and degree are found through the calibration process. Accordingly, the relation between the normalized image point $\tilde{m} = (x_m, y_m, 1)$ and the unit vector P_s in the fisheye (mirror) reference frame can be written as:

$$P_s = g(m) \sim \begin{bmatrix} x_m \\ y_m \\ a_0 + a_2\rho^2 + \ldots + a_N\rho^N \end{bmatrix}, \tag{4.56}$$

where $\rho = \sqrt{x_m^2 + y_m^2}$. As you have probably noticed, the first-order term (i.e., $a_1\rho$) of the polynomial is missing. This follows from the observation that the first derivative of the polynomial calculated at $\rho = 0$ must be null for both catadioptric and fisheye cameras (this is straightforward to verify for catadioptric cameras by differentiating [4.55]). Also observe that because of its polynomial nature, this expression can encompass catadioptric, fisheye, and perspective cameras. This can be done by opportunely choosing the degree of the polynomial. As highlighted by the authors, polynomials of order three or four are able to model very accurately all catadioptric cameras and many types of fisheye cameras available on the market. The applicability of this model to a wide range of commercial cameras is at the origin of its success.

Omnidirectional camera calibration. The calibration of omnidirectional cameras is similar to that for calibrating standard perspective cameras, which we have seen in section 4.2.3.4. Again, the most popular methods take advantage of planar grids that are shown by the user at different positions and orientations. For omnidirectional cameras, it is very important that the calibration images are taken all around the camera and not on a single side only. This in order to compensate for possible misalignments between the camera and mirror.

It is worth to mention three open-source calibration toolboxes currently available for Matlab, which differ mainly for the projection model adopted and the type of calibration pattern.

- The toolbox of Mei uses checkerboard-like images and takes advantage of the projection model of Geyer and Daniilidis discussed earlier. It is particularly suitable for catadioptric cameras using hyperbolic, parabolic, folded mirrors, and spherical mirrors. Mei's toolbox can be downloaded from [349], while the theoretical details can be found in [212].

- The toolbox of Barreto uses line images instead of checkerboards. Like the previous toolbox, it also uses the projection model of Geyer and Daniilidis. It is particularly suitable for parabolic mirrors. The toolbox can be downloaded from [350], while the theoretical details can be found in [67] and [68].

- Finally, the toolbox of Scaramuzza uses checkerboard-like images. Contrary to the previous two, it takes advantage of the unified Taylor model for catadioptric and fisheye cameras developed by the same author. It works with catadioptric cameras using hyperbolic, parabolic, folded mirrors, spherical, and elliptical mirrors. Additionally, it works with a wide range of fisheye lenses available on the market—like Nikon, Sigma, Omnitech-Robotics, and many others—with field of view up to 195 degrees. The toolbox can be downloaded from [351], while the theoretical details can be found in [274] and [275]. Contrary to the other two, this toolbox features an automatic calibration process. In fact, both the center of distortion and the calibration points are detected automatically without any user intervention.

4.2.5 Structure from stereo

4.2.5.1 Introduction

Range sensing is extremely important in mobile robotics, since it is a basic input for successful obstacle avoidance. As we have seen earlier in this chapter, a number of sensors are popular in robotics explicitly for their ability to recover depth estimates: ultrasonic, laser rangefinder, time-of-flight cameras. It is natural to attempt to implement ranging functionality using vision chips as well.

However, a fundamental problem with visual images makes rangefinding relatively difficult. Any vision chip collapses the 3D world into a 2D image plane, thereby losing depth information. If one can make strong assumptions regarding the size of objects in the world, or their particular color and reflectance, then one can directly interpret the appearance of the 2D image to recover depth. But such assumptions are rarely possible in real-world mobile robot applications. Without such assumptions, a single picture does not provide enough information to recover spatial information.

The general solution is to recover depth by looking at *several* images of the scene to gain more information, hopefully enough to at least partially recover depth. The images used must be different, so that taken together they provide additional information. They could differ in camera geometry—such as the focus position or lens iris—yielding depth from focus (or defocus) techniques that we have described in section 4.2.3.1. An alternative is to create different images, not by changing the camera geometry, but by changing the camera viewpoint to a different camera position. This is the fundamental idea behind *structure from stereo* (i.e., *stereo vision*) and *structure from motion* that we will present in the next sections. As we will see, stereo vision processes two distinct images taken at the same time and assumes that the relative pose between the two cameras is known. Structure-from-motion conversely processes two images taken with the same or a different camera at different times and from different unknown positions; the problem consists in recovering both the relative motion between the views and the depth. The 3D scene that we want to reconstruct is usually called *structure*.

4.2.5.2 Stereo vision

Stereopsis (from *stereo* meaning solidity, and *opsis* meaning vision or sight) is the process in visual perception leading to the sensation of depth from the two slightly different projections of the world onto the retinas of the two eyes. The difference in the two retinal images is called horizontal *disparity*, retinal disparity, or binocular disparity. The differences arise from the eyes' different positions in the head. It is the disparity that makes our brain fuse (perceive as a single image) the two retinal images making us perceive the object as a one and solid. To have a clearer understanding of what disparity is, as a simple test, hold your finger vertically in front of you and close each eye alternately. You will see that the finger jumps from left to right. The distance between the left and right appearance of the finger is the disparity. The same phenomenon is visible in the image pair shown in figure 4.48, in which the foreground objects shift left and right relative to the background.

Computational stereopsis, or stereo vision, is the process of obtaining depth information from a pair of images coming from two cameras which look at the same scene from different positions. In stereo vision we can identify two major problems:

1. the correspondence problem
2. 3D reconstruction

The first consists in matching (pairing) points of the two images which are the projection of the same point in the scene. These matching points are called *corresponding points* or *correspondences* (figure 4.45a). This will be clarified later on. Determining the corresponding points is made possible based on the assumption that the two images differ only slightly and therefore a feature in the scene appears similar in both images. Based only of this assumption, however, there might be many possible false matches. As we will see, this problem can be overcome by introducing an additional constraint which makes the correspondence matching feasible. This constraint is called *epipolar constraint* (section 4.2.6.1) and states that the correspondent of a point in an image lies on a line (called *epipolar line*) in the other image (figure 4.45b). Because of this constraint, we will see that the correspondence search becomes one-dimensional instead of two-dimensional.

Knowing the correspondences between the two images, knowing the relative orientation and position of the two cameras, and knowing the intrinsic parameters of the two cameras, it is possible to reconstruct the scene points (i.e., the structure). This process of reconstruction requires the prior calibration of the stereo camera; that is, we need to calibrate the two cameras separately for estimating their extrinsic parameters, but we also need to determine their extrinsic parameters, i.e. the camera relative position.

Figure 4.42 (Left) The STH-MDCS3 form Videre Design uses CMOS sensors, a baseline of 9 cm, an image resolution of 1280×960 at 7.5 frames per second (fps), or 640×480 at 30 fps. (Right) The Bumblebee2 from Point Grey uses CCD sensors, a baseline of 12 cm, an image resolution of 1024×768 at 20 frames per second (fps), or 640×480 at 48 fps.

The theory of stereo vision has been well understood for years, while the engineering challenge of creating a practical stereo-vision sensor has been formidable [21, 43, 44]. Example of commercially available stereo cameras are shown in figure 4.42.

Basic case. First, we consider a simplified case in which two cameras have the same orientation and are placed with their optical axes parallel, at a separation of b (called *baseline*), shown in figure 4.43.

In this figure, a point on the object is described as being at coordinate (x, y, z) with respect to the origin located in the left camera lens. The image coordinate in the left and right image are (u_l, v_l) and (u_r, v_r) respectively. From figure 4.43a and using equations (4.36) and (4.37), we can write

$$\frac{f}{z} = \frac{u_l}{x} , \tag{4.57}$$

$$\frac{f}{z} = \frac{-u_r}{b - x} , \tag{4.58}$$

from which we obtain

$$z = b \frac{f}{u_l - u_r} , \tag{4.59}$$

where the *difference* in the image coordinates, $u_l - u_r$ is called *disparity*. This is an important term in stereo vision, because it is only by measuring disparity that we can recover depth information. Observations from this equation are as follows:

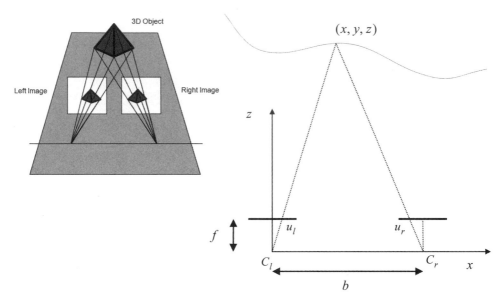

Figure 4.43
Idealized camera geometry for stereo vision. The cameras are assumed be identical (i.e., identical focal lengths and image resolution); furthermore, they are assumed to be perfectly aligned on the horizontal axis.

- Distance is inversely proportional to disparity. The distance to near objects can therefore be measured more accurately than that to distant objects, just as with depth from focus techniques. In general, this is acceptable for mobile robotics, because for navigation and obstacle avoidance closer objects are of greater importance.

- Disparity is proportional to b. For a given disparity error, the accuracy of the depth estimate increases with increasing baseline b.

- As b is increased, because the physical separation between the cameras is increased, some objects may appear in one camera but not in the other. This is due to the field of view of the cameras. Such objects by definition will not have a disparity and therefore will not be ranged.

- If the baseline b is unknown, it is possible to reconstruct the scene point only *up to a scale*. This is the case in structure-from-motion (section 4.2.6).

- A point in the scene visible to both cameras produces a pair of image points known as a *conjugate pair*, or *correspondence pair* (figure 4.44a). Given one member of the conjugate pair, we know that the other member of the pair lies somewhere along a line known

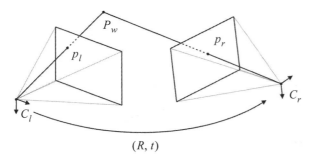

Figure 4.44 Stereo vision: general case.

as *epipolar line*. In the case depicted in figure 4.43a, because the cameras are perfectly aligned with one another, the epipolar lines are horizontal lines (i.e., along the x direction). The concept of epipolar line will be explained later on in this section.

General case. The assumption of perfectly aligned cameras is normally violated in practice. In fact, even the most expensive stereo cameras available in the market do not assume this model. Indeed, two exactly identical cameras do not exist. There will be always differences in the focal length due to manufacturing but, especially, even if such identical cameras could exist, we would never be sure that they are perfectly aligned. The situation is even more complicated by the fact that the internal orientation of the CCD in the camera package is unknown. Ideally it is aligned, but in practice the CCD cannot be considered perfectly aligned. Therefore, the general stereo vision model assumes that the two cameras are different and not aligned (figure 4.44) but requires that the relative position and orientation of the two cameras is known. If the relative position is not known, the stereo camera must be calibrated using the checkerboard-based calibration treated in section 4.2.3.4. Fortunately, the previously mentioned toolbox for calibrating the camera intrinsic parameters [347] allows the user to calibrate stereo cameras as well.

So, let us assume that the two cameras have been previously calibrated. Therefore, the intrinsic parameter matrices A_l and A_r (see equation 4.47) for the left and right camera are known, and the camera extrinsic parameters—i.e. the rotations R_l, R_r and translations t_l, t_r of the two cameras with the respect to the world coordinate system—are also known. In stereo vision, it is a common practice to assume the origin of the world coordinate system in the left camera. Thus, we can write $R_l = I$ and $R_r = R$. This allows us to write the equations of perspective projection for the two cameras as:

$$\lambda_l \tilde{p}_l = A_l [I|0] \tilde{P}_w \text{ (for the left camera),} \tag{4.60}$$

Figure 4.45 A stereo pair. Corresponding points are projections of the same scene point. Because of the epipolar constraint, conjugate points can be searched along the epipolar lines. This heavily reduces the computational cost of the correspondence search: from a two-dimensional search it becomes a one-dimensional search problem.

$$\lambda_r \tilde{p}_r = A_r[R|t]\tilde{P}_w \quad \text{(for the right camera)}, \tag{4.61}$$

where $\tilde{p}_l = [u_l, v_l, 1]^T$ and $\tilde{p}_r = [u_r, v_r, 1]^T$ are the image points (in homogeneous coordinates) corresponding to the world point $\tilde{P}_w = [x, y, z, 1]^T$ (in homogeneous coordinates) in the left and right camera respectively. λ_l and λ_r are the depth factors. Observe that (4.60) and (4.61) actually contribute three equations each. Therefore, we have a system of six equations in five unknowns, three for the world point $P_w = (x, y, z)$ and two for depth factors, i.e. λ_l and λ_r. The system is overdetermined and can be solved either linearly, using least-squares, or nonlinearly by computing the 3D point that minimizes distances between the two light rays passing through $\tilde{p}_l$ and $\tilde{p}_r$. The solution of these two equations is left as an exercise to the reader in section 4.8.

Correspondence problem. Using the preceding equations requires us to have identified the conjugate pair p_l and p_r in the left and right camera images, which originates from the same scene point $\tilde{P}_w$ (figure 4.45a). This fundamental challenge is called the *correspon-*

dence problem. Intuitively, the problem is: given two images of the same scene from different perspectives, how can we identify the same object points in both images? For every such identified object point, we will then be able to recover its 3D position in the scene.

The correspondence search is based on the assumption that the two images of the same scene do not differ too much, that is, a feature in the scene is supposed to appear very similar in both images. Using an opportune image similarity metric (see section 4.3.3), a given point in the first image can be paired with one point in the second image. The problem of *false correspondences* makes the correspondence search challenging. False correspondences occur when a point is paired to another that is not its real conjugate. This is because the assumption of image similarity does not hold very well, for instance if the part of the scene to be paired appears under different illumination or geometric conditions. Other problems that make the correspondence search difficult are:

- *Occlusions*: the scene is seen by two cameras at different viewpoints and therefore there are parts of the scene that appear only in one of the images. This means, there exist points in one image which do not have a correspondent in the other image.

- *Photometric distortion*: there are surfaces in the scene which are nonperfectly *lambertian*, that is, surfaces whose behavior is partly specular. Therefore, the intensity observed by the two cameras is different for the same point in the scene as more as the cameras are farther apart.

- *Projective distortion*: because of the perspective distortion, an object in the scene is projected differently on the two images, as more as the cameras are farther apart.

Some constraints can, however, be exploited for improving the correspondence search, which are:

- *Similarity constraint*: a feature in the image appears similar in the other image.

- *Continuity constraint*: far from the image borders, the depth of the scene points along a continuous surface varies continuously. This constraint clearly limits the *gradient of disparity*.

- *Unicity*: a point of the first image can be paired only with a single point in the other image, and vice versa (it fails in presence of occlusions, specularities, and transparency).

- *Monotonic order*: if point p_l in the left image is the correspondent of p_r in the right image, then the correspondent of a point on the right (left) of p_l can only be found on the right (left) of p_r. This is valid only for points that lie on an opaque object.

- *Epipolar constraint*: the correspondent of a point in the left image can only be found along a line in the right image, which is called *epipolar line* (figure 4.45b). As a matter of fact, this is the most important constraint and will be explained later on.

The methods for searching correspondences can be distinguished into two categories:

- *Area-based*: these algorithms consider a small patch (window) in one image and look for the most similar patch in the second image by means of an appropriate *correlation* measure. This search is done for every pixel and allows us to obtain a *dense* reconstruction. However, in uniform regions—that is, poor texture—these methods fail. There exist different techniques to measure the similarity between image patches for stereo matching. The most used are the *Sum of Absolute Differences* (SAD), the *Sum of Squared Differences* (SSD), the *Normalized Cross Correlation* (NCC), and the *Census Transform*. An overview of some of these algorithms is given in section 4.3.3. Finally, observe that the search for correspondences is a two-dimensional search: the most similar of a patch in the left image must be searched across all rows and columns of the right image. As we will see in the next section, the search can be reduced to only one line, the epipolar line, thus reducing the dimensionality of the search from two to one (figure 4.45b).

- *Feature-based*: these algorithms extract salient features from the images, which are possibly stable with respect to change of view point. The matching process is applied to the attributes associated to the features. Edges, corners, line segments, and blobs are some of the features that can be used. They do not have to correspond necessarily to a well defined geometric entity. An exhaustive overview on feature extraction is given is section 4.5. Feature-based stereo matching algorithms are faster and more robust than area-based methods but provide only sparse depth maps, which then need to be interpolated.

Epipolar geometry. Given a pixel in one image (say the left image), how can we compute its correspondence with the correct pixel in the other image? As we anticipated in the previous section, one way would be to search the correspondences across all pixels of the second image. In the case of stereo matching, however, we have some information available, namely the relative position and the calibration parameters of the two cameras. This information allows us to reduce the search from two dimensions to an only one dimension. Figure 4.46a shows how a pixel point p_l in one image projects to an epipolar line segment in the other image. The segment is bounded at one end by the projection of P_∞ (the original viewing ray at infinity) and at the other end by the projection of C_l into the second camera, which is known as the epipole e_r. By projecting the epipolar line in the second image back into the first image, we get another line which is bounded by the other corresponding epipole e_l. Notice that two corresponding epipolar lines (figure 4.46b) originate from the intersection of the two image planes with the epipolar plane that passes through the camera centers C_l and C_r and the scene point P_w.

To compute the equation of the epipolar line, we must project the optical ray passing through p_l and C_l to the second image. This is straightforward. The equation of the optical

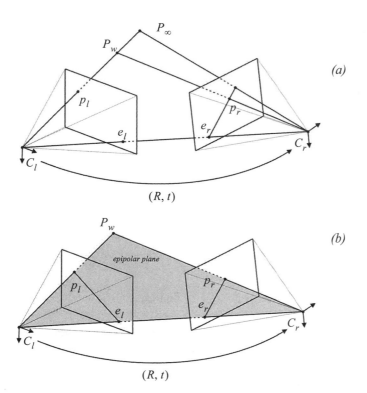

Figure 4.46 Epipolar geometry: (a) epipolar line segment corresponding to one ray; (b) corresponding set of epipolar lines and their epipolar plane.

ray passing through p_l and C_l can be obtained from the perspective projection equation (4.60), which we rewrite here as:

$$\lambda_l \tilde{p}_l = \lambda_l \begin{bmatrix} u_l \\ v_l \\ 1 \end{bmatrix} = A_l [I|0] \tilde{P}_w = A_l P_w = A_l \begin{bmatrix} x \\ y \\ z \end{bmatrix}, \tag{4.62}$$

and therefore the line passing through p_l and C_l has equation:

$$\begin{bmatrix} x \\ y \\ z \end{bmatrix} = \lambda_l A_l^{-1} \begin{bmatrix} u_l \\ v_l \\ 1 \end{bmatrix}, \tag{4.63}$$

which we can rewrite in a more compact form as:

$$P_w = \lambda_l A_l^{-1} \tilde{p}_l. \tag{4.64}$$

Finally, to find the equation of the epipolar line, we just project this line onto the second image using the perspective projection equation (4.61):

$$\lambda_r \tilde{p}_r = A_r [R|t] \tilde{P}_w = A_r R P_w + A_r t, \tag{4.65}$$

and therefore, using (4.64), we obtain the epipolar line

$$\lambda_r \tilde{p}_r = \lambda_l A_r R A_l^{-1} \tilde{p}_l + A_r t, \tag{4.66}$$

where $A_r t$ is actually the epipole e_r in the second image, that is, the projection of the optical center C_l of the left camera into the right image.

By applying equation (4.66) to every image point in the left image, we can compute all the epipolar lines in the right image. The correspondence of one point in the left image will then need to be searched only along its corresponding epipolar line. Note that the epipolar lines pass all through the same epipole. However, observe that in computing equation (4.66) we did not take into account the radial distortion introduced by the lens. Although for some narrow-field-of-view cameras the radial distortion is rather small, it is always opportune to take the radial distortion into account when computing the equation of the epipolar line. The reason is that if the epipolar line is not determined precisely, the correspondence search along a non accurate epipolar line can lead to a larger uncertainty in the computation of the disparity as well as in the reconstruction of the scene point P_w.

Instead of taking into account the radial distortion, a common consolidated procedure in stereo vision is that of undistorting first the two images, that is, remapping the left and right image into new images without distortion. Furthermore, the two images can be remapped in such a way that all epipolar lines in the left and right image are collinear and horizontal (figure 4.47d). The process of transforming a pair of stereo images into a new pair without radial distortion and with horizontal epipolar lines is called *stereo rectification* or *epipolar rectification*. We will briefly explain it in the next section.

Epipolar rectification. Given a pair of stereo images, epipolar rectification is a transformation of each image plane such that all corresponding epipolar lines become collinear and parallel to one of the image axes, for convenience usually the horizontal axis. The resulting rectified images can be thought of as acquired by a new stereo camera obtained by rotating the original cameras about their optical centers. The great advantage of the epipolar recti-

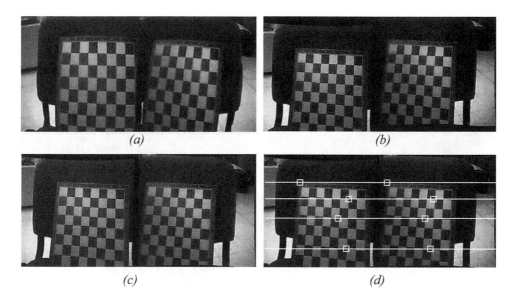

Figure 4.47 Rectification of a stereo pair: (a) original images, (b) compensation of the lens distortion, (c) compensation of rotation and translation, (d) After the epipolar rectification, the epipolar lines appear collinear and horizontal.

fication is the correspondence search becomes simpler and computationally less expensive because the search is done along the horizontal lines of the rectified images. The steps of the epipolar rectification algorithm are illustrated in figure 4.47. Observe that after the rectification, all the epipolar lines in the left and right image are collinear and horizontal (figure 4.47d). The equations for the epipolar rectification algorithm go beyond the scope of this book, but the interested reader can find an easy-to-implement algorithm in [133].

Disparity map. After the calibration of the stereo-rig, the epipolar rectification, and the correspondence search, we can finally reconstruct the scene points in 3D by solving the system of equations (4.60)–(4.61) (see also the problem in section 4.8). Another popular output of stereo vision is the *disparity map*. A disparity map appear as a grayscale image where the intensity of every pixel point is proportional to the disparity of that pixel in the left and right image: objects that are closer to the camera appear lighter, while farther objects appear darker. An example disparity map is shown in figure 4.48. Disparity maps are very useful for obstacle avoidance (figure 4.49). Modern stereo cameras—like those from Videre Design and Point-Grey (figure 4.42)—are able to compute disparity maps directly in hardware.

Left image   Right image

Disparity map

Figure 4.48 An example disparity map computed from the two top images. Every pixel point is proportional to the disparity of that pixel in the left and right image. Objects that are closer to the camera appear lighter, while farther objects appear darker. Image courtesy of Martin Humenberger, AIT Austrian Institute of Technology — http://www.ait.ac.at.

4.2.6 Structure from motion

In the previous section, we described how to recover the structure of the environment from two images of the scene taken from two distinct cameras whose relative position and orientation is known. In this section, we discuss the problem of recovering the structure when the camera relative pose is unknown. This is the case, for instance, when the two images are taken from the same camera but at different positions and at different times,[8] or, alternatively, from different cameras. This implies that both structure and motion must be estimated simultaneously. This problem is known as *Structure from Motion* (SfM). This problem has been studied for long time in the computer vision community, and in this sec-

---

8. For the sake of simplicity, here we assume that the scene is time invariant (i.e., static). One way to deal with dynamic scenes consists in treating moving objects as outliers.

Figure 4.49
A stereo camera from Videre Design on the Shrimp robot developed at the ASL.

tion we provide only the solution to the two-frame structure from motion problem. For an in-depth study of structure from motion, we refer the reader to [21, 22, 29, 36, 53].

Observe that in structure-from-motion, the images do not need to be precalibrated. This allows SfM to work in challenging situations, where, for instance, images are taken by different users using different cameras (for example, images from the Web). The intrinsic parameters can in fact be estimated automatically from SfM itself. A suggestive result of SfM is illustrated in figure 4.50. Here, the scene was reconstructed using dozens of images. Using thousands of images from different viewpoints, SfM can sometimes achieve 3D reconstruction results that are almost comparable in accuracy and density of points to 3D laser rangefinders (page 133). However, this precision is often at the expense of the computation power.

4.2.6.1 Two-view structure-from-motion

Let us start again from the two perspective projection equations (4.60) and (4.61) derived for the stereo vision case, but now remember that R and t denote the relative motion between the first and the second camera position; therefore, we can write:

Figure 4.50 An example of structure-from-motion: salient image points (image features, see section 4.5) are extracted and matched across multiple frames. Wrong data associations (outliers) are removed and the relative motion among the views is determined. Finally, the points are reconstructed by triangulation. The reconstruction of the building was obtained using dozens of images. The camera poses are also displayed. Image courtesy of Friedrich Fraundorfer. Structure from motion also allows dense reconstruction of entire cities and monuments using just images.

$$\lambda_1 \tilde{p}_1 = A_1[I|0]\tilde{P}_w = A_1 P_w \text{ (for the first camera position),} \qquad (4.67)$$

$$\lambda_2 \tilde{p}_2 = A_2[R|t]\tilde{P}_w \text{ (for the second camera position).} \qquad (4.68)$$

In order to simplify our problem, let us make some assumptions. Let us assume that we use the same camera for the first and the second position and that the intrinsic parameters do not change in between; therefore, $A_1 = A_2 = A$. Let us also assume that the camera is calibrated, and that therefore A is known. In this case, it is more convenient to work with *normalized* image coordinates. Let $\tilde{x}_1$ and $\tilde{x}_2$ be the *normalized* coordinates of $\tilde{p}_1$ and $\tilde{p}_2$ respectively, where

$$\tilde{x}_1 = A^{-1}\tilde{p}_1 \text{ and } \tilde{x}_2 = A^{-1}\tilde{p}_2, \qquad (4.69)$$

and $\tilde{x}_1 = (x_1, y_1, 1)$, $\tilde{x}_2 = (x_2, y_2, 1)$. Then, we can rewrite (4.67) and (4.68) as:

$$\lambda_1 \tilde{x}_1 = P_w \text{ (for the first camera position)}, \tag{4.70}$$

$$\lambda_2 \tilde{x}_2 = [R|t]\tilde{P}_w = RP_w + t \text{ (for the second camera position)}. \tag{4.71}$$

As we did before for computing the epipolar lines (page 176), let us map the optical ray corresponding to x_1 into the second image. Thus, by substituting (4.70) into (4.71), we obtain:

$$\lambda_2 \tilde{x}_2 = \lambda_1 R \tilde{x}_1 + t. \tag{4.72}$$

Let us now take the *cross* product of both sides with t. This in order to cancel t on the right-hand side. Then, we obtain:

$$\lambda_2 [t] \times \tilde{x}_2 = \lambda_1 ([t] \times R) \cdot \tilde{x}_1, \tag{4.73}$$

where $[t] \times$ is an antisymmetric matrix defined as

$$[t] \times = \begin{bmatrix} 0 & -t_z & t_y \\ t_z & 0 & -t_x \\ -t_y & t_x & 0 \end{bmatrix}. \tag{4.74}$$

Now, taking the *dot* product of both sides (4.73) with x_2 yields:

$$\lambda_2 \tilde{x}_2^T \cdot ([t] \times \tilde{x}_2) = \lambda_1 \tilde{x}_2^T \cdot ([t] \times R) \cdot \tilde{x}_1. \tag{4.75}$$

Observe that $\tilde{x}_2^T \cdot ([t] \times \tilde{x}_2) = 0$, and therefore from (4.75) we obtain:

$$\tilde{x}_2^T \cdot ([t] \times R) \cdot \tilde{x}_1 = 0, \tag{4.76}$$

which is called *epipolar constraint*. Observe that the epipolar constraint is valid for every pair of conjugate points.

Let us define the *essential matrix* $E = ([t] \times R)$, the epipolar constraint reads as;

$$\tilde{x}_2^T \cdot E \cdot \tilde{x}_1 = 0 \tag{4.77}$$

It can be shown that the essential matrix has two singular values which are equal and another which is zero [29].

Computing the essential matrix. Given this fundamental relationship (4.77), how can we use it to recover the camera motion encoded in the essential matrix E? If we have N corresponding measurements $\{(x_1^i, x_2^i)\}$, we can form N homogeneous equations in the nine elements of $E = [e_{11}, e_{12}, e_{13}, e_{21}, e_{22}, e_{23}, e_{31}, e_{32}, e_{33}]^T$, of the type

$$\tilde{x}_1^i \tilde{x}_2^i e_{11} + \tilde{y}_1^i \tilde{x}_2^i e_{12} + \tilde{x}_2^i e_{13} + \tilde{x}_1^i \tilde{y}_2^i e_{21} + \tilde{y}_1^i \tilde{y}_2^i e_{22} + \tilde{y}_2^i e_{23} + \tilde{x}_1^i e_{31} + \tilde{y}_1^i e_{32} + e_{33} = 0. \quad (4.78)$$

This can be rewritten in a more compact way as:

$$D \cdot E = 0. \quad (4.79)$$

Given $N \geq 8$ such equations, we can compute an estimate (up to a scale) for the entries in E using the *Singular Values Decomposition (SVD)*. The solution of (4.79) will therefore be the eigenvector of D corresponding to the smallest eigenvalue. Because at least eight point correspondences are needed, this algorithm is known as the *eight-point algorithm* [194]. This algorithm is one of the milestones of computer vision. The main advantages of the eight-point algorithm are that it is very easy to implement and that it works also for an uncalibrated camera, that is, when the camera intrinsic parameters are unknown. The drawback is that it does not work for degenerate point configurations such as planar scenes, that is, when all the scene points are coplanar.

In the case of a calibrated camera, at least five point correspondences are required [178]. An efficient algorithm for computing the essential matrix from at least five point correspondences was proposed by Nister [246]. The *five-point algorithm* works only for calibrated cameras but is more complicated to implement. However, in contrast to the eight-point algorithm, it also works for planar scenes.

Decomposing E into R and t. Let us now assume that the essential matrix E has been determined from known point correspondences. How do we determine R and t? Because a complete derivation of the proof is beyond the scope of this book, we will give directly the final expression. The interested reader can find the proof of these equations in [29].

Before decomposing E, we need to enforce the constraint that two of its singular values are equal and the third one is zero. In fact, in presence of image noise this constraint will never be verified in practice. To do this, we compute the closest[9] essential matrix $\hat{E}$ which

---

9. Closest in terms of the Frobenius norm.

satifies this constraint. One popular technique is to use SVD and force the two larger singular values to be equal and the smallest one to be zero. Therefore:

$$[U, S, V] = SVD(E),$$ (4.80)

where $S = diag([S_{11} \ S_{ss} \ S_{33}])$ with $S_{11} \geq S_{22} \geq S_{33}$. Then, the closest essential matrix eo E in the Frobenius norm is given by

$$\hat{E} = U \cdot diag\left(\left[\frac{S_{11} + S_{22}}{2}, \frac{S_{11} + S_{22}}{2}, 0\right]\right) \cdot V^T$$ (4.81)

Then, we replace E with $\hat{E}$. At this point, we can decompose E into R and t.

The decomposition of E returns four solutions for (R,t), two for R and two for t. Let us define

$$B = \begin{bmatrix} 0 & 1 & 0 \\ -1 & 0 & 0 \\ 0 & 0 & 1 \end{bmatrix} \quad \text{and} \quad [U, S, V] = SVD(E),$$ (4.82)

where U, S, and V are such that $U \cdot S \cdot V^T = E$. It can be shown (see [29]) that the two solutions for R are:

$$R_1 = det(U \cdot V^T) \cdot U \cdot B \cdot V^T,$$ (4.83)

$$R_2 = det(U \cdot V^T) \cdot U \cdot B^T \cdot V^T.$$ (4.84)

Now, let us define

$$L = U \cdot \begin{bmatrix} 0 & -1 & 0 \\ 1 & 0 & 0 \\ 0 & 0 & 0 \end{bmatrix} \cdot U^T \quad \text{and} \quad M = -U \cdot \begin{bmatrix} 0 & -1 & 0 \\ 1 & 0 & 0 \\ 0 & 0 & 0 \end{bmatrix} \cdot U^T.$$ (4.85)

The two solutions for t are:

$$t_1 = \frac{\left[L_{32} \ L_{13} \ L_{21}\right]^T}{\left\|\left[L_{32} \ L_{13} \ L_{21}\right]\right\|},$$ (4.86)

$$t_2 = \frac{\left[M_{32} \ M_{13} \ M_{21} \right]}{\left\| \left[M_{32} \ M_{13} \ M_{21} \right]^T \right\|}.$$ (4.87)

These four solutions can be disambiguated using the so-called *cheirality constraint*, which requires that reconstructed point correspondences lie in front of the cameras. In fact, if you analyse the four solutions of the SfM problem, you will always find that three solutions are such that the reconstructed point correspondences appear behind at least one of the two cameras, while only one solution guarantees that they lie in front of both cameras. Thus, testing with a single point correspondence to determine if it is reconstructed in front of both cameras is sufficient to identify the right solution out of the four possible choices. Also, observe that the solution for t is known up to a scale. In fact, with a single camera it is not possible to recover the absolute scale. For the same reason, the recovered structure will also be known up to a scale.

The last step in two-view structure-from-motion is the reconstruction of the scene. Once R and t have been found, the 3D structure can be computed via triangulation of the feature points as done for the stereo camera (page 173).

Free software for multi-view structure from motion. To conclude this section, we would like to point the reader to some interesting, free software to performe structure-from motion from unordered image collections. The most popular is Microsoft Photosynth (http://photosynth.net)—inspired by the research work on Photo Tourism [355]—which is based on the very popular open-source software Bundler (available at http://photo-tour.cs.washington.edu/bundler) and described in [297] and [298].

Very useful and fully open-source tools for on-line processing are: the Parallel Tracking and Mapping (PTAM) tool [358], the Vodoo camera tracker [356], and the ARToolkit [357].

Useful Matlab toolboxes for structure from motion are:

* FIT3D: http://www.fit3d.info

* Structure from Motion toolbox by V. Rabaud: http://code.google.com/p/vincents-structure-from-motion-matlab-toolbox

* Matlab Functions for Multiple View Geometry by A. Zissermann:
  http://www.robots.ox.ac.uk/~vgg/hzbook/code

* Structure and Motion Toolkit by P. Torr:
  http://cms.brookes.ac.uk/staff/PhilipTorr/Code/code_page_4.htm

* Matlab Code for Non-Rigid Structure from Motion using Factorisation by L. Torresani
  http://movement.stanford.edu/learning-nr-shape

Finally, see also the companies 2d3 (http://www.2d3.com) and Vicon (http://www.vicon.com).

4.2.6.2 Visual odometry

Directly linked to structure from motion is *visual odometry*. Visual odometry consists in estimating the motion of a robot or that of a vehicle by using visual input alone. The term "visual odometry" was coined in 2004 by Nister with his homonym landmark paper [245], where he showed successful results on different vehicles (on-road and off-road) using either a single camera or a stereo camera. The basic principle behind visual odometry is a simple iteration of two-view structure from motion that we have seen in the previous section.

Most of the work done about visual odometry has been produced using stereo cameras and can be traced back to 1980 with Moravec's work [236]. Similar work has also been reported elsewhere also (see [160, 174, 181, 244]). Furthermore, stereo visual odometry has also been successfully used on Mars by the NASA rovers since early 2004 [203]. Nevertheless, visual odometry methods for outdoor applications have also been produced, which use a single camera alone (see [107, 244, 278, 279, 307]).

The advantage of using a stereo camera compared to a single camera is that the measurements are directly provided in the absolute scale. Conversely, when using a single camera the absolute scale must be estimated in other ways (e.g., from knowledge of an element in the scene, or the distance between the camera and the ground plane) or using other sensors such as GPS, IMU, wheel odometry, or lasers.

Visual odometry aims at recovering only the trajectory of the vehicle. Nevertheless, it is not uncommon to see results showing also the 3D map of the environment which is usually a simple triangulation of the feature points from the estimated camera poses. An example visual odometry result using a single omnidirectional camera is shown in figure 4.51. Here the scale was obtained by exploiting the nonholonomic constraints of the vehicle as described in [277]. In this figure, visual odometry is performed over a 3 km trajectory. Notice the visible drift toward the end of the trajectory.

All visual odometry algorithms suffer from motion drift due to the integration of the relative displacements between consecutive poses which unavoidably accumulates errors over time. This drift becomes evident usually after a few hundred meters. But the results may vary depending on the abundance of features in the environment, the resolution of the cameras, the presence of moving objects like people or other passing vehicles, and the illumination conditions. Remember that motion drift is also present in wheel odometry, as will be described in section 5.2.4. However, the reason visual odometry is becoming more and more popular in both robotics and automotive is that drift can be canceled if the vehicle revisits a place that has already been observed previously. The possibility of performing location recognition (or place recognition) is one of the main advantages of vision com-

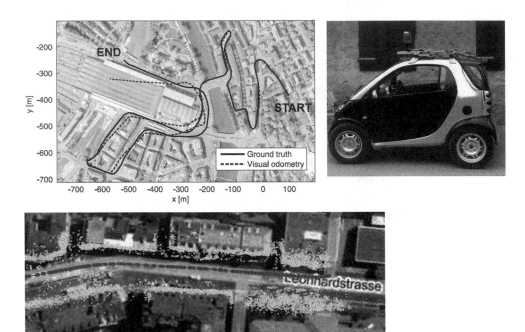

Figure 4.51 (upper left) An example visual odometry result with related map (bottom) obtained using a single omnidirectional camera mounted on the roof of the vehicle (right). The absolute scale was computed automatically by taking advantage of the fact that a wheeled vehicle is constrained to follow a circular coarse, locally, about the instantaneous center of rotation (see Ackerman steering principle in section 3.3.1). This visual odometry result was obtained using the 1-point RANSAC method described in [278]. The advantage of this algorithm compared to the state of the art is that visual odometry runs at 400 frames per second, while standard method work at 20–40 Hz.

pared to other sensor modalities. The most popular computer vision approaches for location recognition will be described in section 4.6. Once a place previously observed is visited a second time by the robot, the accumulated error can be reduced by adding the constraint that the positions of the vehicle at these two places (the previously visited and the revisited one) should actually coincide. This obviously requires an algorithm that modifies ("relaxes") all the previous robot poses until the error between the current and previously visited location is minimized.

The problem of location recognition is also called loop-detection, because a loop is a closed trajectory of a vehicle that returns to a previously-visited point. The problem of minimizing the error at the loop closure is instead called loop-closing. There are several algorithms in the literature to perform loop-closing. Some of them come from the computer vision community and rely on the so called *bundle adjustment*,[10] while others have been

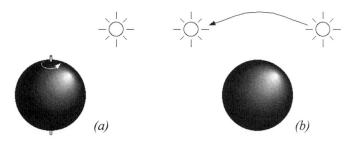

Figure 4.52
Motion of the sphere or the light source here demonstrates that optical flow is not always the same as the motion field.

developed within the robotics community to solve the Simultaneous Localization and Mapping (SLAM) problem (see section 5.8.2). Some of the most popular algorithms can be found in [318] and [352].

4.2.7 Motion and optical flow

A great deal of information can be recovered by recording time-varying images from a fixed (or moving) camera. First, we distinguish between the motion field and optical flow:

- Motion field: this assigns a velocity vector to every point in an image. If a point in the environment moves with velocity v_0, then this induces a velocity v_i in the image plane. It is possible to determine mathematically the relationship between v_i and v_0.

- Optical flow: it can also be true that brightness patterns in the image move as the object that causes them moves (light source). Optical flow is the apparent motion of these brightness patterns.

In our analysis here we assume that the optical flow pattern will correspond to the motion field, although this is not always true in practice. This is illustrated in figure 4.52a, where a sphere exhibits spatial variation of brightness, or shading, in the image of the sphere since its surface is curved. If the surface moves, however, this shading pattern will not move hence the optical flow is zero everywhere even though the motion field is not zero. In figure 4.52b, the opposite occurs. Here we have a fixed sphere with a moving light source. The shading in the image will change as the source moves. In this case the optical

---

10. Given a set of images observing a certain number of 3D points from different viewpoints, bundle adjustment is the problem of simultaneously refining the 3D coordinates of the scene geometry as well as the relative motion and the camera intrinsic parameters. This is done according to an optimality criterion involving the corresponding image projections of all points.

flow is nonzero but the motion field is zero. If the only information accessible to us is the optical flow and we depend on this, we will obtain incorrect results in both cases.

4.2.7.1 Optical flow

There are a number of techniques for attempting to measure optical flow and thereby obtain the scene's motion field. Most algorithms use local information, attempting to find the motion of a local patch in two consecutive images. In some cases, global information regarding smoothness and consistency can help to disambiguate further such *matching* processes. Below we present details for the optical flow constraint equation method. For more details on this and other methods, refer to [69, 151, 316].

Suppose first that the time interval between successive snapshots is so fast that we can assume that the measured intensity of a portion of the same object is effectively constant. Mathematically, let $I(x, y, t)$ be the image irradiance at time t at the image point (x, y). If $u(x, y)$ and $v(x, y)$ are the x and y components of the optical flow vector at that point, we need to search a new image for a point where the irradiance will be the same at time $t + \delta t$, that is, at point $(x + \delta t, y + \delta t)$, where $\delta x = u \delta t$ and $\delta y = v \delta t$. That is,

$$I(x + u\delta t, y + v\delta t, t + \delta t) = I(x, y, t) \tag{4.88}$$

for a small time interval, δt. This will capture the motion of a constant-intensity *patch* through time. If we further assume that the brightness of the image varies smoothly, then we can expand the left-hand side of equation (4.88) as a Taylor series to obtain

$$I(x, y, t) + \delta x \frac{\partial I}{\partial x} + \delta y \frac{\partial I}{\partial y} + \delta t \frac{\partial I}{\partial t} + e = I(x, y, t), \tag{4.89}$$

where e contains second- and higher-order terms in δx, and so on. In the limit as δt tends to zero we obtain

$$\frac{\partial I}{\partial x} \frac{dx}{dt} + \frac{\partial I}{\partial y} \frac{dy}{dt} + \frac{\partial I}{\partial t} = 0, \tag{4.90}$$

from which we can abbreviate

$$u = \frac{dx}{dt} \; ; \quad v = \frac{dy}{dt} \tag{4.91}$$

and

$$I_x = \frac{\partial I}{\partial x} ; \quad I_y = \frac{\partial I}{\partial y} ; \quad I_t = \frac{\partial I}{\partial t} = 0 , \tag{4.92}$$

so that we obtain

$$I_x u + I_y v + I_t = 0 . \tag{4.93}$$

The derivative I_t represents how quickly the intensity changes with time while the derivatives I_x and I_y represent the spatial rates of intensity change (how quickly intensity changes across the image). Altogether, equation (4.93) is known as the *optical flow constraint equation*, and the three derivatives can be estimated for each pixel given successive images.

We need to calculate both u and v for each pixel, but the optical flow constraint equation only provides one equation per pixel, and so this is insufficient. The ambiguity is intuitively clear when one considers that a number of equal-intensity pixels can be inherently ambiguous—it may be unclear which pixel is the resulting location for an equal-intensity originating pixel in the prior image.

The solution to this ambiguity requires an additional constraint. We assume that in general the motion of adjacent pixels will be similar, and that therefore the overall optical flow of all pixels will be smooth. This constraint is interesting in that we know it will be violated to *some* degree, but we enforce the constraint nonetheless in order to make the optical flow computationally tractable. Specifically, this constraint will be violated precisely when different objects in the scene are moving in different directions with respect to the vision system. Of course, such situations will tend to include edges, and so this may introduce a useful visual cue.

Because we know that this smoothness constraint will be somewhat incorrect, we can mathematically define the degree to which we violate this constraint by evaluating the formula

$$e_s = \iint (u^2 + v^2) dx dy , \tag{4.94}$$

which is the integral of the square of the magnitude of the gradient of the optical flow. We also determine the error in the optical flow constraint equation (which in practice will not quite be zero).

$$e_c = \iint (I_x u + I_y v + I_t)^2 dx dy . \tag{4.95}$$

Both of these equations should be as small as possible, so we want to minimize $e_s + \lambda e_c$, where λ is a parameter that weights the error in the image motion equation relative to the departure from smoothness. A large parameter should be used if the brightness measurements are accurate and small if they are noisy. In practice, the parameter λ is adjusted manually and interactively to achieve the best performance.

The resulting problem then amounts to the calculus of variations, and the Euler equations yield

$$\nabla^2 u \;=\; \lambda (I_x u + I_y v + I_t) I_x , \qquad\qquad\qquad\qquad\qquad (4.96)$$

$$\nabla^2 v \;=\; \lambda (I_x u + I_y v + I_t) I_y , \qquad\qquad\qquad\qquad\qquad (4.97)$$

where

$$\nabla^2 \;=\; \frac{\partial^2}{\delta x^2} + \frac{\partial^2}{\delta y^2} , \qquad\qquad\qquad\qquad\qquad\qquad (4.98)$$

which is the Laplacian operator.

Equations (4.96) and (4.97) form a pair of elliptical second-order partial differential equations that can be solved iteratively.

Where occlusions (one object occluding another) occur, discontinuities in the optical flow will occur. This, of course, violates the smoothness constraint. One possibility is to try to find edges that are indicative of such occlusions, excluding the pixels near such edges from the optical flow computation so that smoothness is a more realistic assumption. Another possibility is to make use of these distinctive edges opportunistically. In fact, corners can be especially easy to *pattern-match* across subsequent images and thus can serve as fiducial markers for optical flow computation in their own right.

Optical flow is an important ingredient in vision algorithms that combine cues across multiple algorithms. Obstacle avoidance and navigation control systems for mobile robots (especially flying robots) using optical flow have proved to be broadly effective as long as texture is present [23, 54].

4.2.8 Color tracking

An important aspect of vision sensing is that the vision chip can provide sensing modalities and cues that no other mobile robot sensor provides. One such novel sensing modality is detecting and tracking color in the environment.

Color is an environmental characteristic and represents both a natural cue and an artificial cue that can provide new information to a mobile robot. For example, the annual robot

Figure 4.53
Color markers on the top of EPFL's STeam Engine soccer robots enable a color-tracking sensor to locate the robots and the ball in the soccer field.

soccer events (RoboCup) make extensive use of color both for environmental marking and for robot localization (see figure 4.53).

Color sensing has two important advantages. First, detection of color is a straightforward function of a single image, therefore no correspondence problem needs be solved in such algorithms. Second, because color sensing provides a new, independent environmental cue, if it is combined (i.e., *sensor fusion*) with existing cues, such as data from stereo vision or laser rangefinding, we can expect significant information gains.

Efficient color-tracking sensors are also available commercially—such as the CMUcam from Carnegie Mellon University—but they can also be implemented straightforwardly using a standard camera. The simplest way of doing this is using *constant thresholding*: a given pixel point is selected if and only if its RGB values (r, g, b) fall simultaneously in the chosen R, G, and B ranges, which are defined by six thresholds $[R_{min}, R_{max}]$, $[G_{min}, G_{max}]$, $[B_{min}, B_{max}]$. Therefore

$$R_{min} < r < R_{max} \text{ and } G_{min} < g < G_{max} \text{ and } B_{min} < b < B_{max} . \tag{4.99}$$

If we represent the RGB color space as a three-dimensional Euclidean space, the aforementioned method selects those pixels whose color components belong to the cube specified by the given thresholds. Alternatively, a sphere could be used. In this case, a pixel would be selected only if its RGB components are within a certain distance from a given point in the RGB space.

Alternatively to RGB, the YUV color space can be used. While R, G, and B values encode the intensity of each color, YUV separates the color (or *chrominance*) measure

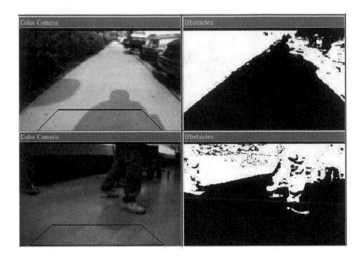

Figure 4.54
Examples of adaptive floor plane extraction. The trapezoidal polygon identifies the floor sampling region.

from the brightness (or *luminosity*) measure. Y represents the image's luminosity while U and V together capture its chrominance. Thus, a bounding box expressed in YUV space can achieve greater stability with respect to changes in illumination than is possible in RGB space.

A popular application of color segmentation in robotics is floor plane extraction (figure 4.54). In this case, color segmentation techniques more complex than color thresholding are used, like adaptive thresholding, or *k-means clustering*[11] [18]. Floor plane extraction is a vision approach for identifying the traversable portions of the ground. Because it makes use of edges (section 4.3.2) and color in a variety of implementations, such obstacle detection systems can easily detect obstacles in cases that are difficult for traditional ranging devices. As is the case with all vision algorithms, floor plane extraction succeeds only in environments that satisfy several important assumptions:

- Obstacles differ in appearance from the ground.

- The ground is flat, and its angle to the camera is known.

---

11. In statistics and machine learning, *k-means clustering* is a method of cluster analysis that aims to partition n observations into k clusters in which each observation belongs to the cluster with the nearest mean. The k-means clustering algorithm is commonly used in computer vision as a form of image segmentation.

- There are no overhanging obstacles.

The first assumption is a requirement in order to discriminate the ground from obstacles using its appearance. A stronger version of this assumption, sometimes invoked, states that the ground is uniform in appearance and different from all obstacles. The second and third assumptions allow floor-plane-extraction algorithms to estimate the robot's distance to obstacles detected.

4.3   Fundamentals of Image Processing

Image processing is a form of signal processing where the input signal is an image (such as a photo or a video) and the output is either an image or a set of parameters associated with the image. Most image-processing techniques treat the image as a two-dimensional signal $I(x, y)$ where x and y are the *spatial* image coordinates and the amplitude of I at any pair of coordinates (x, y) is called *intensity* or *gray level* of the image at that point.

Image processing is a huge field and typical operations, among many others, are:

- Filtering, image enhancing, edge detection

- Image restoration and reconstruction

- Wavelets and multiresolution processing

- Image compression (e.g., JPEG)

- Euclidean geometry transformations such as enlargement, reduction, and rotation

- Color corrections such as brightness and contrast adjustments, quantization, or color translation to a different color space

- Image registration (the alignment of two or more images)

- Image recognition (for example, extracting a face from the image by using some face recognition algorithm)

- Image segmentation (partitioning the image in characteristic regions according to color, edges, or other features)

Because a review of all these techniques goes beyond the scope of this book, here we focus only on the most important image processing operations that are relevant for robotics. In particular, we describe image filtering operations such as smoothing and edge detection. We will then describe some image similarity measures for finding point correspondences between images, which are helpful in structure from stereo and structure from motion. For an in-depth study of image processing in general, we refer the reader to [26].

4.3.1 Image filtering

Image filtering is one of the principal tools in image processing. The word *filter* comes from frequency domain processing, where "filtering" refers to the process of accepting or rejecting certain frequency components. For example, a filter that passes low frequencies is called a *lowpass* filter. The effect produced by a lowpass filter is to blur (smooth) an image, which has the main effect of reducing image noise. Conversely, a filter that passes high frequencies is called *highpass* filter and is typically used for edge detection. Image filters can be implemented both in the frequency domain and in the spatial domain. In the latter case, the filter is called *mask* or *kernel*. In this section, we will review the fundamentals of spatial filtering.

In figure 4.55, the basic principle of spatial filtering is explained. A spatial filter consists of (1) a neighborhood of the pixel under examination, (typically a small rectangle), and (2) a predefined operation T that is performed on the image pixels encompassed by the neighborhood. Let S_{xy} denote the set of coordinates of a neighborhood centered on an arbitrary point (x,y) in an image I. Spatial filtering generates a corresponding pixel at the same coordinates in an output image I' where the value of that pixel is determined by a specified operation on the pixels in S_{xy}. For example, suppose that the specified operation is to compute the average value of the pixels in a rectangular window of size $m \times n$ centered on (x,y). The locations of pixels in this region constitute the set S_{xy}. Figure 4.55a–b illustrates the process. We can express this operation in equation form as

$$I'(x, y) = \frac{1}{mn} \sum_{(r, c) \in S_{xy}} I(r, c) \tag{4.100}$$

where r and c are the row and column coordinates of the pixels in the set S_{xy}. The new image I' is created by varying the coordinates (x, y) so that the center of the window moves from pixel to pixel in image I. For instance the image in figure 4.55d was created in this manner using a window of size 21×21 applied on the image in figure 4.55c.

The filter used to illustrate the example above is called *averaging filter*. More generally, the operation performed on the image pixels can be linear or nonlinear. In these cases, the filter is called either a *linear* or *nonlinear* filter. Here, we concentrate on linear filters. In general, linear spatial filtering of an image with a filter w of size $m \times n$ is given by the expression

$$I'(x, y) = \sum_{s = -a}^{a} \sum_{t = -b}^{b} w(s, t) \cdot I(x + s, y + t), \tag{4.101}$$

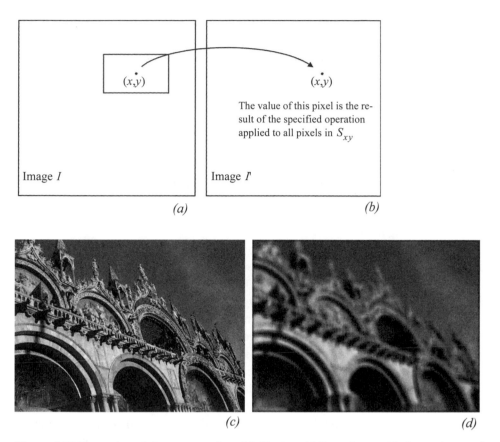

(a)   *(b)*

(c)   *(d)*

Figure 4.55 Illustration of the concept of spatial filtering. (c) Input image. (d) Output image after application of average filter.

where $m = 2a + 1$ and $n = 2b + 1$ are usually assumed odd integers. The filter w is also called *kernel*, *mask*, or *window*. As observed in (4.101), linear filtering is the process of moving a filter mask over the entire image and computing the sum of products at each location. In signal processing, this particular operation is also called *correlation* with the kernel w. It is, however, opportune to specify that an equivalent linear filtering operation is the *convolution*

$$I'(x, y) = \sum_{s=-a}^{a} \sum_{t=-b}^{b} w(s, t) \cdot I(x-s, y-t) \tag{4.102}$$

where the only difference with the correlation is the presence of the minus sign, meaning that the image must be flipped. Observe that for symmetric filters *convolution* and *correlation* return the same result and the two terms can therefore be used interchangeably. The operation of convolution with the kernel *w* can be written in a more compact way as

$$I'(x, y) = w(x, y)*I(x, y),$$ (4.103)

where * denotes the convolution operator.

Generating linear spatial filters requires that we specify the *mn* coefficients of the kernel. These coefficients are chosen based on what the filter is supposed to do. In the next section, we will see how to select these coefficients.

4.3.1.1 Smoothing filters

Smoothing filters are used for blurring and for noise reduction. Blurring is used in tasks such as removal of small details or filling of small gaps in lines or curves. Both blurring and noise reduction can be accomplished via linear or nonlinear filters. Here, we review some linear filters.

The output of a smoothing filter is simply the weighted average of the pixels contained in the filter mask. These filters are sometimes called *averaging filters* or *lowpass filters*. As we explained before, every pixel in an image is replaced by the average of the intensity of the pixels in the neighborhood defined by the filter mask. This process results in a new image with reduced sharp transitions. Accordingly, image noise gets reduced. As a side effect, however, edges—which are usually a desirable feature of an image—also get blurred. This side effect can be limited by choosing the filter coefficients appropriately. Finally observe that a *nonlinear* averaging filter can also be easily implemented by taking the *median* of the pixels contained in the mask. Median filters are particularly useful to remove *salt and pepper* noise.[12]

In the previous section, we already saw an example of constant averaging filter (figure 4.55) that simply yields the standard average of the pixels in the mask. Assuming a 3×3 mask, the filter can be written as

$$w = \frac{1}{9}\begin{bmatrix} 1 & 1 & 1 \\ 1 & 1 & 1 \\ 1 & 1 & 1 \end{bmatrix},$$ (4.104)

---

12. Salt and pepper noise represents itself as randomly occurring white and black pixels and is a typical form of noise in images.

where all the coefficients sum to 1. This normalization is important to keep the same value as the original image if the region by which the filter is multiplied is uniform. Also note that, instead of being 1/9, the coefficients of the filter are all 1s. The idea is that the pixels are first summed up and the result is then divided by 9. Indeed, this is computationally more efficient than multiplying each element by 1/9.

Many image-processing algorithms make use of the second derivative of the image intensity. Because of the susceptibility of such high-order derivative algorithms to changes in illumination in the basic signal, it is important to smooth the signal so that changes in intensity are due to real changes in the luminosity of objects in the scene rather than random variations due to imaging noise. A standard approach is the use of a Gaussian averaging filter whose coefficients are given by

$$G_\sigma(x, y) = e^{-\frac{x^2 + y^2}{2\sigma^2}}. \tag{4.105}$$

To generate, say, a 3×3 filter mask from this function, we sample it about its center. For example, with $\sigma = 0.85$, we get

$$G = \frac{1}{16}\begin{bmatrix} 1 & 2 & 1 \\ 2 & 4 & 2 \\ 1 & 2 & 1 \end{bmatrix}, \tag{4.106}$$

where, again, the coefficients were rescaled so that they sum to 1. Also notice that the coefficients are all powers of 2, which makes it extremely efficient to compute. This filter is actually very popular. Such a lowpass filter effectively removes high-frequency noise, and this in turn causes the first derivative and especially the second derivative of intensity to be far more stable. Because of the importance of gradients and derivatives to image processing, such Gaussian smoothing preprocessing is a popular first step of virtually all computer vision algorithms.

4.3.2 Edge detection

Figure 4.56 shows an image of a scene containing a part of a ceiling lamp as well as the edges extracted from this image. Edges define regions in the image plane where a *significant* change in the image brightness takes place. As shown in this example, edge detection significantly reduces the amount of information in an image, and is therefore a useful potential feature during image interpretation. The hypothesis is that edge contours in an image correspond to important scene contours. As figure 4.56b shows, this is not entirely true. There is a difference between the output of an edge detector and an ideal line drawing.

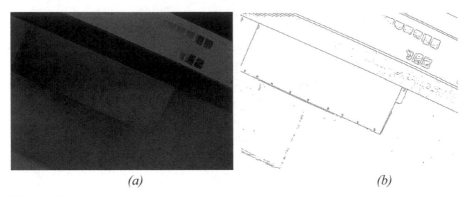

(a)   *(b)*

Figure 4.56
(a) Photo of a ceiling lamp. (b) Edges computed from (a).

Typically, there are missing contours, as well as noise contours, that do not correspond to anything of significance in the scene.

The basic challenge of edge detection is visualized in figure 4.57. The top left portion shows the 1D section of an ideal edge. But the signal produced by a camera will look more like figure 4.57 (top right) because of noise. The location of the edge is still at the same x value, but a significant level of high-frequency noise affects the signal quality.

A naive edge detector would simply differentiate, since an edge by definition is located where there are large transitions in intensity. As shown in figure 4.57 (bottom right), differentiation of the noisy camera signal results in subsidiary peaks that can make edge detection very challenging. A far more stable derivative signal can be generated simply by preprocessing the camera signal using the Gaussian smoothing function described above. Below, we present several popular edge detection algorithms, all of which operate on this same basic principle, that the derivative(s) of intensity, following some form of smoothing, comprises the basic signal from which to extract edge features.

Optimal edge detection: the Canny edge detector. The current reference edge detector throughout the vision community was invented by John Canny in 1983 [91]. This edge detector was born out of a formal approach in which Canny treated edge detection as a signal-processing problem in which there are three explicit goals:

- Maximizing the signal-to-noise ratio;

- Achieving the highest precision possible on the location of edges;

- Minimizing the number of edge responses associated with each edge.

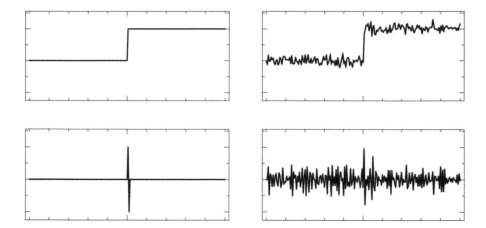

Figure 4.57
Step function example of second derivative shape and the impact of noise.

The Canny edge extractor smooths the image I via Gaussian convolution and then looks for maxima in the (rectified) derivative. In practice, the smoothing and differentiation are combined into one operation because

$$(G*I)' = G'*I \quad^{13} \tag{4.107}$$

Thus, smoothing the image by convolving with a Gaussian G_σ and then differentiating is equivalent to convolving the image with G'_σ, which is the first derivative of G_σ (figure 4.58b).

We wish to detect edges in any direction. Since G' is directional, this requires application of two perpendicular filters (figure 4.59). We define the two filters as $f_V(x, y) = G'_\sigma(x)G_\sigma(y)$ and $f_H(x, y) = G'_\sigma(y)G_\sigma(x)$. The result is a basic algorithm for detecting edges at arbitrary orientations:

The algorithm for detecting edge pixels at an arbitrary orientation is as follows:

1. Convolve the image $I(x, y)$ with $f_V(x, y)$ and $f_H(x, y)$ to obtain the gradient components $R_V(x, y)$ and $R_H(x, y)$, respectively.

2. Define the square of the gradient magnitude $R(x, y) = R_V^2(x, y) + R_H^2(x, y)$.

3. Mark those peaks in $R(x, y)$ that are above some predefined threshold T.

---

13. This is a known property of convolution.

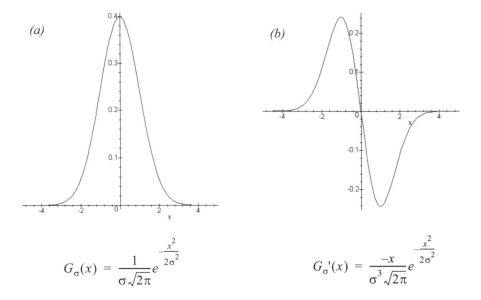

$$G_\sigma(x) = \frac{1}{\sigma\sqrt{2\pi}} e^{-\frac{x^2}{2\sigma^2}} \qquad\qquad G_\sigma'(x) = \frac{-x}{\sigma^3\sqrt{2\pi}} e^{-\frac{x^2}{2\sigma^2}}$$

Figure 4.58
(a) A Gaussian function. (b) The first derivative of a Gaussian function.

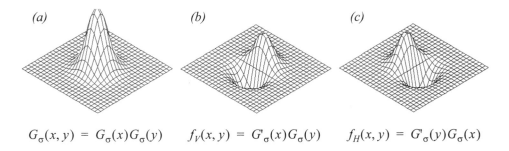

$$G_\sigma(x, y) = G_\sigma(x)G_\sigma(y) \qquad f_V(x, y) = G_\sigma'(x)G_\sigma(y) \qquad f_H(x, y) = G_\sigma'(y)G_\sigma(x)$$

Figure 4.59
(a) Two-dimensional Gaussian function. (b) Vertical filter. (c) Horizontal filter.

Once edge pixels are extracted, the next step is to construct complete edges. A popular next step in this process is *nonmaxima suppression*. Using edge direction information, the process involves revisiting the gradient value and determining whether or not it is at a local maximum. If not, then the value is set to zero. This causes only the maxima to be preserved, and thus reduces the thickness of all edges to a single pixel (figure 4.60).

Figure 4.60
(a) Example of an edge image; (b) Nonmaxima suppression of (a).

Finally, we are ready to go from edge pixels to complete edges. First, find adjacent (or connected) sets of edges and group them into ordered lists. Second, use thresholding to eliminate the weakest edges.

Gradient edge detectors. On a mobile robot, computation time must be minimized to retain the real-time behavior of the robot. Therefore simpler, discrete kernel operators are commonly used to approximate the behavior of the Canny edge detector. One such early operator was developed by Roberts in 1965 [43]. He used two 2×2 masks to calculate the gradient across the edge in two diagonal directions. Let r_1 be the value calculated from the first mask and r_2 that from the second mask. Roberts obtained the gradient magnitude $|G|$ with the equation

$$|G| \cong \sqrt{r_1^2 + r_2^2} \; ; \quad r_1 = \begin{bmatrix} -1 & 0 \\ 0 & 1 \end{bmatrix} \; ; \quad r_2 = \begin{bmatrix} 0 & -1 \\ 1 & 0 \end{bmatrix} \tag{4.108}$$

Prewitt (1970) [43] used two 3×3 masks oriented in the row and column directions. Let p_1 be the value calculated from the first mask and p_2 the value calculated from the second mask. Prewitt obtained the gradient magnitude $|G|$ and the gradient direction θ taken in a clockwise angle with respect to the column axis shown in the following equation.

$$|G| \cong \sqrt{p_1^2 + p_2^2} \; ;$$

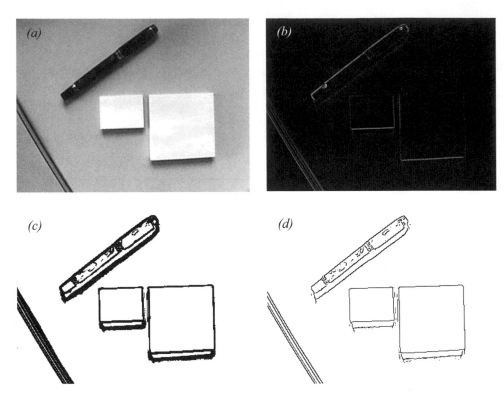

Figure 4.61
Example of visual feature extraction with the different processing steps: (a) raw image data; (b) filtered image using a Sobel filter; (c) thresholding, selection of edge pixels (d) nonmaxima suppression.

$$\theta \cong \operatorname{atan}\left(\frac{p_1}{p_2}\right) \; ; \quad p_1 = \begin{bmatrix} -1 & -1 & -1 \\ 0 & 0 & 0 \\ 1 & 1 & 1 \end{bmatrix} \; ; \quad p_2 = \begin{bmatrix} -1 & 0 & 1 \\ -1 & 0 & 1 \\ -1 & 0 & 1 \end{bmatrix} \tag{4.109}$$

In the same year, Sobel [43] used, like Prewitt, two 3×3 masks oriented in the row and column direction. Let s_1 be the value calculated from the first mask and s_2 the value calculated from the second mask. Sobel obtained the same results as Prewitt for the gradient magnitude $|G|$ and the gradient direction θ taken in a clockwise angle with respect to the column axis. Figure 4.61 shows application of the Sobel filter to a visual scene.

$$|G| \cong \sqrt{s_1^2 + s_2^2} \ ;$$

$$\theta \cong \mathrm{atan}\!\left(\frac{s_1}{s_2}\right) \ ; \qquad s_1 = \begin{bmatrix} -1 & -2 & -1 \\ 0 & 0 & 0 \\ 1 & 2 & 1 \end{bmatrix} \ ; \qquad s_2 = \begin{bmatrix} -1 & 0 & 1 \\ -2 & 0 & 2 \\ -1 & 0 & 1 \end{bmatrix} . \qquad (4.110)$$

Dynamic thresholding. Many image-processing algorithms have generally been tested in laboratory conditions or by using static image databases. Mobile robots, however, operate in dynamic real-world settings where there is no guarantee regarding optimal or even stable illumination. A vision system for mobile robots has to adapt to the changing illumination. Therefore a constant threshold level for edge detection is not suitable. The same scene with different illumination results in edge images with considerable differences. To adapt the edge detector dynamically to the ambient light, a more adaptive threshold is required, and one approach involves calculating that threshold based on a statistical analysis of the image about to be processed.

To do this, a histogram of the gradient magnitudes of the processed image is calculated (figure 4.62). With this simple histogram it is easy to consider only the n pixels with the highest gradient magnitude for further calculation steps. The pixels are counted backward, starting at the highest magnitude. The gradient magnitude of the point where n is reached will be used as the temporary threshold value.

The motivation for this technique is that the n pixels with the highest gradient are expected to be the most relevant ones for the processed image. Furthermore, for each image, the same number of relevant edge pixels is considered, independent of illumination. It is important to pay attention to the fact that the number of pixels in the edge image delivered by the edge detector is not n. Because most detectors use nonmaxima suppression, the number of edge pixels will be further reduced.

Straight edge extraction: Hough transform. In mobile robotics, the straight edge is often extracted as a specific feature. Straight vertical edges, for example, can be used as clues to the location of doorways and hallway intersections. The Hough transform is a simple tool for extracting edges of a particular shape [21, 28]. Here we explain its application to the problem of extracting straight edges.

Suppose a pixel (x_p, y_p) in the image I is part of an edge. Any straight-line edge including point (x_p, y_p) must satisfy the equation: $y_p = m_1 x_p + b_1$. This equation can only be satisfied with a constrained set of possible values for m_1 and b_1. In other words, this equation is satisfied only by lines through I that pass through (x_p, y_p).

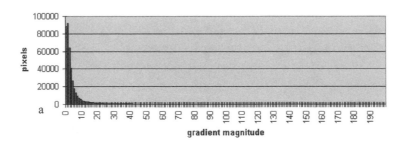

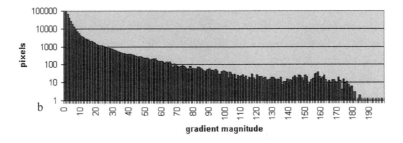

Figure 4.62
(a) Number of pixels with a specific gradient magnitude in the image of figure 4.61b. (b) Same as (a), but with logarithmic scale

Now consider a second pixel, (x_q, y_q) in I. Any line passing through this second pixel must satisfy the equation: $y_q = m_2 x_q + b_2$. What if $m_1 = m_2$ and $b_1 = b_2$? Then the line defined by both equations is one and the same: it is the line that passes through both (x_p, y_p) and (x_q, y_q).

More generally, for all pixels that are part of a single straight line through I, they must all lie on a line defined by the *same* values for m and b. The general definition of this line is, of course, $y = mx + b$. The Hough transform uses this basic property, creating a mechanism so that each edge pixel can "vote" for various values of the (m, b) parameters. The lines with the most votes at the end are straight edge features:

- Create a 2D array A with axes that tessellate the values of m and b.

- Initialize the array to zero: $A[m, b] = 0$ for all values of m, b.

- For each edge pixel (x_p, y_p) in I, loop over all values of m and b:
  if $y_p = m x_p + b$ then $A[m, b] += 1$.

- Search the cells in A to identify those with the largest value. Each such cell's indices (m, b) correspond to an extracted straight-line edge in I.

4.3.3 Computing image similarity

In this section, we review the three most popular image similarity measures used for solving the correspondence problem in structure from stereo (section 4.2.5) and structure from motion (section 4.2.6). The methods we are about to describe are all *area-based* (page 174). Suppose that we want to compare a $m \times n$ patch in image I_1 centered on (u,v) with another patch of the same size centered on (u', v') in image I_2 . We assume that these are odd integers, therefore $m = 2a+1$ and $n = 2b+1$ .The similarity is then computed between the gray intensity levels of the two patches. Some of the most popular criteria are:

Sum of Absolute Differences (SAD)

$$SAD = \sum_{k=-a}^{a} \sum_{l=-b}^{b} \left| I_1(u+k, v+l) - I_2(u'+k, v'+l) \right| . \tag{4.111}$$

Sum of Squared Differences (SSD)

$$SSD = \sum_{k=-a}^{a} \sum_{l=-b}^{b} [I_1(u+k, v+l) - I_2(u'+k, v'+l)]^2 . \tag{4.112}$$

Normalized Cross Correlation (NCC)

$$NCC = \frac{\displaystyle\sum_{k=-a}^{a} \sum_{l=-b}^{b} [I_1(u+k, v+l) - \mu_1] \cdot [I_2(u'+k, v'+l) - \mu_2]}{\sqrt[2]{\displaystyle\sum_{k=-a}^{a} \sum_{l=-b}^{b} [I_1(u+k, v+l) - \mu_1]^2 \sum_{k=-a}^{a} \sum_{l=-b}^{b} [I_2(u'+k, v'+l) - \mu_2]^2}} , \tag{4.113}$$

where

$$\mu_1 = \frac{1}{mn} \sum_{k=-a}^{a} \sum_{l=-b}^{b} I_1(u+k, v+l) \tag{4.114}$$

$$\mu_2 = \frac{1}{mn} \sum_{k=-a}^{a} \sum_{l=-b}^{b} I_2(u'+k, v'+l) \tag{4.115}$$

are the mean values of the two image patches.

The SAD is the simplest among these similarity measures. It is calculated by subtracting pixels between the reference image I_1 and the target image I_2 followed by the aggregation of absolute differences within the patch. The SSD has a higher computational complexity compared to the SAD, since it involves numerous multiplication operations (i.e., squared). Notice that if the left and right images match perfectly, the resultant of SAD and SSD will be zero.

The NCC is even more complex than both SAD and SSD algorithms, since it involves numerous multiplication, division, and square root operations; however, it but provides more distinctiveness than SSD and SAD and also invariance to affine intensity changes (see also figure 4.69). Finally, note that the value of NCC ranges between -1 and 1 where 1 corresponds to maximum similarity between the two image patches.

4.4   Feature Extraction

An autonomous mobile robot must be able to determine its relationship to the environment by making measurements with its sensors and then using those measured signals. A wide variety of sensing technologies are available, as shown in section 4.1. But every sensor we have presented is imperfect: measurements always have error and, therefore, uncertainty associated with them. Therefore, sensor inputs must be used in a way that enables the robot to interact with its environment successfully in spite of measurement uncertainty.

There are two strategies for using uncertain sensor input to guide the robot's behavior. One strategy is to use each sensor measurement as a raw and individual value. Such raw sensor values could, for example, be tied directly to robot behavior, whereby the robot's actions are a function of its sensor inputs. Alternatively, the raw sensor values could be used to update an intermediate model, with the robot's actions being triggered as a function of this model rather than of the individual sensor measurements.

The second strategy is to extract information from one or more sensor readings first, generating a higher-level *percept* that can then be used to inform the robot's model and perhaps the robot's actions directly. We call this process *feature extraction*, and it is this next, optional step in the perceptual interpretation pipeline (figure 4.63) that we will now discuss.

In practical terms, mobile robots do not necessarily use feature extraction and scene interpretation for every activity. Instead, robots will interpret sensors to varying degrees depending on each specific functionality. For example, in order to guarantee emergency stops in the face of immediate obstacles, the robot may make direct use of raw forward-facing range readings to stop its drive motors. For local obstacle avoidance, raw ranging sensor strikes may be combined in an occupancy grid model, enabling smooth avoidance of obstacles meters away. For map-building and precise navigation, the range sensor values and even vision-sensor measurements may pass through the complete perceptual pipeline,

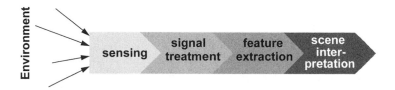

Figure 4.63
The perceptual pipeline: from sensor readings to knowledge models.

being subjected to feature extraction followed by scene interpretation to minimize the impact of individual sensor uncertainty on the robustness of the robot's mapmaking and navigation skills. The pattern that thus emerges is that, as one moves into more sophisticated, long-term perceptual tasks, the feature-extraction and scene-interpretation aspects of the perceptual pipeline become essential.

Feature definition. Features are recognizable structures of elements in the environment. They can usually be extracted from measurements and mathematically described. Good features are always perceivable and easily detectable from the environment. We distinguish between *low-level features* (*geometric primitives*) such as lines, points, corners, blobs, circles, or polygons and *high-level features* (*objects*) such as doors, tables, or trash cans. At one extreme, raw sensor data provide a large volume of data, but with low distinctiveness of each individual quantum of data. Making use of raw data has the potential advantage that every bit of information is fully used, and thus there is a high conservation of information. Low-level features are abstractions of raw data, and as such they provide a lower volume of data while increasing the distinctiveness of each feature. The hope, when one incorporates low-level features, is that the features are filtering out poor or useless data, but of course it is also likely that some valid information will be lost as a result of the feature-extraction process. High-level features provide maximum abstraction from the raw data, thereby reducing the volume of data as much as possible while providing highly distinctive resulting features. Once again, the abstraction process has the risk of filtering away important information, potentially lowering data utilization.

Although features must have some spatial locality, their geometric extent can range widely. For example, a corner feature inhabits a specific coordinate location in the geometric world. In contrast, a visual "fingerprint" identifying a specific room in an office building applies to the entire room, but it has a location that is spatially limited to the one particular room.

In mobile robotics, features play an especially important role in the creation of environmental models. They enable more compact and robust descriptions of the environment, helping a mobile robot during both map-building and localization. When designing a mobile robot, a critical decision revolves around choosing the appropriate features for the robot to use. A number of factors are essential to this decision.

Target environment. For geometric features to be useful, the target geometries must be readily detected in the actual environment. For example, line features are extremely useful in office building environments due to the abundance of straight wall segments, while the same features are virtually useless when navigating on Mars. Conversely, point features (such as corners and blobs) are more likely to be found in any textured environment. As an example, consider that the two NASA Mars explorations rovers, Spirit and Opportunity, used corner features (section 4.5) for visual odometry [203].

Available sensors. Obviously, the specific sensors and sensor uncertainty of the robot impacts the appropriateness of various features. Armed with a laser rangefinder, a robot is well qualified to use geometrically detailed features such as corner features owing to the high-quality angular and depth resolution of the laser scanner. In contrast, a sonar-equipped robot may not have the appropriate tools for corner feature extraction.

Computational power. Visual feature extraction can effect a significant computational cost, particularly in robots where the vision sensor processing is performed by one of the robot's main processors.

Environment representation. Feature extraction is an important step toward scene interpretation, and by this token the features extracted must provide information that is consonant with the representation used for the environmental model. For example, nongeometric visual features are of little value in purely geometric environmental models but can be of great value in topological models of the environment. Figure 4.64 shows the application of two different representations to the task of modeling an office building hallway. Each approach has advantages and disadvantages, but extraction of line and corner features has much more relevance to the representation on the left. Refer to chapter 5, section 5.5 for a close look at map representations and their relative trade-offs.

In sections 4.5–4.7, we present specific feature extraction techniques based on the two most popular sensing modalities of mobile robotics: vision and range sensing.

Visual interpretation is, as we have mentioned before, an extremely challenging problem to fully solve. Significant research effort has been dedicated over the past several decades to inventing algorithms for understanding a scene based on 2D images, and the research efforts have slowly produced fruitful results.

In section 4.2 we saw vision ranging and color-tracking sensors that are commercially available for mobile robots. These specific vision applications have witnessed commercial

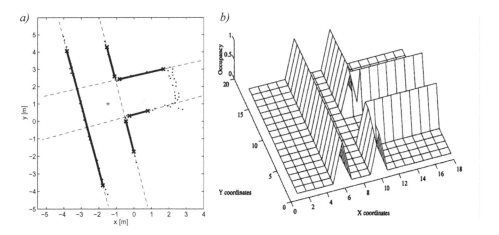

Figure 4.64
Environment representation and modeling: (a) feature-based (continuous metric); (b) occupancy grid (discrete metric). Courtesy of Sjur Vestli.

solutions primarily because the challenges are in both cases relatively well focused and the resulting, problem-specific algorithms are straightforward. But images contain much more than implicit depth information and color blobs. We would like to solve the more general problem of extracting a large number of feature types from images.

The next section presents some point feature extraction techniques that are relevant to mobile robotics along these lines. Two key requirements must be met for a visual feature extraction technique to have mobile robotic relevance. First, the method must operate in real time. Mobile robots move through their environment, and so the processing simply cannot be an offline operation. Second, the method must be robust to the real-world conditions outside of a laboratory. This means that carefully controlled illumination assumptions and carefully painted objects are unacceptable requirements.

Throughout the following descriptions, keep in mind that vision interpretation is primarily about the challenge of *reducing information*. A sonar unit produces perhaps fifty bits of information per second. By contrast, a CCD camera can output 240 *million* bits per second! The sonar produces a tiny amount of information from which we hope to draw broader conclusions. But the CCD chip produces too much information, and this overabundance of information mixes together relevant and irrelevant information haphazardly. For example, we may intend to measure the color of a landmark. The CCD camera does not simply report its color, but also measures the general illumination of the environment, the direction of illumination, the defocusing caused by optics, the side effects imposed by nearby objects with different colors, and so on. Therefore, the problem of visual feature extraction is largely one of removing the majority of irrelevant information in an image so that the remaining information unambiguously describes specific features in the environment.

4.5   Image Feature Extraction: Interest Point Detectors

In this section, we define the concept of the local feature and review some of the most consolidated feature extractors. As the computer vision literature in this field is very large, we will only describe in detail the two most popular feature detectors, namely Harris and SIFT, and will briefly introduce the others by explaining the main advantages and disadvantages and domain of application. For the interested reader, a comprehensive survey on local feature detectors can be found in [320].

4.5.1   Introduction

A local feature is an image pattern that differs from its immediate neighborhood in terms of intensity, color, and texture. Local features can be small image patches (such as regions of uniform color), edges, or points (such as corners originated from line intersections). In the modern terminology, local features are also called interest points, interest regions, or keypoints.

Depending on their semantic content, local features can be divided into three different categories. In the first category are features that have a semantic interpretation such as, for instance, edges corresponding to lanes of the road or blobs corresponding to blood cells in medical images. This is the case in most automotive applications, airborne images, and medical image processing. Furthermore, these were also the first applications for which local feature detectors have been proposed. In the second category are features that do not have a semantic interpretation. Here, what the features actually represent is not relevant. What matters is that their location can be determined accurately and robustly over time. Typical applications are feature tracking, camera calibration, 3D reconstruction, image mosaicing, and panorama stitching. Finally, in the third category are features that still do not have a semantic interpretation if taken individually, but that can be used to recognize a scene or an object if taken all together. For instance, a scene could be recognized counting the number of feature matches between the observed scene and the query image. In this case, the location of the feature is not important; only the number of matches is relevant. Application domains include texture analysis, scene classification, video mining, and image retrieval (see, for instance, Google Images, Microsoft Bing Images, Youtube, or Tineye.com). This principle is the basis of the *visual-word*-based place recognition that will be described in section 4.6.

4.5.2 Properties of the ideal feature detector

In this section we summarize the properties that an ideal feature detector should have. Let us start with a concrete example from digital image photography. Most of today's digital consumer cameras come with software for automatic stitching of panoramas from multiple photos. An example is shown in figure 4.65. The user simply takes several shots of the scene with little overlap between adjacent pictures and the software automatically aligns and fuses them all together into a cylindrical panorama (figure 4.65a). The key challenge is to identify corresponding regions between overlapping images. As the reader may perceive, one way to solve this problem is to extract feature points from adjacent pictures, find corresponding pairs according to some similarity measure (figure 4.65b), and compute the transformation (e.g., homography) to align them (figure 4.65c). The first problem is how to detect the same points independently in both images. In figure 4.65d, for instance, the features from the left image are not redetected in the right image. Because of this, we need a "repeatable" feature detector. The second problem is: for each point in the first image we need to correctly recognize the corresponding one in the second image. Thus, the detected features should be very distinctive (i.e., highly distinguishable).

"Repeatability" is probably the most important property of a good feature detector. Given two images of the same scene taken under different viewing and illumination conditions, it is desirable that a high percentage of the features of the first image can be redetected in the second image. This requires the feature be invariant to view point changes, such as camera rotation or zoom (i.e., scale), and illumination changes.

The second important property is "distinctiveness," that is, the information carried by the patch surrounding the feature point should be as distinctive as possible so that the features can be distinguished and matched. For instance, the corners of a chessboard are not distinctive because they cannot be distinguished from each other. As we will see later, distinctiveness is also the main difference between Harris and SIFT features. Harris privileges corners (such as edge intersections), while SIFT privileges image patches with highly informative content (i.e., not corners).

Other important properties of a good feature detector are:

- Localization accuracy: the detected features should be accurately localized, both in image position and scale. Accuracy is especially important in camera calibration, 3D reconstruction from images ("structure from motion"), and panorama stitching.

- Quantity of features: the ideal number of detected features depends on the application. For most of the tasks like object or scene recognition, image retrieval, and 3D reconstruction it is important to have a sufficiently large number of features, this in order to increase the recognition rate or the accuracy of the reconstruction. However, if the feature had a semantic interpretation, then a small number of features would be enough to

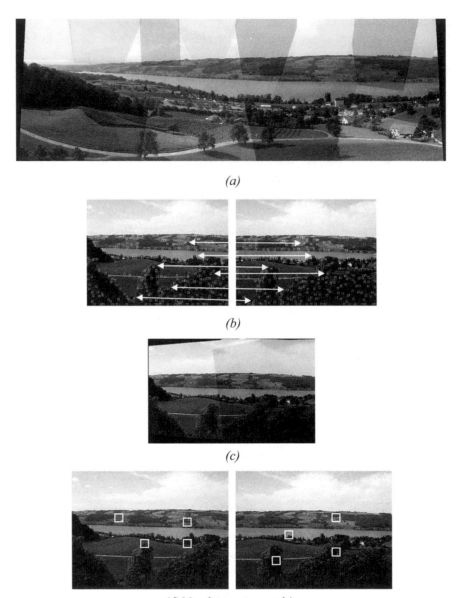

(a)

(b)

(c)

(d) No chance to match!

Figure 4.65 (a) Panorama built from multiple overlapping images using Autostitch software. (b) First step: select salient features in both images and match corresponding ones. (c) Second step: compute the transformation between the two corresponding sets and align the images. (d) Two example images where features were not redetected and therefore there is no chance to match them.

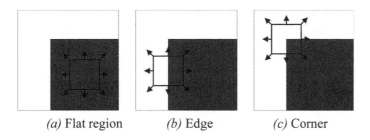

(a) Flat region   *(b)* Edge   *(c)* Corner

Figure 4.66 (a) "Flat" region: no change in all directions. (b) "Edge": no change along the edge direction. (c) "Corner": significant change in all directions.

recognize a scene (as an example, some semantic "high level" features could be individual objects or objects parts such as table, chair, table leg, door, and so on).

- Invariance: good features should be invariant to changes of camera viewpoint, environment illumination, and scale (like zoom or camera translation). Invariance can be achieved within a certain range when these changes can be modeled as mathematical transformations (see section 4.5.4). A successful result in this direction has been successfully demonstrated by some of the recent detectors like SIFT (section 4.5.5.1).

- Computational efficiency: it is also desirable that features can be detected and matched very efficiently. In the framework of the project undertaken by Google Images of organizing all the images on the web, computation efficiency is a critical component as its database—nowadays composed of billions of images—grows more and more every year. This is even important in robotics, where most of the applications need to work in real-time. However, the time of detection and matching of a feature is strictly related to the degree of invariance desired: the higher the level of invariance, the more image transformations to check, and, thus, the longer the computation time.

- Robustness: the detected features should be robust to image noise, discretization effects, compression artifacts, blur, deviations from the mathematical model used to obtain invariance, and so on.

4.5.3 Corner detectors

A corner in an image can be defined as the intersection of two or more edges. Corners are features with high repeatability.

The basic concept of corner detection. One of the earliest corner detectors was invented by Moravec [234, 235]. He defined a corner as a point where there is a large intensity variation in every direction. An intuitive explanation of his corner detection algorithm is given in figure 4.66. Intuitively, one could recognize a corner by looking through a small window centered on the pixel. If the pixel lies in a "flat" region (i.e., a region of uniform intensity), then the adjacent windows will look similar. If the pixel is along an edge, then adjacent windows in the direction perpendicular to the edge will look different, but adjacent windows in a direction parallel to the edge will result only in a small change. Finally, if the pixel lies on a corner, then none of the adjacent windows will look similar. Moravec used the Sum of Squared Differences (SSD, section 4.3.3) as a measure of the similarity between two patches. A low SSD indicates more similarity. If this number is locally maximal, then a corner is present.

4.5.3.1 The Harris corner detector

Harris and Stephens [146] improved Moravec's corner detector by considering the partial derivatives of the SSD instead of using shifted windows.

Let I be a grayscale image. Consider taking an image patch centered on (u, v) and shifting it by (x, y). The Sum of Squared Differences SSD between these two patches is given by:

$$SSD(x, y) = \sum_u \sum_v ((I(u, v)) - I(u + x, v + y))^2 . \tag{4.116}$$

$I(u + x, v + y)$ can be approximated by a first-order Taylor expansion. Let I_x and I_y be the partial derivatives of I, such that

$$I(u + x, v + y) \approx I(u, v) + I_x(u, v)x + I_y(u, v)y . \tag{4.117}$$

This produces the approximation

$$SSD(x, y) \approx \sum_u \sum_v (I_x(u, v)x + I_y(u, v)y)^2 , \tag{4.118}$$

which can be written in matrix form:

$$SSD(x, y) \approx \begin{bmatrix} x & y \end{bmatrix} M \begin{bmatrix} x \\ y \end{bmatrix} , \tag{4.119}$$

where M is the second moment matrix

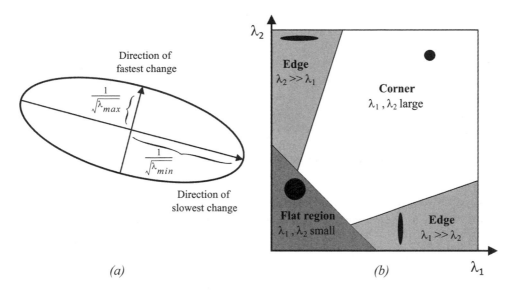

Figure 4.67 (a) This ellipse is built from the second moment matrix and visualizes the directions of fastest and lowest intensity change. (b) The classification of corner and edges according to Harris and Stephens.

$$M = \sum_u \sum_v \begin{bmatrix} I_x^2 & I_x I_y \\ I_x I_y & I_y^2 \end{bmatrix} = \begin{bmatrix} \sum\sum I_x^2 & \sum\sum I_x I_y \\ \sum\sum I_x I_y & \sum\sum I_y^2 \end{bmatrix}. \tag{4.120}$$

And since M is symmetric, we can rewrite M as

$$M = R^{-1} \begin{bmatrix} \lambda_1 & 0 \\ 0 & \lambda_2 \end{bmatrix} R, \tag{4.121}$$

where λ_1 and λ_2 are the eigenvalues of M.

As mentioned before, a corner is characterized by a large variation of SSD in all directions of the vector (x, y). The Harris detector analyses the eigenvalues of M to decide if we are in presence of a corner or not. Let us first give an intuitive explanation before showing the mathematical expression.

Using equation (4.119) we can visualize M as an ellipse (figure 4.67a) of equation:

$$\begin{bmatrix} x & y \end{bmatrix} M \begin{bmatrix} x \\ y \end{bmatrix} = const. \tag{4.122}$$

The axis lengths of this ellipse are determined by the eigenvalues of M and the orientation is determined by R.

Based on the magnitudes of the eigenvalues, the following inferences can be made based on this argument:

- If both λ_1 and λ_2 are small, SSD is almost constant in all directions (i.e., we are in presence of a flat region).

- If either $\lambda_1 \gg \lambda_2$ or $\lambda_2 \gg \lambda_1$, we are in presence of an edge: SSD has a large variation only in one direction, which is the one perpendicular to the edge.

- If both λ_1 and λ_2 are large, SSD has large variations in all directions and then we are in presence of a corner.

The three situations mentioned above are pictorially summarized in figure 4.67b.

Because the calculation of the eigenvalues is computationally expensive, Harris and Stephens suggested the use of the following "cornerness function" instead:

$$C = \lambda_1 \lambda_2 - \kappa(\lambda_1 + \lambda_2)^2 = det(M) - \kappa \cdot trace^2(M), \tag{4.123}$$

where κ is a tunable sensitivity parameter. This way, instead of computing the eigenvalues of M, we just need to evaluate the determinant and trace of M. The value of κ has to be determined empirically. In the literature, values are often reported in the range 0.04–0.15.

The last step of the Harris corner detector consists in extracting the local maxima of the cornerness function, using *nonmaxima suppression*[14]. Finally only the local maxima which are above a given threshold are retained. The processing steps are illustrated in figure 4.68.

Figure 4.68c shows the corners detected for the two example images. Notice that the images are related by a rotation and a slight change of illumination. As can be seen, many of the features detected in the left image have also been redetected in the right image. This means that the repeatability of the Harris detector under rotations and small changes of illumination is high. In the next section, we will point out the properties of the Harris detector as well as its drawbacks.

---

14. *Nonmaxima suppression* involves revisiting every pixel of the cornerness function and determining whether or not it is at a local maximum. If not, then the value is set to zero. This causes only the maxima to be preserved.

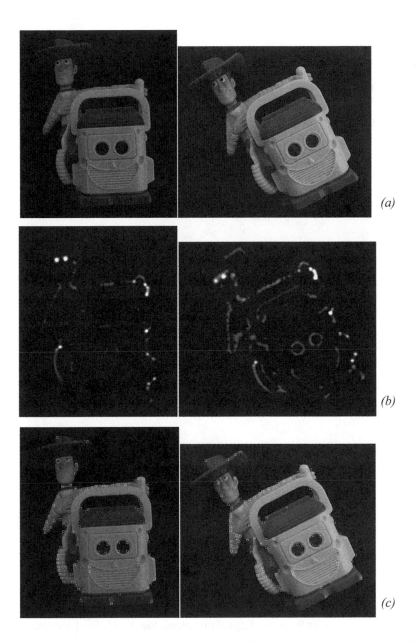

(a)

(b)

(c)

Figure 4.68 Extraction of the Harris corners. (a) Original image. (b) Cornerness function. (c) Harris corners are identified as local maxima of the cornerness function (only local maxima larger than a given threshold are retained). Two images of the same object are shown, which differ by illumination and orientation.

Geometric change

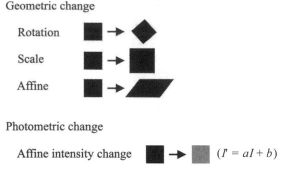

Photometric change

Affine intensity change   ■ ➜ ■ $(I' = aI + b)$

Figure 4.69 Models of image changes.

4.5.4 Invariance to photometric and geometric changes

As observed in section 4.5.2, in general we want features to be detected despite geometric and photometric changes in the image: if we have two transformed versions of the same image, features should be detected in corresponding locations. Image transformations can affect the geometric or the photometric properties of an image. Consolidated models of image transformations are the following (see also figure 4.69):

- Geometric changes:

    - 2D Rotation

    - Scale (uniform rescaling)

    - Affine

- Photometric changes

    - Affine intensity

Observe that we did not mention changes of camera viewpoint (i.e., perspective distortions). In this case the transformation valid only for planar objects would be a homography. However, when the viewpoint changes are small and the object is locally planar, the affine transformation is a good approximation of the homography. Observe also that 2D rotation occurs only if the camera rotates purely about its optical axis. Uniform rescaling, instead, appears when the camera zooms (in or out) or translates along the direction of its optical axis, but the latter is valid only for locally planar objects.

As an example, let us now examine the invariance of the Harris detector to the above mentioned transformations. We can observe that the Harris detector is invariant to 2D image rotations. This can be explained by observing that the eigenvalues of the second moment matrix do not change under pure rotation. Indeed, the ellipse rotates, but its shape

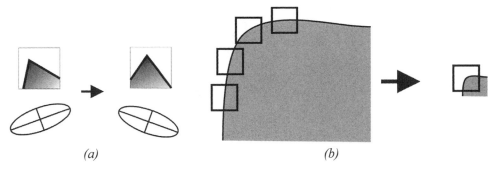

Figure 4.70 (a) Harris detector is invariant to image rotations: the ellipse rotates but its shape (i.e. eigenvalues) remains the same. (b) Conversely, it is not invariant to image scale: at a large scale (left) all points along the corner would be classified as edges, while at a smaller scale case (right) the point would be classified as corner.

(i.e., the eigenvalues) remains the same (see figure 4.70a). As a matter of fact, observe that to make the Harris detector isotropic (i.e., uniform for all rotations), the second moment matrix should be computed in a circular region rather in a squared window. This is usually done by averaging equation (4.120) with a circular symmetric Gaussian function.

The Harris detector is also invariant to affine intensity changes. In this case, the eigenvalues, and so the cornerness function, are rescaled by a constant factor, but the position of the local maxima of the cornerness function remains the same. Conversely, the Harris detector is not invariant to geometric affine transformations or scale changes. Intuitively, an affine transformation distorts the neighborhood of the feature along the x and y directions and, accordingly, a corner can get reduced or increased its curvature. Regarding scale changes, this is immediately clarified as observed in figure 4.70b. In this figure, the corner would be classified as an edge at a high scale and as a corner at a smaller scale.

The performance of the Harris detector against scale changes is shown in figure 4.71 according to a comparative study made in [216]. In this figure, the repeatability rate is plotted versus the scale factor. The repeatability rate between two images is computed as the ratio between the number of found correspondences and the number of all possible correspondences. As observed, after rescaling the image by a factor 2, only 20% of the possible correspondences are redetected.

Although it is not invariant to scale changes, the Harris detector, as in its original implementation by Harris and Stephens, is still widely used and can be found in the well-known Intel open-source computer vision library (OpenCV [343]). Furthermore, in a comparative study of different interest point detectors, Schmid et al. [285] showed that the Harris corners are among the most repeatable and most informative features. As we will see in the next sections, some modifications to the original implementation have made the Harris detector also invariant to scale and affine changes. Additionally, the location accuracy of

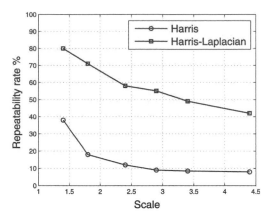

Figure 4.71 Repeatability rate of Harris detector. Comparison with Harris-Laplacian. This plot is the result of a comparative study presented in [216].

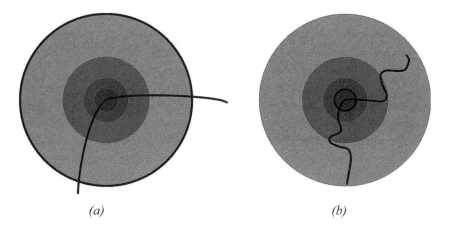

(a)   *(b)*

Figure 4.72 To achieve scale-invariant detection, the image is analyzed at different scales. This means, the Harris detector is applied several times on the image, each time with a different window size.

the Harris corners can be improved up to subpixel precision. This can be achieved by approximating the cornerness function in the neighborhood of a local maximum through a quadratic function.

4.5.4.1 Scale-invariant detection

In this section, we will describe the modifications that have been devised to make the Harris detector invariant to scale changes. If we look at figure 4.72, we will notice that one way to detect the corner at higher scale is to use a multiscale Harris detector. This means that

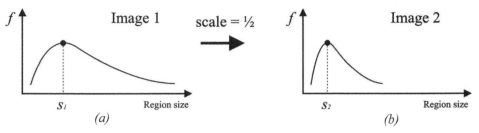

Figure 4.73 Average intensity as a function of the region size. (a) Original image. (b) Resized image.

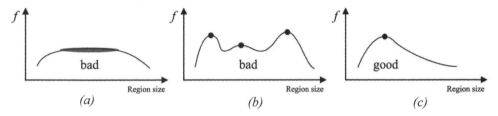

Figure 4.74 (a)-(b) are bad scale invariant functions. (c) is a good.

the same detector is applied several times on the image, each time with a different window (circle) size. Note, an efficient implementation of multiscale detection uses the so called scale-space pyramid: instead of varying the window size of the feature detector, the idea is to generate upsampled or downsampled versions of the original image (i.e., pyramid). Using multiscale Harris detector, we can be sure that at some point the corner of image 4.72a will get detected. Once the point has been detected in both images a question arises as: how do we select the corresponding scale? In other words, how do we choose corresponding circles independently in each image?

In computer vision, the correct-scale selection is usually done by following the approach proposed in 1998 by Lindeberg [193]: the appropriate scale of a local feature can be chosen as the one for which a given function reaches a maximum or minimum over scales. Let us give an intuitive explanation. For every circle in the two images in figure 4.72, let us plot the average intensity of the pixels within the circle as a function of the circle size. For these images, we will get the two plots shown in figure 4.73. As expected, these two functions look the same up to a rescaling in the x-axis. The solution to our problem is therefore to take as corresponding scales the circle sizes for which these functions reach their maximum. Depending of the chosen function, we might take the minimum instead of the maximum.

The problem is how to design a good function. The first observation we can make is that a "good" function for scale detection should have one stable sharp peak like that in figure 4.74. Despite the one used in our example, the average intensity is not good because it can

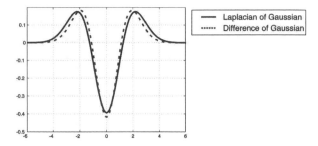

Figure 4.75 Comparison between Laplacian of Gaussian and Difference of Gaussian

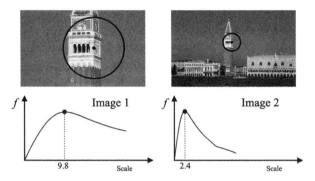

Figure 4.76 Response of the LOG operator over two corresponding points from two images taken at different scales.

return multiple peaks or even no peaks at all. As a matter of fact, it turns out that for usual images a good function is one that corresponds to contrast, that is, sharp local intensity changes. The Laplacian of Gaussian (LoG) (figure 4.75) is a good operator for identifying sharp intensity changes and is currently the one used for scale selection by the Harris corner detector. In a comparative study presented in [216, 101], the LoG operator has been shown to give the best results with respect to other functions. The response of the LoG over two image features is shown in figure 4.76.

The multiscale Harris detector is known as Harris-Laplacian and was implemented by Mikolajczyk and Schmid [217]. The comparison between the standard Harris and the Harris-Laplacian over scale is shown in figure 4.71.

4.5.4.2 Affine invariant detection
As mentioned in section 4.5.4, affine transformation is a good approximation of perspective distortion of locally planar patches under small viewpoint changes. In the previous section, we considered the problem of detection under uniform rescaling. The problem now is how

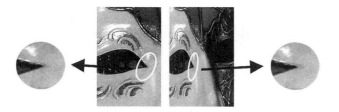

Figure 4.77 Computation of the affine invariant ellipse in two images related by an affine transformation.

to detect the same features under affine transformation, which can be seen as a nonuniform rescaling. The procedure consists in the following steps:

- First, the features are identified using the scale invariant Harris-Laplacian detector.

- Then, the second moment matrix (4.120) is used to identify the two directions of slowest and fastest change of intensity around the feature.

- Out of these two directions, an ellipse is computed to the same size as the scale computed with the LoG operator.

- The region inside the ellipse is normalized to a circular one.

- The initial detected ellipse and the resulting normalized circular shape are shown in figure 4.77.

The affine invariant Harris detector is known as Harris-Affine and was devised by Mikolajczyk and Schmid [218].

4.5.4.3 Other corner detectors

The Shi-Tomasi corner detector. This is also sometimes referred to as the Kanade-Tomasi corner detector [284]. This detector is strongly based on the Harris corner detector. The authors show that for image patches undergoing affine transformations, $min(\lambda_1, \lambda_2)$ is a better measure than the cornerness function C (4.123).

The SUSAN corner detector. SUSAN stands for Smallest Univalue Segment Assimilating Nucleus and, besides being used for corner detection, it is also used for edge detection and noise suppression. The SUSAN corner detector has been introduced by Smith and Brady [296]. Its working principle is different from the Harris detector. As we have seen, Harris is based on local image gradients, which are computationally expensive to compute. Conversely, SUSAN is based on a morphological approach, which is computationally much more efficient than Harris.

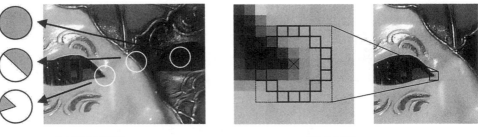

(a) SUSANT corners   *(b)* FAST corners

Figure 4.78 (a) SUSAN detector compares pixels within a circular region, while FAST (b) compares them only on a circle.

The working principle of SUSAN is very simple (see also figure 4.78a). For each pixel in the image, SUSAN considers a circular window of fixed radius centered on it. Then, all the pixels within this window are divided into two categories, depending on whether they have "similar" or "different" intensity values as the center pixel. Accordingly, on uniform intensity regions of the image, most pixels within the window will have a similar brightness as the center pixel. Near edges, the fraction of pixels with similar intensity will drop to 50%, while near corners it will decrease further to about 25%. Thus, SUSAN corners are identified as image locations where the number of pixels with similar brightness in a local neighborhood attains a local minimum and is below a specified threshold. As a final step, nonmaxima suppression (page 218) is used to identify local minima.

The SUSAN corners show a high repeatability, however they are heavily sensitive to noise. Indeed, many of the features are often located on edges than on real corners.

The FAST corner detector. The FAST (Features from Accelerated Segment Test) detector, was introduced by Rosten and Drummond [267, 268]. This detector builds upon the SUSAN detector. As we have seen, SUSAN computes the fraction of pixels within a circular window, which have similar intensity as the center pixel. Conversely, FAST compares pixels only on a circle of 16 pixels around the candidate corner (see figure 4.78b). This results in a very efficient detector that is up to thirty times faster than Harris: FAST takes only 1–2 milliseconds on a 2GHz Dual Core laptop and is currently the most computationally efficient feature detector available. However, like the SUSAN, it is not robust at high levels of noise.

4.5.4.4 Discussion about corner detectors

The Harris detector, with its scale and affine invariant extensions, is a convenient tool for extracting a large number of corners. Additionally, it has been identified as the most stable corner detector, as reported in several evaluations [28, 219, 285]. Alternatively, the

SUSAN or the FAST detectors can be used. They are much more efficient but also more sensitive to noise.

Shi-Tomasi, SUSAN, and FAST can also be made scale invariant like the Harris-Laplacian by analyzing the image at multiple scales, as seen in section 4.5.4.1. However, the scale estimation of corners is less accurate than blobs (e.g., SIFT, MSER, or SURF) due to the multiscale nature of corners: by definition, a corner is found at the intersection of edges, therefore its appearance changes very little at adjacent scales.

Finally, it is important to remind that the affine transformation model holds only for small viewpoint changes and in case of locally planar regions, that is, assuming the camera is relatively far from the object.

In the next section, we will describe the SIFT detector. Despite being a blob detector, the SIFT features incorporate all the properties of the scale-affine-invariant Harris but they are much more distinctive and robust to image noise, small illumination changes, and large changes of camera viewpoint.

4.5.5 Blob detectors

A blob is an image pattern that differs from its immediate neighborhood in terms of intensity, color, and texture. It is not an edge, nor a corner. The location accuracy of a blob is typically smaller than that of a corner, but its scale and shape are better defined. To be clearer, a corner can be localized by a single point (e.g., the intersection of two edges), while a blob can only be localized by its boundary. On the other hand, a corner is less accurately localized over the scale because, as we pointed out before, a corner is found at the intersection of edges and therefore its appearance changes very little at adjacent scales. Conversely, a blob is more accurately localized over the scale because the boundary of a blob defines immediately its size and so its scale.

Using the new terminology, blob detectors can also be referred to as interest point operators, or alternatively interest region operators. Some examples of bloblike features are shown in figure 4.79. In this figure you can see two feature types that will be described in this section, namely SIFT and MSER. As observed, MSER privileges regions with uniform intensity, while SIFT does not.

4.5.5.1 SIFT features

SIFT stands for Scale Invariant Feature Transform and is a method to detect and match robust keypoints, which was invented in 1999 by Lowe [196, 197]. The uniqueness of SIFT is that these features are extremely distinctive and can be successfully matched between images with very different illumination, rotation, viewpoint, and scale changes. Its high repeatability and high matching rate in very challenging conditions have made SIFT the best feature detector so far. It has found many applications in object recognition, robotic

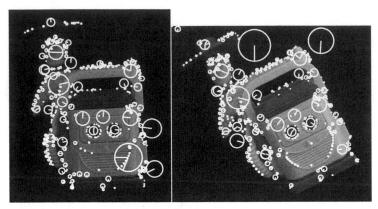

(a) SIFT features

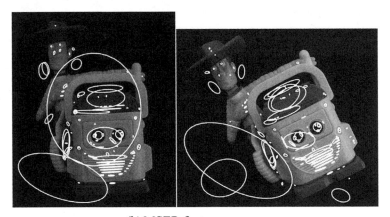

(b) MSER features

Figure 4.79 Extraction of SIFT and MSER features from the same sample image used for the Harris detector in figure 4.68. Observe that both SIFT and MSER avoid corners. Furthemore, MSER privileges regions with uniform intensity. Both these feature detectors are robust to large changes of intensity, scale, and viewpoint.

mapping and navigation, image stitching (e.g. panoramas, mosaics), 3D modeling, gesture recognition, video tracking, and face recognition.

The main advantage of the SIFT features in comparison to all previously explained methods is that a "descriptor" is computed from the region around the interest point, which distinctively describes the information carried by the feature. As we will see, this descriptor is a vector that represents the local distribution of the image gradients around the interest point. As proven by its inventor, it is actually this descriptor that makes SIFT robust to rotation and small changes of illumination, scale, and viewpoint.

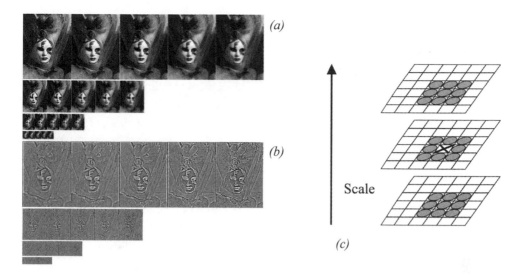

Figure 4.80 (a) Gaussian blurred images at different scales. (b) Difference of Gaussian images. (c) Keypoint selection as local maxima or minima of the DoG images across adjacent scales.

In the following, we will analyze the main steps of the SIFT algorithm, which are:

- Identification of keypoint location and scale

- Orientation assignment

- Generation of keypoint descriptor

Identification of keypoint location and scale. The first step toward the identification of SIFT keypoints is the generation of the so-called Difference of Gaussian (DoG) images. This is done by first blurring the original image with Gaussian filters at different scales (i.e., different sigma) and then by taking the difference of successive Gaussian-blurred images. This process is shown in figure 4.80a: the original image (top left) is blurred with four Gaussian filters with different sigma, and this is repeated after downsampling the image of a factor 2. Finally, DoG images are computed by simply taking the difference between successive blurred images 4.80b.

The second step is the selection of the keypoints. SIFT keypoints are identified as local maxima or minima of the DoG images across scales. In particular, each pixel in the DoG images is compared to its eight neighbors at the same scale, plus the nine neighbors at adjacent scales (figure 4.80c). If the pixel is a local maximum or minimum, it is selected as a candidate keypoint.

The third step consists in refining the location, in both space and scale, of the keypoints by interpolation of nearby data. Finally, keypoints with low contrast or along edges are removed because of their low distinctiveness and due to their instability to image noise.

Note that another way of generating DoG images consists in convolving the image with a DoG operator, which is nothing but the difference between to Gaussian filters (figure 4.75). As shown in figure 4.75, the DoG function is actually a very good approximation of the Laplacian of Gaussian (LoG). However, DoG images are more efficient to compute, and therefore they have been used in SIFT in lieu of LoG. At this point the attentive reader will recognize that the scale extrema selection of the SIFT is very similar to the scale extrema selection of the Harris-Laplacian. Indeed, the main difference with the Harris-Laplacian is the identification of the keypoint location. While in Harris the keypoint is identified in the image plane as local maximum of the cornerness function, in SIFT the keypoint is a local minimum or maximum of the DoG image in both position and scale. To recap, in SIFT the DoG operator is used to identify both position and scale of the keypoints.

Orientation assignment. This step consists in assigning each keypoint a specific orientation in order to make it invariant to image rotation.

To determine the keypoint orientation, a gradient orientation histogram is computed in the neighborhood of the keypoint. In other words, for every pixel in the neighborhood of the keypoints, the intensity gradient (magnitude and orientation) is computed. Then a histogram of orientations is built such that the contribution of each pixel is weighted by the gradient magnitude.

Peaks in the histogram correspond to dominant orientations (figure 4.81a). Once the histogram is filled, the orientation corresponding to the highest peak is assigned to the keypoint. In the case of multiple peaks that are within 80% of the highest peak, an additional keypoint is created for each additional orientation, having the same location and scale as the original keypoint. All the properties of the keypoint will be measured relative to the keypoint orientation. This provides invariance to rotation.

Final keypoints with selected orientation and scale are shown in figure 4.81b.

Generation of keypoint descriptor. In the previous steps, we have described how to detect SIFT keypoints in both location and scale spaces and how to assign orientations to them. The last step of the SIFT algorithm is to compute descriptor vectors for these keypoints such that the descriptors are highly distinctive and partially invariant to illumination and viewpoint.

The descriptor is based on gradient orientation histograms. In order to achieve orientation invariance, the gradient orientations are rotated relative to the keypoint orientation. The neighboring region of the keypoint is then divided into 4×4 smaller regions, and a gradient histogram with eight orientation bins is computed within each of these regions.

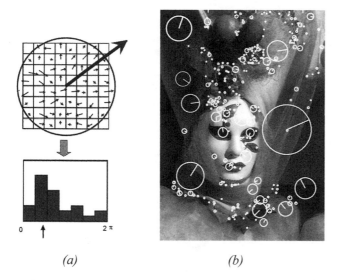

(a)   *(b)*

Figure 4.81 (a) Orientation assignment. (b) Some SIFT features with detected orientation and scale.

Finally, the descriptor is built by stacking all the orientation histogram entries. Therefore, the final length of the descriptor vector is $4 \times 4 \times 8 = 128$ elements. To achieve partial illumination invariance, the descriptor vector is finally normalized to have unit norm.

Observe, lower dimension descriptors could be built by using smaller region partitions or less histogram bins. However according to the literature, the 128 element vector is the one for which the best results in terms of robustness to image variations were reported.

In [197], it was also shown that feature matching accuracy is above 50% for viewpoint changes of up to 50 degrees (see figure 4.82). Therefore SIFT descriptors are invariant to minor viewpoint changes. To evaluate the distinctiveness of the SIFT descriptor, many tests were performed by counting the number of correct matches in a database with a varying number of keypoints. These tests revealed that matching accuracy of SIFT descriptor decreases only very little for very large database sizes. This implies that SIFT features are highly distinctive.

Because of its high repeatability and distinctiveness, SIFT has demonstrated in the last ten years to be the best feature detectors in a wide range of applications, although it is outperformed in efficiency by SURF (section 4.5.5.2). Excellent results have been achieved in robot navigation, 3D object recognition, place recognition, SLAM, panorama stitching, image retrieval, and many others.

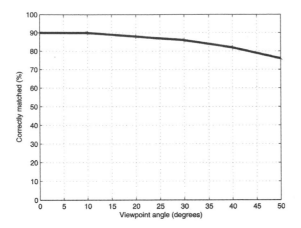

Figure 4.82 Fraction of SIFT keypoints correctly matched as a function of the viewpoint angle.

4.5.5.2 Other blob detectors

The MSER detector. Maximally Stable Extremal Regions (MSER) have been proposed by Matas et al. [210] for matching features that are robust under large viewpoint changes. A *maximally stable extremal region* is a connected component of pixels that have either higher or lower intensity than all the pixels on its outer boundary (figure 4.79b). These *extremal regions* are selected using an appropriate intensity thresholding and have a number of desirable properties. First, they are completely invariant to monotonic changes of intensity. Second, they are invariant to affine image transformations.

The SURF detector. SURF stands for Speeded Up Robust Features and have been proposed by Bay et al. [71]. This scale-invariant feature detector is strongly inspired by SIFT but is several times faster. Basically, it uses *Haar wavelets*[15] to approximate DoG filters and *integral images*[16] for convolution, which make the filtering process much more efficient at the expense of a minor robustness with respect to SIFT.

4.5.5.3 Summary on features detectors
Table 4.2 gives an overview of the most important properties for the feature detectors described in the previous sections. The highest repeatability and localization accuracy is

---

15. A Haar wavelet is a piecewise constant function.
16. Integral image is an algorithm for quickly and efficiently generating the sum of values in a rectangular subset of a grid.

Table 4.2
Comparison of feature detectors: properties and performance.

	Corner detector	Blob detector	Rotation invariant	Scale invariant	Affine invariant	Repeatability	Localization accuracy	Robustness	Efficiency
Harris	x		x			+++	+++	++	++
Shi-Tomasi	x		x			+++	+++	++	++
Harris-Laplacian	x	x	x	x		+++	+++	++	+
Harris-Affine	x	x	x	x	x	+++	+++	++	++
SUSAN	x		x			++	++	++	+++
FAST	x		x			++	++	++	++++
SIFT		x	x	x	x	+++	++	+++	+
MSER		x	x	x	x	+++	+	+++	+++
SURF		x	x	x	x	++	++	++	++

obtained by the Harris detector and its scale and affine invariant versions. The SUSAN and FAST detectors avoid computation of image derivatives and are therefore more efficient than Harris but the absence of smoothing makes them more sensitive to noise. The original Harris, Shi-Tomasi, SUSAN, and FAST are not scale-invariant, however some literature exists on how achieving scale invariance using the approach described in section 4.5.4.1.

In contrast to the original Harris, Harris-Laplace attains scale invariance; however, its scale estimation is less accurate than SIFT, MSER, or SURF due to the multiscale nature of corners. Finally, the SURF detector shows high repeatability, scale, and viewpoint invariance. However, it was devised for efficiency, and therefore it does not perform as well as SIFT.

GPU and FPGA implementations. Some of these feature detectors have been implemented to take advantage of the parallelism offered by modern Graphics Processing Units (GPUs) and Field Programmable Gate Arrays (FPGA). GPU implementations of SIFT are described in [150, 292]. An FPGA implementation of the Harris-Affine feature detector is discussed in [89] and of the SIFT detector in [286]. The availability of these algorithms for GPUs and FPGA make computer vision algorithm able to work at high frame rates.

Open source software: Web resources.

- Most of the feature detectors described in this section (Harris, MSER, FAST, SURF, and several others) are available as ready-to-use code in the Intel open-source computer vision library (OpenCV): http://opencv.willowgarage.com/wiki

- SUSAN, original source code: http://users.fmrib.ox.ac.uk/~steve/susan

- FAST, original source code: http://mi.eng.cam.ac.uk/~er258/work/fast.html

- SIFT, the original executable from David Lowe: http://people.cs.ubc.ca/~lowe/key-points

- A reimplementation of SIFT, MSER and other featured detectors by Andrea Vedaldi. Source code in C and Matlab: http://www.vlfeat.org

- 3D object recognition toolkit, based on SIFT. Developed at the Autonomous Systems Lab at the ETH Zurich: http://robotics.ethz.ch/~ortk

- ERSP Vision Tool. An exiting demo for live object recognition by Evolution Robotics: http://www.evolution.com/product/oem/download/?ch=Vision (it becomes download-able after user registration)

- SURF, precompiled software (GPU implementation also available): http://www.vision.ee.ethz.ch/~surf

4.6   Place Recognition

4.6.1   Introduction

Location recognition (or *place recognition*) describes the capability of naming discrete places in the world. A requirement is that it is possible to obtain a discrete partitioning of the environment into places and a representation of the place and that the places with the corresponding representations are stored in a database. The location recognition process then works by computing a representation from the current sensor measurements of the robot and searching the database for the most similar representation stored. The retrieved representation then tells us the location of the robot.

   Location recognition is the natural form of robot localization in a topological environment map as described by many authors [108, 131, 138, 208, 214, 325]. Visual sensors (i.e., cameras) are perfectly suited to create a rich representation that is both descriptive and discriminative. Most visual representations proposed so far can be divided into *global* representations and *local* representations. Global representations use the whole camera image as a representation of the place, most in a domain-transformed way, for instance, as PCA transformed image [159], Fourier-transformed image [213], image histograms, image fingerprints [182], GIST descriptors [253, 254], and so on. Local representations instead identify salient regions of the image first and create the representation out of this only. This approach largely depends on the detection of salient regions using interest point or interest region detectors, which we have seen in section 4.5. With the development of many effective interest point detectors, local methods have proven to be practical and are nowadays applied in many systems. We will therefore present this method first as the preferred way to location recognition. However, in the last two sections, we will also review some of the earliest approaches to place recognition using image histograms and fingerprints. In fact, although largely outperformed by the local *visual-word*-based approaches, these methods are still used in some robot applications.

4.6.2 From bag of features to visual words

A representation of an image by a set of interest points only is usually called a bag of features. For each interest, point a descriptor is usually computed in a manner that is invariant to rotation, scale, intensity, and viewpoint change (section 4.5.4). A popular way is to use gradient histograms, e.g. SIFT (section 4.5.5.1) or SURF (section 4.5.5.2). This set of descriptors is the new representation of the image. It is called a *bag of features* because the original spatial relation between the interest points is removed and only the descriptors are remembered. The similarity between two sets of descriptors can be computed by counting the number of common feature descriptors. For this, a matching function needs to be defined, which allows us to determine whether two feature descriptors are the same. This matching function usually depends on the type of feature descriptor. But in general a feature descriptor is a high-dimensional vector, and matching features can be found by computing the distance using the L_2 norm. *Visual words* are a 1-dimensional representation of the high-dimensional feature descriptor. This means that the visual word for a 128-dimensional SIFT descriptor is just a single integer number. The conversion to visual words creates a bag of visual words instead of a bag of features. For this conversion, the high-dimensional descriptor space is divided into nonoverlapping cells. This division is computed by *k-means clustering* [18]. For the clustering, a large number of feature descriptors is necessary. The computed cluster borders form the cell divisions of the feature space. Each of the cells is now assigned a number that will be assigned to any feature descriptor within the cell. This number is referred to as visual word. Similar feature descriptors will be then sorted into the same cell and therefore get the same visual word assigned. This is illustrated in figure 4.83, which is a very efficient method of finding matching-feature descriptors. The visual words created by the partitioning is called *visual vocabulary*.

For *quantization*, a prototype vector for each cell is stored, which is the mean descriptor vector of all training descriptors from the cell. To assign a feature descriptor to its cell it needs to be compared to all prototype vectors. For a large number of cells this can be a very expensive operation. It can be sped up by creating a hierarchical splitting of the feature space called *vocabulary tree* [243].

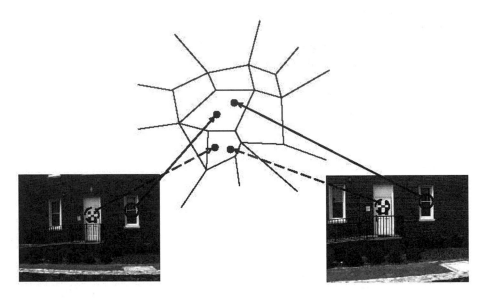

Figure 4.83 Partition of the descriptor feature space. Each cell stands for a visual word. Similar feature descriptors will be sorted into the same cell and therefore get the same visual word assigned.

4.6.3   Efficient location recognition by using an inverted file

Feature quantization into visual words is one key ingredient for efficient location recognition. Another one is the use of an *inverted file* for the database and a voting scheme for similarity computations. The database organized as an inverted file consists of a list of all possible visual words. Each element of this list points to another list that holds all the image identifiers in which this particular visual word appeared. This is illustrated in figure 4.84.

The voting scheme to find the most similar set of visual words in the database to a given query set works as follows. A voting array is initialized, which has as many cells as images in the database. A visual word from the query image is taken, and the list of image identifiers attached to this visual word is processed. For all the image identifiers in the list, a vote is cast by increasing the value at the corresponding position in the voting array. The most similar image to the query image is then the one with the highest vote. This voting scheme can exactly compute the L_2 norm if the descriptor vectors are correctly normalized [243].

This algorithm not only gives the most similar image in the database but also creates a ranking of all images in the database by similarity without any additional computational cost. This can be used to robustify place recognition.

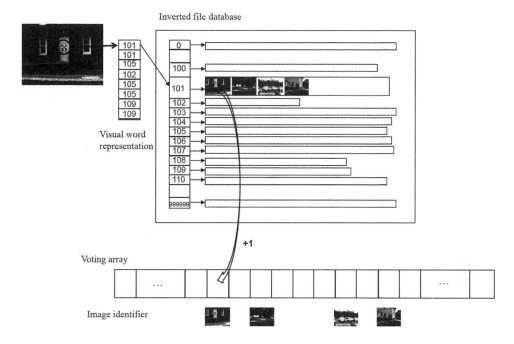

Figure 4.84 Visual word based location recognition using an inverted file system. An image in the database gets a vote if the same visual word is present in the query image.

4.6.4 Geometric verification for robust place recognition

The set of visual words does not contain the spatial relations anymore, thus an image that has the same visual words but in a different spatial arrangement would also have high similarity. The spatial relations however can be enforced again by a final geometric verification. For this, the k most similar images in a query are tested for geometric consistency. The geometric consistency test computes geometric transformations using the x and y image coordinates of matching visual words. Transformations used are affine transformations, homographies, or the essential matrix between images (section 4.2.6, page 184). The computation is performed in a robust way using RANSAC [128] (section 4.7.2.4), and the number of inliers to the transformation is counted. The image that achieves the largest number of inliers with the query image is then reported as the final match. This returns the desired location in place recognition.

4.6.5 Applications

This method for place recognition has already been used successfully in several applications. It was used for topological localization and mapping in [131]. It is also the core algorithm of FABMAP [108], for which it was extended by a probabilistic formulation. Other methods that use this scheme are described in [56, 276].

Available source code on the Web:

- Vocabulary-tree-based image search by F. Fraundorfer et al. [131]. It is a framework for fast image retrieval and place recognition very popular in robotics (useful for loop detection in visual SLAM):
  http://www1.ethz.ch/cvg/people/postgraduates/fraundof/vocsearch

- FABMAB, by M. Cummins et al. [108], is another framework for fast image retrieval and place recognition also very popular in robotics: http://www.robots.ox.ac.uk/~mobile/wikisite/pmwiki/pmwiki.php?n=Software.FABMAP

- Bag of features: another powerful tool for image retrieval and visual recognition using vocabulary trees: http://www.vlfeat.org/~vedaldi/code/bag/bag.html

These algorithms are all very useful for loop detection in the problem of simultaneous localization and mapping, which we will see in section 5.8.

4.6.6 Other image representations for place recognition

In this section, we review two of the early and most successful approaches to place recognition before the advent of the visual words based methods described above. The first method uses image histograms, while the second one uses image fingerprints.

4.6.6.1 Image histograms

A single visual image provides so much information regarding a robot's immediate surroundings that an alternative to searching the image for spatially localized features is to make use of the information captured by the entire image (i.e., all the image pixels) to extract a *whole-image feature* or *global image feature*. Whole-image features are not designed to identify specific spatial structures such as obstacles or the position of specific landmarks. Rather, they serve as compact representations of the entire local region. From the perspective of robot localization, the goal is to extract one or more features from the image that are correlated well with the robot's position. In other words, small changes in robot position should cause only small changes to whole-image features, while large changes in robot position should cause correspondingly large changes to whole-image features.

A logical first step in designing a vision sensor for this purpose is to maximize the field of view of the camera. As the field of view increases, a small-scale structure in the robot's environment occupies a smaller proportion of the image, thereby mitigating the impact of individual scene objects on image characteristics. A catadioptric camera system, nowadays very popular in mobile robotics, offers an extremely wide field of view (section 4.2.4).

A catadioptric image is a 360-degree image warped onto a 2D image surface. Because of this, it offers another critical advantage in terms of sensitivity to small-scale robot motion. If the camera is mounted vertically on the robot so that the image represents the environment surrounding the robot (i.e., its horizon; figure 4.37a), then rotation of the camera and robot simply results in image rotation. In short, the catadioptric camera can be invariant to rotation of the field of view.

Of course, mobile robot rotation will still change the image; that is, pixel positions will change, although the new image will simply be a rotation of the original image. But we intend to extract image features via histogramming. Because histogramming is a function of the set of pixel values and not of the position of each pixel, the process is pixel position-invariant. When combined with the catadioptric camera's field of view invariance, we can create a system that is invariant to robot rotation and insensitive to small-scale robot translation.

A color camera's output image generally contains useful information along multiple *bands*: r, g, and b values as well as hue, saturation, and luminance values. The simplest histogram extraction strategy is to build separate 1D histograms characterizing each band. Given a color camera image, G, the first step is to create mappings from G to each of the n available bands. We use G_i to refer to an array storing the values in band i for all pixels in G. Each band-specific histogram H_i is calculated as before:

- As preprocessing, smooth G_i using a Gaussian smoothing operator.

- Initialize H_i with n levels: $H[j] = 0$ for $j = 1, \ldots, n$.

- For every pixel (x,y) in G_i, increment the histogram: $H_i[G_i[x, y]]\mathrel{+}=1$.

Given the image shown in figure 4.37a, the image histogram technique extracts six histograms (for each of r, g, b, hue, saturation, and luminance) as shown in figure 4.85. In order to make use of such histograms as whole-image features, we need ways to compare to histograms to quantify the likelihood that the histograms map to nearby robot positions. The problem of defining useful histogram distance metrics is itself an important subfield within the image retrieval field. For an overview refer to [270]. One of the most successful distance metrics encountered in mobile robot localization is the *Jeffrey divergence*. Given two histograms H and K, with h_i and k_i denoting the histogram entries, the Jeffrey divergence $d(H, K)$ is defined as

$$d(H, K) = \sum_i \left(h_i \log \frac{2h_i}{h_i + k_i} + k_i \log \frac{2k_i}{h_i + k_i} \right). \tag{4.124}$$

Using measures such as the Jeffrey divergence, mobile robots have used whole-image histogram features to identify their position in real time against a database of previously

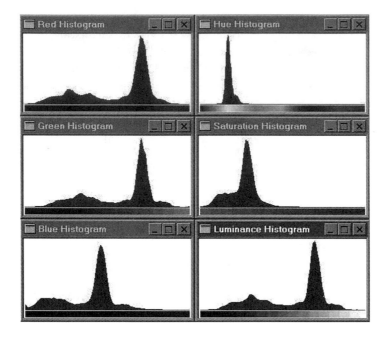

Figure 4.85
Six 1D histograms of the image above. A 5×5 smoothing filter was convolved with each band before histogramming.

recorded images of locations in their environment. Using this whole-image extraction approach, a robot can readily recover the particular hallway or particular room in which it is located [325].

Finally, note that in the last decade another global image descriptor—known as GIST—has been devised. The image is represented by a 320 dimensional vector per color band. The feature vector corresponds to the mean response to steerable filters at different scales and orientations computed over 4×4 sub-windows. Because a complete explanation of the GIST descriptor goes beyond the scope of this book, for an in-depth study we refer the reader to [253, 254] and [238].

4.6.6.2 Image fingerprints
This method is similar to the *visual word* approach with the difference that here the features are not interest points but rather morphological features, lines and colored blobs. Although outperformed by the new *visual word* based place recognition methods, this approach is still quite used in many applications of mobile robotics in both indoor and outdoor.

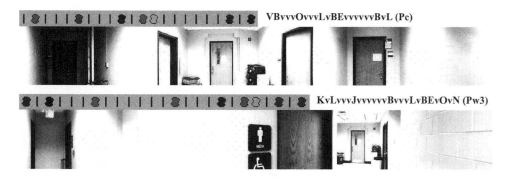

Figure 4.86
Two panoramic images and their associated fingerprint sequences [182].

We describe one particular implementation of this approach called the image *finger-print*, which was developed first in [182]. Such as the previous method, the system makes use of a 360-degree panoramic image. The first extraction tier searches the panoramic image for spatially localized features: vertical edges and sixteen discrete hues of color. The vertical edge detector is a straightforward gradient approach implementing a horizontal difference operator. Vertical edges are "voted upon" by each edge pixel just as in a vertical edge Hough transform. An adaptive threshold is used to reduce the number of edges. Suppose the Hough table's tallies for each candidate vertical line have a mean μ and a standard deviation σ. The chosen threshold is simply $\mu + \sigma$.

Vertical color bands are identified in largely the same way, identifying statistics over the occurrence of each color, then filtering out all candidate color patches except those with tallies greater than $\mu + \sigma$. Figure 4.86 shows two sample panoramic images and their associated fingerprints. Note that each fingerprint is converted to an ASCII string representation.

Just as with histogram distance metrics in the case of image histogramming, we need a quantifiable measure of the distance between two fingerprint strings. String-matching algorithms are yet another large field of study, with particularly interesting applications today in the areas of genetics [55]. Note that we may have strings that differ not just in a single element value, but even in their overall length. For example, figure 4.87 depicts three actual sequences generated using the preceding algorithm. The top string should match *Place 1*, but note that there are deletions and insertions between the two strings.

The technique used in the fingerprinting approach for string differencing is known as a *minimum energy algorithm*. Taken from the stereo vision community, this optimization algorithm will find the minimum energy required to "transform" one sequence into another sequence. The result is a distance metric that is relatively insensitive to the addition or sub-

Place x: vvBEvvCvvvMvOBvvvvv

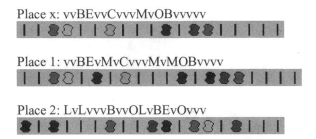

Place 1: vvBEvMvCvvvMvMOBvvvv

Place 2: LvLvvvBvvOLvBEvOvvv

Figure 4.87
Three actual string sequences. The top two are strings extracted by the robot at the same position [182].

traction of individual local features while still able to identify robustly the correct matching string in a variety of circumstances.

It should be clear to the reader that the image histogram and image fingerprint place representations are straightforward to implement. For this reason these methods became very popular, although recently the *visual-word*-based approaches for a greater variety of applications have outperformed them.

4.7   Feature Extraction Based on Range Data (Laser, Ultrasonic)

Most of today's features extracted from ranging sensors are geometric primitives such as line segments or circles. The main reason for this is that for most other geometric primitives the parametric description of the features becomes too complex and no closed-form solution exists. In this section, we will focus on line extraction, since line segments are the simplest features to extract. As we will see in chapter 5, lines are used to match laser scans for performing tasks like robot localization or automatic map building.

There are three main problems in line extraction in unknown environments:

- How many lines are there?

- Which points belong to which line?

- Given the points that belong to a line, how to estimate the line model parameters?

For answering these questions, we will present the description of the six most popular line extraction algorithms for 2D range scans. Our selection is based on their performance and popularity in both mobile robotics, especially feature extraction, and computer vision. Only basic versions of the algorithms are given, even though their details may vary in different applications and implementations. The interested reader should refer to the indicated references for more details. Our implementation follows closely the pseudocode described below in most cases, otherwise it will be stated.

Before describing the six algorithms, we will first explain the line fitting problem, which answers the third question: "Given the points that belong to a line, how to estimate the line model parameters?" In describing line fitting, we will demonstrate how the uncertainty models presented in section 4.1.3 can be applied to the problem of combining multiple sensor measurements. Then, we will answer the first two questions by describing six line-extraction algorithms from noisy range measurements. Finally, we will briefly present other very successful features of indoor mobile robots using range data, the corner and the plane features, and demonstrate how these features can be combined into a single representation.

4.7.1 Line fitting

Geometric feature fitting is usually the process of comparing and matching measured sensor data against a predefined description, or template, of the expected feature. Usually, the system is overdetermined in that the number of sensor measurements exceeds the number of feature parameters to be estimated. Since the sensor measurements all have some error, there is no perfectly consistent solution and, instead, the problem is one of optimization. One can, for example, fit the feature that minimizes the discrepancy with all sensor measurements used (e.g.,. *least-squares estimation*).

In this section we present an optimization solution to the problem of extracting a line feature from a set of uncertain sensor measurements. For greater detail than what is presented below, refer to [17, pages 15 and 221].

4.7.1.1 Probabilistic line fitting from uncertain range sensor data

Our goal is to fit a line to a set of sensor measurements as shown in figure 4.88. There is uncertainty associated with each of the noisy range sensor measurements, and so there is no single line that passes through the set. Instead, we wish to select the best possible match, given some optimization criterion.

More formally, suppose n ranging measurement points in polar coordinates $x_i = (\rho_i, \theta_i)$ are produced by the robot's sensors. We know that there is uncertainty associated with each measurement, and so we can model each measurement using two random variables $X_i = (P_i, Q_i)$. In this analysis we assume that the uncertainty with respect to the actual value of P and Q is independent. Based on equation (4.11), we can state this formally:

$$E[P_i \cdot P_j] = E[P_i]E[P_j] \qquad \forall\, i,j = 1, \ldots, n \mid i \neq j. \tag{4.125}$$

$$E[Q_i \cdot Q_j] = E[Q_i]E[Q_j] \qquad \forall\, i,j = 1, \ldots, n \mid i \neq j. \tag{4.126}$$

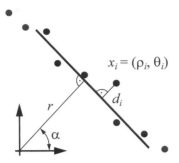

Figure 4.88
Estimating a line in the least-squares sense. The model parameters r (length of the perpendicular) and α (its angle to the abscissa) uniquely describe a line.

$$E[P_i \cdot Q_j] = E[P_i]E[Q_j] \qquad \forall \ i,j = 1, ..., n .\tag{4.127}$$

Furthermore, we assume that each random variable is subject to a Gaussian probability density curve, with a mean at the true value and with some specified variance:

$$P_i \sim N(\rho_i, \sigma^2_{\rho_i}) .\tag{4.128}$$

$$Q_i \sim N(\theta_i, \sigma^2_{\theta_i}) .\tag{4.129}$$

Given some measurement point (ρ, θ), we can calculate the corresponding Euclidean coordinates as $x = \rho\cos\theta$ and $y = \rho\sin\theta$. If there were no error, we would want to find a line for which all measurements lie on that line:

$$\rho\cos\theta\cos\alpha + \rho\sin\theta\sin\alpha - r = \rho\cos(\theta - \alpha) - r = 0 .\tag{4.130}$$

Of course, there is measurement error, and so this quantity will not be zero. When it is nonzero, this is a measure of the error between the measurement point (ρ, θ) and the line, specifically in terms of the minimum orthogonal distance between the point and the line. It is always important to understand how the error that shall be minimized is being measured. For example, a number of line-extraction techniques do not minimize this orthogonal point-line distance, but instead the distance parallel to the y-axis between the point and the line. A good illustration of the variety of optimization criteria is available in [25], where several algorithms for fitting circles and ellipses are presented that minimize algebraic and geometric distances.

For each specific (ρ_i, θ_i), we can write the orthogonal distance d_i between (ρ_i, θ_i) and the line as

$$\rho_i \cos(\theta_i - \alpha) - r = d_i .$$ (4.131)

If we consider each measurement to be equally uncertain, we can sum the square of all errors together, for all measurement points, to quantify an overall fit between the line and all of the measurements:

$$S = \sum_i d_i^2 = \sum_i (\rho_i \cos(\theta_i - \alpha) - r)^2 .$$ (4.132)

Our goal is to minimize S when selecting the line parameters (α, r). We can do so by solving the nonlinear equation system

$$\frac{\partial S}{\partial \alpha} = 0 \qquad \frac{\partial S}{\partial r} = 0 .$$ (4.133)

This formalism is considered an *unweighted least-squares* solution because no distinction is made from among the measurements. In reality, each sensor measurement may have its own, unique uncertainty based on the geometry of the robot and environment when the measurement was recorded. For example, we know with regard to vision stereo ranging that uncertainty and, therefore, variance increase as a square of the distance between the robot and the object. To make use of the variance σ_i^2 that models the uncertainty regarding distance ρ_i of a particular sensor measurement, we compute an individual weight w_i for each measurement using the formula[17]

$$w_i = 1/\sigma_i^2 .$$ (4.134)

Then, equation (4.132) becomes

$$S = \sum w_i d_i^2 = \sum w_i (\rho_i \cos(\theta_i - \alpha) - r)^2 .$$ (4.135)

---

17. The issue of determining an adequate weight when σ_i is given (and perhaps some additional information) is complex in general and beyond the scope of this text. See [11] for a careful treatment.

It can be shown that the solution to equation (4.133) in the *weighted* least-squares sense[18] is

$$\alpha = \frac{1}{2}\text{atan}\left(\frac{\sum w_i\rho_i^2\sin 2\theta_i - \frac{2}{\sum w_i}\sum\sum w_iw_j\rho_i\rho_j\cos\theta_i\sin\theta_j}{\sum w_i\rho_i^2\cos 2\theta_i - \frac{1}{\sum w_i}\sum\sum w_iw_j\rho_i\rho_j\cos(\theta_i + \theta_j)}\right). \tag{4.136}$$

$$r = \frac{\sum w_i\rho_i\cos(\theta_i - \alpha)}{\sum w_i}. \tag{4.137}$$

In practice, equation (4.136) uses the four-quadrant arc tangent (atan2).[19]

Let us demonstrate equations (4.136) and (4.137) with a concrete example. The seventeen measurements (ρ_i, θ_i) in table 4.3 have been taken with a laser range sensor installed on a mobile robot. The measurements are shown in figure 4.89. The measurement uncertainty is usually considered proportional to the measured distance, but, to simplify the calculation, in this case we assume that the uncertainties of all measurements are equal. We also assume that the measurements are uncorrelated, and that the robot was static during the measurement process. Direct application of this solution equations yields the line defined by $\alpha = 37.36$ and $r = 0.4$. This line represents the best fit in a least-squares sense and is shown visually in figure 4.89.

4.7.1.2 Propagation of uncertainty during line fitting

Returning to the subject of section 4.1.3, we would like to understand how the uncertainties of specific range sensor measurements propagate to govern the uncertainty of the extracted line. In other words, how does uncertainty in ρ_i and θ_i propagate in equations (4.136) and (4.137) to affect the uncertainty of α and r?

This requires direct application of equation (4.15) with A and R representing the random output variables of α and r respectively. The goal is to derive the 2×2 output covariance matrix

---

18. We follow here the notation of [17] and distinguish a weighted least-squares problem if C_X is diagonal (input errors are mutually independent) and a generalized least-squares problem if C_X is non-diagonal.

19. Atan2 computes $\tan(x/y)^{-1}$ but uses the signs of both x and y to determine the quadrant in which the resulting angles lies. For example $\text{atan2}(-2, -2) = -135°$, whereas $\text{atan2}(2, 2) = 45°$, a distinction which would be lost with a single-argument arc tangent function.

Table 4.3 Measured values

pointing angle of sensor θ_i [deg]	range ρ_i [m]
0	0.5197
5	0.4404
10	0.4850
15	0.4222
20	0.4132
25	0.4371
30	0.3912
35	0.3949
40	0.3919
45	0.4276
50	0.4075
55	0.3956
60	0.4053
65	0.4752
70	0.5032
75	0.5273
80	0.4879

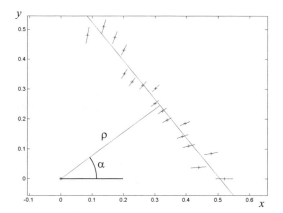

Figure 4.89
Extracted line from laser range measurements (+). The small lines at each measurement point represent the measurement uncertainty σ that is proportional to the measured distance.

$$C_{AR} = \begin{bmatrix} \sigma_A^2 & \sigma_{AR} \\ \sigma_{AR} & \sigma_R^2 \end{bmatrix}, \tag{4.138}$$

given the $2n \times 2n$ input covariance matrix

$$C_X = \begin{bmatrix} C_P & 0 \\ 0 & C_Q \end{bmatrix} = \begin{bmatrix} diag(\sigma_{\rho_i}^2) & 0 \\ 0 & diag(\sigma_{\theta_i}^2) \end{bmatrix} \tag{4.139}$$

and the system relationships [equations (4.136) and (4.137)]. Then by calculating the Jacobian,

$$F_{PQ} = \begin{bmatrix} \dfrac{\partial \alpha}{\partial P_1} & \dfrac{\partial \alpha}{\partial P_2} & \cdots & \dfrac{\partial \alpha}{\partial P_n} & \dfrac{\partial \alpha}{\partial Q_1} & \dfrac{\partial \alpha}{\partial Q_2} & \cdots & \dfrac{\partial \alpha}{\partial Q_n} \\ \dfrac{\partial r}{\partial P_1} & \dfrac{\partial r}{\partial P_2} & \cdots & \dfrac{\partial r}{\partial P_n} & \dfrac{\partial r}{\partial Q_1} & \dfrac{\partial r}{\partial Q_2} & \cdots & \dfrac{\partial r}{\partial Q_n} \end{bmatrix}, \tag{4.140}$$

we can instantiate the uncertainty propagation equation (4.15) to yield C_{AR}:

$$C_{AR} = F_{PQ} C_X F_{PQ}^T \tag{4.141}$$

Thus we have calculated the probability C_{AR} of the extracted line (α, r) based on the probabilities of the measurement points. For more details about this method, refer to [8, 59].

4.7.2 Six line-extraction algorithms

The previous section described how to fit a line feature given a set of range measurements. Unfortunately, the feature extraction process is significantly more complex than that. A mobile robot does indeed acquire a set of range measurements, but in general the range measurements are not all part of one line. Rather, only some of the range measurements should play a role in line extraction and, further, there may be more than one line feature represented in the measurement set. This more realistic scenario is shown in figure 4.90.

The process of dividing up a set of measurements into subsets that can be interpreted one by one is termed *segmentation* and is the most important step of line extraction. In the fol-

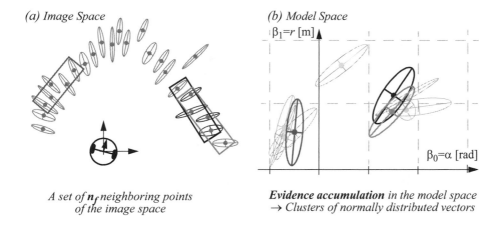

(a) Image Space

(b) Model Space

A set of **n_f** neighboring points
of the image space

Evidence accumulation *in the model space*
→ Clusters of normally distributed vectors

Figure 4.90
Clustering: finding neighboring segments of a common line [59].

Algorithm 1: *Split-and-Merge*

---

1. Initial: set s_1 consists of N points. Put s_1 in a list L

2. Fit a line to the next set s_i in L

3. Detect point P with maximum distance d_P to the line

4. If d_P is less than a threshold, continue (go to step 2)

5. Otherwise, split s_i at P into s_{i1} and s_{i2}, replace s_i in L by s_{i1} and s_{i2}, continue (go to 2)

6. When all sets (segments) in L have been checked, merge collinear segments.

---

lowing, we describe six popular line-extraction (segmentation) algorithms. For both an overview and a comparison among these algorithms, we refer the reader to [247].

4.7.2.1 Algorithm 1: Split-and-merge

Split-and-Merge is the most popular line extraction algorithm. This algorithm has originated from computer vision [257] and has been studied and used in many works [96, 121, 287, 78, 336]. The algorithm is outlined in algorithm 1.

Notice that this algorithm can be slightly modified on line 3 to make it more robust to noise. Indeed, sometimes the splitting position can be the result of a point which still belongs to the same line but which, because of noise, appears too far away from this line.

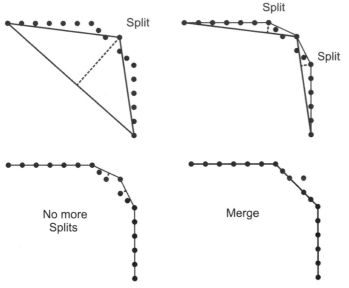

Figure 4.91 Split-and-merge implemented in the Iterative-End-Point-Fit fashion. In this case, the line is not fitted to the points but is constructed by connecting the first and last points.

In this case, we scan for a splitting position where two adjacent points P_1 and P_2 are on the same side of the line and both have distances to the line greater than the threshold. If we find only one such point, then we automatically discard it as a noisy point.

Observe that in line 2 one can use for line fitting the least-squares method described in section 4.7.1. Alternatively, one can construct the line by simply connecting the first and the last points. In this case, the algorithm is named *Iterative-End-Point-Fit* [19, 287, 78, 336] and is a well consolidated approach to implement split-and-merge. This procedure is illustrated in figure 4.91.

Finally, an application of split-and-merge to a 2D laser scan is shown in figure 4.92.

4.7.2.2 Algorithm 2: Line regression

This algorithm was proposed in [59]. It uses a sliding window of size N_f. At every step, a line is fitted to the N_f points within the window. The window is then shifted one point forward (this is why it is called sliding window), and the line-fitting operation is repeated again. The goal is to find adjacent line segments and merge them together. To do this, at every step the *Mahalanobis*[20] distance between the last two windows is computed and is stored in a *fidelity* array. When all the points have been analyzed, the fidelity array is scanned for consecutive similar elements. This is done by using an appropriate clustering

---

20. The Mahalanobis distance is defined in section "Matching" on page 334.

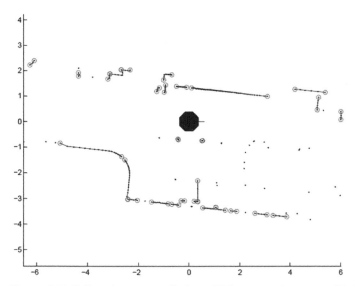

Figure 4.92 Split-and-merge applied to a 2D laser scan (courtesy of B. Jensen).

Algorithm 2: *Line-Regression*

---

    1. Initialize sliding window size N_f

    2. Fit a line to every N_f consecutive points

    3. Compute a line fidelity array. Each element of the array contains the sum of Mahalanobis distances between every three adjacent windows

    4. Construct line segments by scanning the fidelity array for consecutive elements having values less than a threshold

    5. Merge overlapped line segments and recompute line parameters for each segment

---

algorithm. At the end, the clustered consecutive line segments are merged together using again line regression. This algorithm is outlined in algorithm 2, while the main steps are depicted in figure 4.93.

    Notice that the sliding window size N_f is very dependent on the environment and has a strong influence on the algorithm performance. In typical applications, $N_f = 7$ is used.

4.7.2.3 Algorithm 3: Incremental

This algorithm is straightforward to implement and has been used in many applications [24, 328, 308]. The algorithm is outlined in algorithm 3. At the beginning, the set consists of

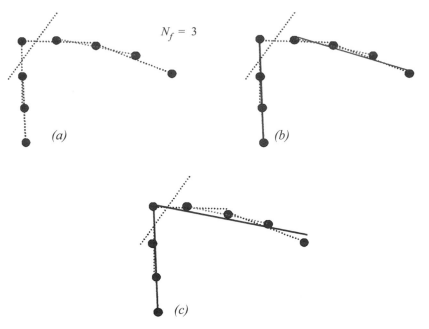

Figure 4.93 A sliding window with size $N_f = 3$ is used in this example. (a) A line is fitted to every set of consecutive three points. (b) Consecutive similar segments are then merged together. (c) The set of all segments is then checked again, and the remaining consecutive similar segments are merged with those generated from the previous step.

two points. Next, an extra point is added to this set and a line is constructed. If the line satisfies a predefined line condition, then a new point is added to the set and the procedure is repeated. If the line does not verify the line condition, the new point is put back and the line is computed from all previous visited points. At this point, the procedure starts again with the two new following points.

Observe that to speed up the incremental process, one can also add multiple points at each step instead of just one point. In [247], five points were added at each step. When the line does not satisfy the predefined line condition, the last five points are put back, and the algorithm switches back to adding individual points at a time.

4.7.2.4 Algorithm 4: RANSAC

RANSAC (*Random Sample Consensus* [128]) is an algorithm to estimate robustly the parameters of a model from a given data in the presence of *outliers*. Outliers are data that do not fit the model. Such outliers can be due to high noise in the data, wrong measurements, or they can more simply be points which come from other objects for which our mathematical model does not apply. For example, a typical laser scan in indoor environ-

---

Algorithm 3: *Incremental*

---

1. Start by the first 2 points, construct a line

2. Add the next point to the current line model

3. Recompute the line parameters by line fitting

4. If it satisfies the line condition, continue (go to step 2)

5. Otherwise, put back the last point, recompute the line parameters, return the line

6. Continue with the next two points, go to step 2

---

ments may contain distinct lines from the surrounding walls but also points from other static and dynamic objects (like chairs or humans). In this case, an outlier is any entity which does not belong to a line (i.e., chair, human, and so on).

RANSAC is an iterative method and is nondeterministic in that the probability to find a line free of outliers increases as more iterations are used. RANSAC is not restricted to line extraction from laser data but it can be generally applied to any problem where the goal is to identify the inliers which satisfy a predefined mathematical model. Typical applications in robotics are: line extraction from 2D range data (sonar or laser); plane extraction from 3D laser point clouds; and structure-from-motion (section 4.2.6), where the goal is to identify the image correspondences which satisfy a rigid body transformation.

Let us see how RANSAC works for the simple case of line extraction from 2D laser scan points. The algorithm starts by randomly selecting a sample of two points from the dataset. Then a line is constructed from these two points and the distance of all other points to this line is computed. The inliers set comprises all the points whose distance to the line is within a predefined threshold d. The algorithm then stores the inliers set and starts again by selecting another minimal set of two points at random. The procedure is iterated until a set with a maximum number of inliers is found, which is chosen as a solution to the problem. The algorithm is outlined in algorithm 4, while figure 4.94 illustrates its working principle.

Because we cannot know in advance if the observed set contains the maximum number of inliers, the ideal would be to check all possible combinations of 2 points in a dataset of N points. The number of combinations is given by $N \cdot (N-1)/2$, which makes it computationally unfeasible if N is too large. For example, in a laser scan of 360 points we would need to check all $360 \cdot 359/2 = 64{,}620$ possibilities!

At this point, a question arises: Do we really need to check all possibilities, or can we stop RANSAC after k iterations? The answer is that indeed we do not need to check all combinations but just a subset of them if we have a rough estimate of the percentage of inliers in our dataset. This can be done by thinking in a probabilistic way.

Algorithm 4: *RANSAC*

---

1. Initial: let A be a set of N points

2. **repeat**

3. Randomly select a sample of 2 points from A

4. Fit a line through the 2 points

5. Compute the distances of all other points to this line

6. Construct the inlier set (i.e. count the number of points with distance to the line $< d$)

7. Store these inliers

8. **until** Maximum number of iterations k reached

9. The set with the maximum number of inliers is chosen as a solution to the problem

---

Let p be the probability of finding a set of points free of outliers. Let w be the probability of selecting an inlier from our dataset of N points. Hence, w expresses the fraction of inliers in the data, that is, w = number of inliers/N. If we assume that the two points needed for estimating a line are selected independently, w^2 is the probability that both points are inliers and $1 - w^2$ is the probability that at least one of these two points is an outlier. Now, let k be the number of RANSAC iterations executed so far, then $(1 - w^2)^k$ will be the probability that RANSAC never selects two points that are both inliers. This probability must be equal to $1 - p$. Accordingly,

$$1 - p = (1 - w^2)^k,$$ (4.142)

and therefore

$$k = \frac{\log(1 - p)}{\log(1 - w^2)}.$$ (4.143)

This expression tells us that knowing the fraction w of inliers, after k RANSAC iterations we will have a probability p of finding a set of points free of outliers. For example, if we want a probability of success equal to 99% and we know that the percentage of inliers in the dataset is 50%, then according to (4.143) we could stop RANSAC after 16 iterations, which is much less than the number of all possible combinations that we had to check in the previous example! Also observe that in practice we do not need a precise knowledge of

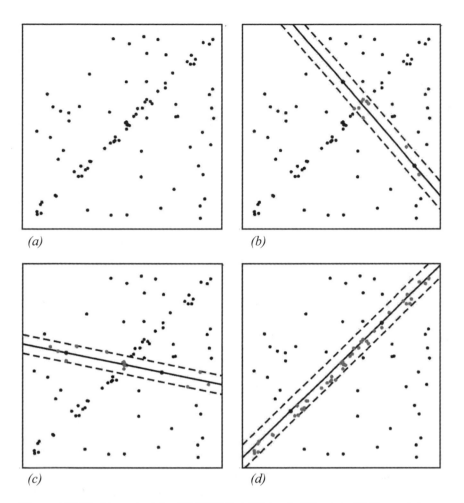

Figure 4.94 Working principle of RANSAC. (a) Dataset of N points. (b) Two points are randomly selected, a line is fitted through them, and the points within a predefined distance to it are identified. (c) The procedure is repeated (iterated) several times. (d) The set with the maximum number of inliers is chosen as a solution to the problem.

the fraction of inliers but just a rough estimate. More advanced implementations of RANSAC estimate the fraction of inliers by changing it adaptively iteration after iteration.

The main advantage of RANSAC is that it is a generic extraction method and can be used with many types of features once we have the feature model. Because of this, it is very popular in computer vision [29]. It is also simple to implement. Another advantage is its ability to cope with large amount of outliers, even more than 50%. Clearly, if we want to extract multiple lines, we need to run RANSAC several time and remove sequentially all

---

Algorithm 5: *Hough Transform*

---

    1. Initial: let A be a set of N points

    2. Initialize the accumulator array by setting all elements to 0

    3. Construct values for the array

    4. Choose the element with max. votes V_{max}

    5. If V_{max} is less than a threshold, terminate

    6. Otherwise, determine the inliers

    7. Fit a line through the inliers and store the line

    8. Remove the inliers from the set, go to step 2

---

the lines extracted so far. A disadvantage of RANSAC is that when the maximum number of iterations k is reached, the solution obtained may not be the optimal one (i.e., the one with the maximum number of inliers). Furthermore, this solution may not even be the one that fits the data in the best way.

4.7.2.5 Algorithm 5: Hough Transform (HT)

This algorithm was already described for straight edge detection in intensity images (page 205) but it can be applied without any modification to 2D range images. The algorithm is outlined in algorithm 5. Although it has been developed within the computer vision community, it has been brought into robotics for extracting lines from scan images [158] and [261]. In fact, 2D scan images are nothing but binary images.

Typical drawbacks with the Hough transform are that it is usually difficult to choose an appropriate grid size and the fact that this transform does not take noise and uncertainty into account when estimating the line parameters. To overcome the second problem, in line 7 one can use the line fitting method described in section 4.7.1, which takes into account feature uncertainty.

4.7.2.6 Algorithm 6: Expectation maximization (EM)

Expectation Maximization (EM), is a probabilistic method commonly used in missing variable problems. EM has been used as a line extraction tool in computer vision [24] and robotics [261]. There are some drawbacks with the EM algorithm. First, it can fall into local minima. Second, it is difficult to choose a good initial value. The algorithm is outlined in algorithm 6. For a detailed implementation of this algorithm for extracting lines, we refer the reader to [24].

Algorithm 6: *Expectation Maximization*

---

1. Initial: let A be a set of N points

2. **repeat**

3.   Randomly generate parameters for a line

4.   Initialize weights for remaining points

5.   **repeat**

6.   *E-Step*: Compute the weights of the points from the line model

7.   *M-Step*: Recompute the line model parameters

8.   **until** Maximum number of steps reached or convergence

9.   **until** Maximum number of trials reached or found a line

10. If found, store the line, remove the inliers, go to step 2

11 Otherwise, terminate

4.7.2.7 Implementation details

Clustering. In most cases, 2D laser scans present some agglomerations of a few sparse points (figure 4.92). These points can be caused for instance by small objects or moving people. In this case, a simple clustering algorithm is usually used for preprocessing: it divides the raw points into groups of close points and discards groups consisting of too few points. Basically, this algorithm scans for big jumps in radial differences of consecutive points and puts breakpoints in those positions. As a result, the scan is segmented into contiguous clusters of points. Clusters having too few number of points are removed.

Merging. Due to occlusions, a line may be observed and extracted as several segments. When this happens, it is likely good to merge collinear line segments into a single line segment. This merging routine should be applied at the output end of each previously seen algorithm, after segments have been extracted. To decide if two consecutive line segments have to be merged, the Mahalanobis distance[21] between each pair of line segments is typically used. If the two line segments have Mahalanobis distance less than a predefined

---

21. The Mahalanobis distance depends on the covariance matrix of the parameters of each line segment as explained on page 334.

threshold, then they are merged. Using line fitting, the new line parameters are finally recomputed from the raw scan points that constitute the two segments.

4.7.2.8 A comparison of line extraction algorithms

These six algorithms can be divided into two categories: deterministic and nondeterministic methods:

  1. Deterministic: Split-and-Merge, Incremental, Regression, Hough transform.

  2. Nondeterministic: RANSAC, EM.

RANSAC and EM are nondeterministic because their results can be different at every run. This is because these two algorithms generate random hypotheses.

A comparison between all six algorithms has been done by Nguyen et al. [247]. They evaluated four quality measures: complexity, speed, correctness (false positives), and precision. The results of that study are shown in table 4.4. The terminology used is explained as follows:

- N: Number of points in the input scan (e.g., 722)

- S: Number of line segments extracted (e.g., 7 in average depending on the algorithm)

- N_f: Sliding window size for Line-Regression (e.g., 9)

- N_{Trials}: Number of trials for RANSAC (e.g., 1000)

- N_C, N_R: Number of columns, rows respectively for the Hough accumulator array (N_C = 401, N_R = 671 for resolutions of 1 cm and 0.9 degrees)

- N_1, N_2: Number of trials and convergence iterations, respectively, for EM (e.g. N_1 = 50, N_2 = 200).

Observe that the values shown in parentheses are typical numbers used in practical implementations.

As shown in the third column (Speed) of table 4.4, Split-and-Merge, Incremental, and Line-Regression perform much faster than the others. The Split-and-Merge algorithm takes the lead. The reason why these three algorithms are much faster is mainly because they are deterministic and, especially, because they take advantage of the sequential ordering of the raw scan points (the points are not captured randomly but according to the rotation direction of the laser beam). If these three algorithms were applied on randomly distributed points (e.g., general binary images), they would not be able to segment all lines, while RANSAC, EM, and Hough would. Indeed, these last three algorithms are popular for their ability to extract lines in binary images which obviously present a large number of outliers.

The Incremental algorithm seems to perform the best in terms of correctness. In fact, it has a very low number of false positives, which is very important for localization, mapping,

Table 4.4 Comparison of algorithms for line extraction from 2D laser data.

	Complexity	Speed [Hz]	False positives	Precision
Split-and-Merge	$N \cdot \log N$	1500	10%	+++
Incremental	$S \cdot N^2$	600	6%	+++
Line-Regression	$N \cdot N_f$	400	10%	+++
RANSAC	$S \cdot N \cdot N_{Trials}$	30	30%	++++
Hough-Transform	$S \cdot N \cdot N_C + S \cdot N_R \cdot N_C$	10	30%	++++
Expectation Maximization	$S \cdot N_1 \cdot N_2 \cdot N$	1	50%	++++

and SLAM (section 5.8). Conversely, RANSAC, HT, and EM seem to produce many more false positives. This is due to the fact that they do not take advantage of the sequential ordering of the scan points and therefore they often try to fit lines falsely across the scan map. Their behavior could be improved by increasing the minimum number of points per line segment, but the drawback of this would then be that short segments might be left out.

Despite their bad correctness, as observed in the fourth column of table 4.4, RANSAC, HT, and EM produce more precise lines than the other algorithms. This is due to their ability to get rid of outliers or largely noisy inliers. For instance, with RANSAC the probability of extracting a stable line increases with the number of iterations, while with HT the outlier (or a largely noise inlier) would vote another grid cell than that representing the line.

In conclusion, Split-and-Merge and Incremental are the best choice in terms of correctness and efficiency and are therefore the best candidates for 2D laser-based robot localization and mapping. However, the right choice depends highly on the type of application and the desired precision.

4.7.3 Range histogram features

A histogram is a simple way to combine characteristic elements of an image. An angle histogram, as presented in figure 4.95, plots the statistics of lines extracted by two adjacent range measurements. First, a 360-degree scan of the room is taken with the range scanner, and the resulting "hits" are recorded in a map. Then the algorithm measures the relative angle between any two adjacent hits (see figure 4.95b). After compensating for noise in the readings (caused by the inaccuracies in position between adjacent hits), the angle histogram

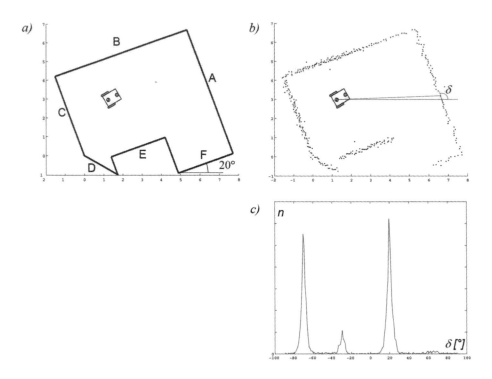

Figure 4.95
Angle histogram [329].

shown in figure 4.95c can be built. The uniform direction of the main walls are clearly visible as peaks in the angle histogram. Detection of peaks yields only two main peaks: one for each pair of parallel walls. This algorithm is very robust with regard to openings in the walls, such as doors and windows, or even cabinets lining the walls.

4.7.4 Extracting other geometric features
Line features are of particular value for mobile robots operating in man-made environments, where, for example, building walls and hallway walls are usually straight. In general, a mobile robot makes use of multiple features simultaneously, comprising a *feature set* that is most appropriate for its operating environment. For indoor mobile robots, the line feature is certainly a member of the optimal feature set.

In addition, other geometric kernels consistently appear throughout the indoor manmade environment. *Corner* features are defined as a point feature with an orientation. *Step discontinuities*, defined as a step change perpendicular to the direction of hallway travel,

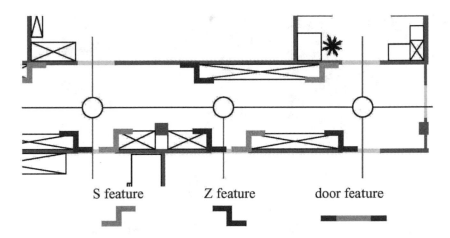

Figure 4.96
Multiple geometric features in a single hallway, including doorways and discontinuities in the width of the hallway.

are characterized by their form (convex or concave) and step size. *Doorways*, defined as openings of the appropriate dimensions in walls, are characterized by their width.

Thus, the standard segmentation problem is not so simple as deciding on a mapping from sensor readings to line segments, but rather it is a process in which features of different types are extracted based on the available sensor measurements. Figure 4.96 shows a model of an indoor hallway environment along with both indentation features (i.e., step discontinuities) and doorways.

Note that different feature types can provide quantitatively different information for mobile robot localization. The line feature, for example, provides two degrees of information, angle and distance. But the step feature provides 2D relative position information as well as angle.

The set of useful geometric features is essentially unbounded, and as sensor performance improves we can only expect greater success at the feature extraction level. For example, an interesting improvement upon the line feature described above relates to the advent of successful vision ranging systems (e.g., stereo cameras and time-of-flight cameras) and 3D laser rangefinder. Because these sensor modalities provide a full 3D set of range measurements, one can extract plane features in addition to line features from the resulting data set. Plane features are valuable in man-made environments due to the flat walls, floors, and ceilings of our indoor environments. Thus they are promising as another highly informative feature for mobile robots to use for mapping and localization. Some

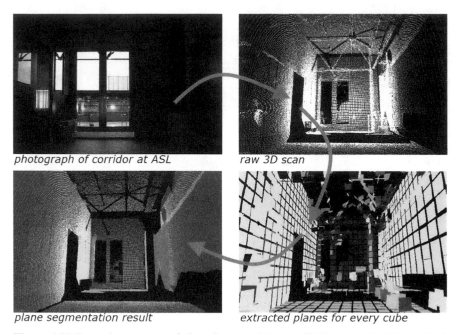

photograph of corridor at ASL   raw 3D scan

plane segmentation result   extracted planes for every cube

Figure 4.97 Extraction process of plane features: (Upper left) Photograph of the original environment. (Upper right) Raw 3D scan. (Bottom right) Plane feature segmentation and fitting. (Bottom left) final plane segmentation result. Image courtesy of J. Weingarten [331].

experiments using plane features have been done at the ASL (ETH Zurich) [331], the plane feature extraction process is illustrated in figure 4.97.

4.8   Problems

1. Consider an omnidirectional robot with a ring of eight 70 KHz sonar sensors that are fired sequentially. Your robot is capable of accelerating and decelerating at 50 cm/s^2. It is moving in a world filled with sonar-detectable fixed (nonmoving) obstacles that can only be detected at 5 meters and closer. Given the bandwidth of your sonar sensors, compute your robot's appropriate maximum speed to ensure no collisions.

2. Design an optical triangulation system with the best possible resolution for the following conditions: specify b (as in figure 4.15):
(a) the system must have sensitivity of 1 cm at a range of 2 meters.
(b) The PSD has a sensitivity of 0.1 mm.
(c) f = 10 cm.

3. Identify a specific digital CMOS-based camera on the market. Using product specifications for this camera, collect and compute the following values. Show your derivations:
- Dynamic range
- Resolution (of a single pixel)
- Bandwidth

4. Stereo vision. Solve the system given by equations (4.60) and (4.61) and find the optimal point (x, y, z) that minimizes the distance between the optical rays passing through $\tilde{p}_l$ and $\tilde{p}_r$. For doing this, observe that these two equations define two distinct lines in the 3D space. The problems consists in rewriting these two equations as the difference between 3D points along these two lines. Then, impose that the partial derivatives of this distance with respect to λ_l and λ_r equal zero. From this, you will obtain the two points along the two lines at minimum distance between each other. The optimal point (x, y, z) can then be found as the middle point between those points.

5. Challenge Question.
Implement a basic two-view structure-from-motion algorithm from scratch:
(a) Implement the basic Harris corner detector in Matlab.
(b) Take two images of the same scene from different view points.
(c) Extract and match Harris features using SSD.
(d) Implement the 8-point algorithm to compute the essential matrix.
(e) Compute rotation and translation up to a scale from the essential matrix. Disambiguate the four solutions using the cheirality constraint.

5 Mobile Robot Localization

5.1 Introduction

Navigation is one of the most challenging competences required of a mobile robot. Success in navigation requires success at the four building blocks of navigation: *perception* (the robot must interpret its sensors to extract meaningful data); *localization* (the robot must determine its position in the environment, figure 5.1); *cognition* (the robot must decide how to act to achieve its goals); and *motion control* (the robot must modulate its motor outputs to achieve the desired trajectory).

Of these four components (figure 5.2), localization has received the greatest research attention in the past decade, and as a result, significant advances have been made on this front. In this chapter, we explore the successful localization methodologies of recent years. First, section 5.2 describes how sensor and effector uncertainty is responsible for the difficulties of localization. Section 5.3 describes two extreme approaches to dealing with the challenge of robot localization: avoiding localization altogether, and performing explicit map-based localization. The remainder of the chapter discusses the question of representa-

Figure 5.1
Where am I?

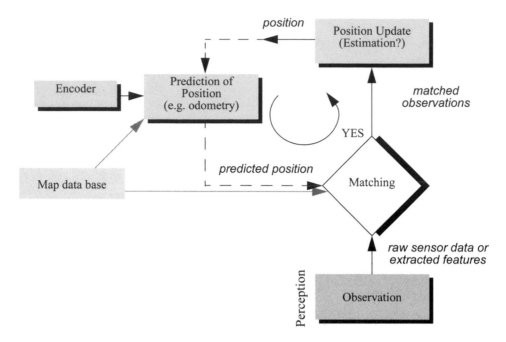

Figure 5.2
General schematic for mobile robot localization.

tion, then presents case studies of successful localization systems using a variety of representations and techniques to achieve mobile robot localization competence.

5.2   The Challenge of Localization: Noise and Aliasing

If one could attach an accurate GPS (global positioning system) sensor to a mobile robot, much of the localization problem would be obviated. The GPS would inform the robot of its exact position, indoors and outdoors, so that the answer to the question, "Where am I?" would always be immediately available. Unfortunately, such a sensor is not currently practical. The existing GPS network provides accuracy to within several meters, which is unacceptable for localizing human-scale mobile robots as well as miniature mobile robots such as desk robots and the body-navigating nanorobots of the future. Furthermore, GPS technologies cannot function indoors or in obstructed areas and are thus limited in their workspace.

But, looking beyond the limitations of GPS, localization implies more than knowing one's absolute position in the Earth's reference frame. Consider a robot that is interacting with humans. This robot may need to identify its absolute position, but its relative position with respect to target humans is equally important. Its localization task can include identifying humans using its sensor array, then computing its relative position to the humans. Furthermore, during the *cognition* step a robot will select a strategy for achieving its goals. If it intends to reach a particular location, then localization may not be enough. The robot may need to acquire or build an environmental model, a *map*, that aids it in planning a path to the goal. Once again, localization means more than simply determining an absolute pose in space; it means building a map, then identifying the robot's position relative to that map.

Clearly, the robot's sensors and effectors play an integral role in all the these forms of localization. It is because of the inaccuracy and incompleteness of these sensors and effectors that localization poses difficult challenges. This section identifies important aspects of this sensor and effector suboptimality.

5.2.1 Sensor noise

Sensors are the fundamental robot input for the process of *perception*, and therefore the degree to which sensors can discriminate the world state is critical. *Sensor noise* induces a limitation on the consistency of sensor readings in the same environmental state and, therefore, on the number of useful bits available from each sensor reading. Often, the source of sensor noise problems is that some environmental features are not captured by the robot's representation and are thus overlooked.

For example, a vision system used for indoor navigation in an office building may use the color values detected by its color CCD camera. When the sun is hidden by clouds, the illumination of the building's interior changes because of the windows throughout the building. As a result, hue values are not constant. The color CCD appears noisy from the robot's perspective as if subject to random error, and the hue values obtained from the CCD camera will be unusable, unless the robot is able to note the position of the sun and clouds in its representation.

Illumination dependence is only one example of the apparent noise in a vision-based sensor system. Picture jitter, signal gain, blooming, and blurring are all additional sources of noise, potentially reducing the useful content of a color video image.

Consider the noise level (i.e., apparent random error) of ultrasonic range-measuring sensors (e.g., sonars) as discussed in section 4.1.2.3. When a sonar transducer emits sound toward a relatively smooth and angled surface, much of the signal will coherently reflect away, failing to generate a return echo. Depending on the material characteristics, a small amount of energy may return nonetheless. When this level is close to the gain threshold of the sonar sensor, then the sonar will, at times, succeed and, at other times, fail to detect the object. From the robot's perspective, a virtually unchanged environmental state will result in two different possible sonar readings: one short and one long.

The poor signal-to-noise ratio of a sonar sensor is further confounded by interference between multiple sonar emitters. Often, research robots have between twelve and forty-eight sonars on a single platform. In acoustically reflective environments, multipath interference is possible between the sonar emissions of one transducer and the echo detection circuitry of another transducer. The result can be dramatically large errors (i.e., underestimation) in ranging values due to a set of coincidental angles. Such errors occur rarely, less than 1% of the time, and are virtually random from the robot's perspective.

In conclusion, sensor noise reduces the useful information content of sensor readings. Clearly, the solution is to take multiple readings into account, employing temporal fusion or multisensor fusion to increase the overall information content of the robot's inputs.

5.2.2   Sensor aliasing

A second shortcoming of mobile robot sensors causes them to yield little information content, further exacerbating the problem of perception and, thus, localization. The problem, known as *sensor aliasing*, is a phenomenon that humans rarely encounter. The human sensory system, particularly the visual system, tends to receive unique inputs in each unique local state. In other words, every different place looks different. The power of this unique mapping is only apparent when one considers situations where this fails to hold. Consider moving through an unfamiliar building that is completely dark. When the visual system sees only black, one's localization system quickly degrades. Another useful example is that of a human-sized maze made from tall hedges. Such mazes have been created for centuries, and humans find them extremely difficult to solve without landmarks or clues because, without visual uniqueness, human localization competence degrades rapidly.

In robots, the nonuniqueness of sensor readings, or *sensor aliasing*, is the norm and not the exception. Consider a narrow-beam rangefinder such as an ultrasonic or infrared rangefinder. This sensor provides range information in a single direction without any additional data regarding material composition such as color, texture, and hardness. Even for a robot with several such sensors in an array, there are a variety of environmental states that would trigger the same sensor values across the array. Formally, there is a many-to-one mapping from environmental states to the robot's perceptual inputs. Thus, the robot's percepts cannot distinguish from among these many states. A classic problem with sonar-based robots involves distinguishing between humans and inanimate objects in an indoor setting. When facing an apparent obstacle in front of itself, should the robot say "Excuse me" because the obstacle may be a moving human, or should the robot plan a path around the object because it may be a cardboard box? With sonar alone, these states are aliased, and differentiation is impossible.

The problem posed to navigation because of sensor aliasing is that, even with noise-free sensors, the amount of information is generally insufficient to identify the robot's position from a single-percept reading. Thus, techniques must be employed by the robot programmer that base the robot's localization on a series of readings and, thus, sufficient information to recover the robot's position over time.

5.2.3 Effector noise

The challenges of localization do not lie with sensor technologies alone. Just as robot sensors are noisy, limiting the information content of the signal, so robot effectors are also noisy. In particular, a single action taken by a mobile robot may have several different possible results, even though from the robot's point of view the initial state before the action was taken is well known.

In short, mobile robot effectors introduce uncertainty about future state. Therefore, the simple act of moving tends to increase the uncertainty of a mobile robot. There are, of course, exceptions. Using *cognition,* the motion can be carefully planned so as to minimize this effect, and indeed sometimes to actually result in more certainty. Furthermore, when the robot's actions are taken in concert with careful interpretation of sensory feedback, it can compensate for the uncertainty introduced by noisy actions using the information provided by the sensors.

First, however, it is important to understand the precise nature of the effector noise that impacts mobile robots. It is important to note that, from the robot's point of view, this error in motion is viewed as an error in odometry, or the robot's inability to estimate its own position over time using knowledge of its kinematics and dynamics. The true source of error generally lies in an incomplete model of the environment. For instance, the robot does not model the fact that the floor may be sloped, the wheels may slip, and a human may push the robot. All of these unmodeled sources of error result in inaccuracy between the physical motion of the robot, the intended motion of the robot, and the proprioceptive sensor estimates of motion.

In odometry (wheel sensors only) and dead reckoning (also heading sensors) the position update is based on *proprioceptive* sensors. The movement of the robot, sensed with wheel encoders or heading sensors or both, is integrated to compute position. Because the sensor measurement errors are integrated, the position error accumulates over time. Thus, the position has to be updated from time to time by other localization mechanisms. Otherwise the robot is not able to maintain a meaningful position estimate in the long run.

In the following we concentrate on odometry based on the wheel sensor readings of a differential-drive robot only (see also [5, 99, 102]). Using additional heading sensors (e.g., gyroscope) can help to reduce the cumulative errors, but the main problems remain the same.

There are many sources of odometric error, from environmental factors to resolution:

- Limited resolution during integration (time increments, measurement resolution, etc.);

- Misalignment of the wheels (deterministic);

- Uncertainty in the wheel diameter and in particular unequal wheel diameter (deterministic);

- Variation in the contact point of the wheel;

- Unequal floor contact (slipping, nonplanar surface, etc.).

Some of the errors might be *deterministic* (systematic); thus, they can be eliminated by proper calibration of the system. However, there are still a number of *nondeterministic* (random) errors that remain, leading to uncertainties in position estimation over time. From a geometric point of view, one can classify the errors into three types:

1. Range error: integrated path length (distance) of the robot's movement
   $\rightarrow$ sum of the wheel movements

2. Turn error: similar to range error, but for turns
   $\rightarrow$ difference of the wheel motions

3. Drift error: difference in the error of the wheels leads to an error in the robot's angular orientation

Over long periods of time, turn and drift errors far outweigh range errors, since their contribution to the overall position error is nonlinear. Consider a robot whose position is initially perfectly wellknown, moving forward in a straight line along the x-axis. The error in the y-position introduced by a move of d meters will have a component of $d \sin \Delta \theta$, which can be quite large as the angular error $\Delta \theta$ grows. Over time, as a mobile robot moves about the environment, the rotational error between its internal reference frame and its original reference frame grows quickly. As the robot moves away from the origin of these reference frames, the resulting linear error in position grows quite large. It is instructive to establish an error model for odometric accuracy and see how the errors propagate over time.

5.2.4 An error model for odometric position estimation
Generally the pose (position) of a robot is represented by the vector

$$p = \begin{bmatrix} x \\ y \\ \theta \end{bmatrix}. \tag{5.1}$$

For a differential-drive robot (figure 5.3) the position can be estimated starting from a known position by integrating the movement (summing the incremental travel distances). For a discrete system with a fixed sampling interval Δt, the incremental travel distances $(\Delta x; \Delta y; \Delta \theta)$ are

$$\Delta x = \Delta s \cos(\theta + \Delta \theta / 2), \tag{5.2}$$

$$\Delta y = \Delta s \sin(\theta + \Delta \theta / 2), \tag{5.3}$$

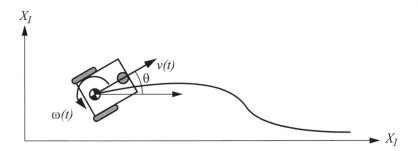

Figure 5.3
Movement of a differential-drive robot.

$$\Delta\theta = \frac{\Delta s_r - \Delta s_l}{b} \, , \tag{5.4}$$

$$\Delta s = \frac{\Delta s_r + \Delta s_l}{2} \, , \tag{5.5}$$

where

$(\Delta x; \Delta y; \Delta\theta)$ = path traveled in the last sampling interval;

$\Delta s_r; \Delta s_l$ = traveled distances for the right and left wheel respectively;

b = distance between the two wheels of differential-drive robot.

Thus we get the updated position p' :

$$p' = \begin{bmatrix} x' \\ y' \\ \theta' \end{bmatrix} = p + \begin{bmatrix} \Delta s \cos(\theta + \Delta\theta/2) \\ \Delta s \sin(\theta + \Delta\theta/2) \\ \Delta\theta \end{bmatrix} = \begin{bmatrix} x \\ y \\ \theta \end{bmatrix} + \begin{bmatrix} \Delta s \cos(\theta + \Delta\theta/2) \\ \Delta s \sin(\theta + \Delta\theta/2) \\ \Delta\theta \end{bmatrix} . \tag{5.6}$$

By using the relation for $(\Delta s; \Delta\theta)$ of equations (5.4) and (5.5) we further obtain the basic equation for odometric position update (for differential drive robots):

$$p' = f(x, y, \theta, \Delta s_r, \Delta s_l) = \begin{bmatrix} x \\ y \\ \theta \end{bmatrix} + \begin{bmatrix} \dfrac{\Delta s_r + \Delta s_l}{2} \cos\left(\theta + \dfrac{\Delta s_r - \Delta s_l}{2b}\right) \\ \dfrac{\Delta s_r + \Delta s_l}{2} \sin\left(\theta + \dfrac{\Delta s_r - \Delta s_l}{2b}\right) \\ \dfrac{\Delta s_r - \Delta s_l}{b} \end{bmatrix}. \tag{5.7}$$

As we discussed earlier, odometric position updates can give only a very rough estimate of the actual position. Owing to integration errors of the uncertainties of p and the motion errors during the incremental motion $(\Delta s_r; \Delta s_l)$, the position error based on odometry integration grows with time.

In the next step we will establish an error model for the integrated position p' to obtain the covariance matrix $\Sigma_{p'}$ of the odometric position estimate. To do so, we assume that at the starting point the initial covariance matrix Σ_p is known. For the motion increment $(\Delta s_r; \Delta s_l)$ we assume the following covariance matrix Σ_Δ:

$$\Sigma_\Delta = covar(\Delta s_r, \Delta s_l) = \begin{bmatrix} k_r |\Delta s_r| & 0 \\ 0 & k_l |\Delta s_l| \end{bmatrix}, \tag{5.8}$$

where Δs_r and Δs_l are the distances traveled by each wheel, and k_r, k_l are error constants representing the nondeterministic parameters of the motor drive and the wheel-floor interaction. As you can see, in equation (5.8) we made the following assumptions:

• The two errors of the individually driven wheels are independent,[22]

• The variance of the errors (left and right wheels) are proportional to the absolute value of the traveled distances $(\Delta s_r; \Delta s_l)$.

These assumptions, while not perfect, are suitable and will thus be used for the further development of the error model. The *motion errors* are due to imprecise movement because of deformation of wheel, slippage, unequal floor, errors in encoders, and so on. The values for the error constants k_r and k_l depend on the robot and the environment and should be experimentally established by performing and analyzing representative movements.

If we assume that p and $\Delta_{rl} = [\Delta s_r, \Delta s_l]^T$ are uncorrelated and the derivation of f (equation [5.7]) is reasonably approximated by the first-order Taylor expansion (linearization), we conclude, using the error propagation law (see section 4.1.3.2),

---

22. If there is more knowledge regarding the actual robot kinematics, the correlation terms of the covariance matrix could also be used.

$$\Sigma_{p'} = \nabla_p f \cdot \Sigma_p \cdot \nabla_p f^T + \nabla_{\Delta_{rl}} f \cdot \Sigma_\Delta \cdot \nabla_{\Delta_{rl}} f^T. \tag{5.9}$$

The covariance matrix Σ_p is, of course, always given by the $\Sigma_{p'}$ of the previous step, and can thus be calculated after specifying an initial value (e.g., 0).

Using equation (5.7) we can develop the two *Jacobians*, $F_p = \nabla_p f$ and $F_{\Delta_{rl}} = \nabla_{\Delta_{rl}} f$:

$$F_p = \nabla_p f = \nabla_p(f^T) = \left[\frac{\partial f}{\partial x}\ \frac{\partial f}{\partial y}\ \frac{\partial f}{\partial \theta}\right] = \begin{bmatrix} 1 & 0 & -\Delta s \sin(\theta + \Delta\theta/2) \\ 0 & 1 & \Delta s \cos(\theta + \Delta\theta/2) \\ 0 & 0 & 1 \end{bmatrix}, \tag{5.10}$$

$$F_{\Delta_{rl}} = \begin{bmatrix} \frac{1}{2}\cos\left(\theta + \frac{\Delta\theta}{2}\right) - \frac{\Delta s}{2b}\sin\left(\theta + \frac{\Delta\theta}{2}\right) & , \frac{1}{2}\cos\left(\theta + \frac{\Delta\theta}{2}\right) + \frac{\Delta s}{2b}\sin\left(\theta + \frac{\Delta\theta}{2}\right) \\ \frac{1}{2}\sin\left(\theta + \frac{\Delta\theta}{2}\right) + \frac{\Delta s}{2b}\cos\left(\theta + \frac{\Delta\theta}{2}\right) & , \frac{1}{2}\sin\left(\theta + \frac{\Delta\theta}{2}\right) - \frac{\Delta s}{2b}\cos\left(\theta + \frac{\Delta\theta}{2}\right) \\ \frac{1}{b} & -\frac{1}{b} \end{bmatrix}$$

$$\tag{5.11}$$

The details for arriving at equation (5.11) are

$$F_{\Delta_{rl}} = \nabla_{\Delta_{rl}} f = \left[\frac{\partial f}{\partial \Delta s_r}\ \frac{\partial f}{\partial \Delta s_l}\right] = \dots \tag{5.12}$$

$$\begin{bmatrix} \frac{\partial \Delta s}{\partial \Delta s_r}\cos\left(\theta + \frac{\Delta\theta}{2}\right) + \frac{\Delta s}{2}-\sin\left(\theta + \frac{\Delta\theta}{2}\right)\frac{\partial \Delta\theta}{\partial \Delta s_r} & , \frac{\partial \Delta s}{\partial \Delta s_l}\cos\left(\theta + \frac{\Delta\theta}{2}\right) + \frac{\Delta s}{2}-\sin\left(\theta + \frac{\Delta\theta}{2}\right)\frac{\partial \Delta\theta}{\partial \Delta s_l} \\ \frac{\partial \Delta s}{\partial \Delta s_r}\sin\left(\theta + \frac{\Delta\theta}{2}\right) + \frac{\Delta s}{2}\cos\left(\theta + \frac{\Delta\theta}{2}\right)\frac{\partial \Delta\theta}{\partial \Delta s_r} & , \frac{\partial \Delta s}{\partial \Delta s_l}\sin\left(\theta + \frac{\Delta\theta}{2}\right) + \frac{\Delta s}{2}\cos\left(\theta + \frac{\Delta\theta}{2}\right)\frac{\partial \Delta\theta}{\partial \Delta s_l} \\ \frac{\partial \Delta\theta}{\partial \Delta s_r} & \frac{\partial \Delta\theta}{\partial \Delta s_l} \end{bmatrix}$$

$$\tag{5.13}$$

and with

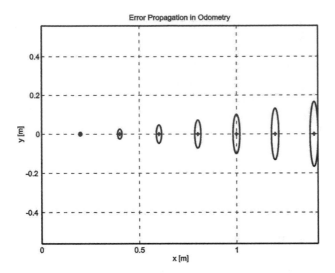

Figure 5.4
Growth of the pose uncertainty for straight-line movement: Note that the uncertainty in y grows much faster than in the direction of movement. This results from the integration of the uncertainty about the robot's orientation. The ellipses drawn around the robot positions represent the uncertainties in the x,y direction (e.g. 3σ). The uncertainty of the orientation θ is not represented in the picture, although its effect can be indirectly observed.

$$\Delta s = \frac{\Delta s_r + \Delta s_l}{2} \quad ; \quad \Delta\theta = \frac{\Delta s_r - \Delta s_l}{b} \tag{5.14}$$

$$\frac{\partial \Delta s}{\partial \Delta s_r} = \frac{1}{2} \quad ; \quad \frac{\partial \Delta s}{\partial \Delta s_l} = \frac{1}{2} \quad ; \quad \frac{\partial \Delta\theta}{\partial \Delta s_r} = \frac{1}{b} \quad ; \quad \frac{\partial \Delta\theta}{\partial \Delta s_l} = -\frac{1}{b}, \tag{5.15}$$

we obtain equation (5.11).

Figures 5.4 and 5.5 show typical examples of how the position errors grow with time. The results have been computed using the error model presented earlier.

Once the error model has been established, the error parameters must be specified. One can compensate for deterministic errors properly calibrating the robot. However the error parameters specifying the nondeterministic errors can only be quantified by statistical (repetitive) measurements. A detailed discussion of odometric errors and a method for calibration and quantification of deterministic and nondeterministic errors can be found in [6]. A method for on-the-fly odometry error estimation is presented in [205].

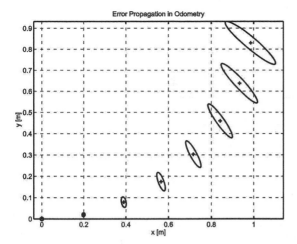

Figure 5.5
Growth of the pose uncertainty for circular movement (r = const): Again, the uncertainty perpendicular to the movement grows much faster than that in the direction of movement. Note that the main axis of the uncertainty ellipse does not remain perpendicular to the direction of movement.

5.3   To Localize or Not to Localize: Localization-Based Navigation Versus Programmed Solutions

Figure 5.6 depicts a standard indoor environment that a mobile robot navigates. Suppose that the mobile robot in question must deliver messages between two specific rooms in this environment: rooms A and B. In creating a navigation system, it is clear that the mobile robot will need sensors and a motion control system. Sensors are absolutely required to avoid hitting moving obstacles such as humans, and some motion control system is required so that the robot can deliberately move.

It is less evident, however, whether or not this mobile robot will require a *localization system*. Localization may seem mandatory in order to navigate successfully between the two rooms. It is through localizing on a map, after all, that the robot can hope to recover its position and detect when it has arrived at the goal location. It is true that, at the least, the robot must have a way of detecting the goal location. However, explicit localization with reference to a map is not the only strategy that qualifies as a goal detector.

An alternative, espoused by the behavior-based community, suggests that, since sensors and effectors are noisy and information-limited, one should avoid creating a geometric map for localization. Instead, this community suggests designing sets of behaviors that together result in the desired robot motion. Fundamentally, this approach avoids explicit reasoning about localization and position, and thus generally avoids explicit path planning as well.

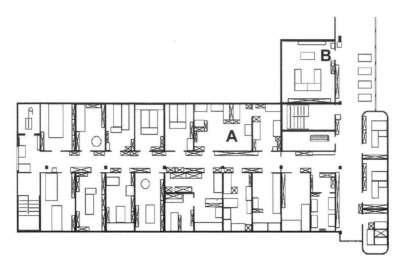

Figure 5.6
A sample environment.

This technique is based on a belief that there exists a procedural solution to the particular navigation problem at hand. For example, in figure 5.6, the behavioralist approach to navigating from room *A* to room *B* might be to design a left-wall following behavior and a detector for room *B* that is triggered by some unique queue in room *B*, such as the color of the carpet. Then the robot can reach room *B* by engaging the left-wall follower with the room *B* detector as the termination condition for the program.

The architecture of this solution to a specific navigation problem is shown in figure 5.7. The key advantage of this method is that, when possible, it may be implemented very quickly for a single environment with a small number of goal positions. It suffers from some disadvantages, however. First, the method does not directly scale to other environments or to larger environments. Often, the navigation code is location-specific, and the same degree of coding and debugging is required to move the robot to a new environment.

Second, the underlying procedures, such as *left-wall-follow,* must be carefully designed to produce the desired behavior. This task may be time-consuming and is heavily dependent on the specific robot hardware and environmental characteristics.

Third, a behavior-based system may have multiple active behaviors at any one time. Even when individual behaviors are tuned to optimize performance, this fusion and rapid switching between multiple behaviors can negate that fine-tuning. Often, the addition of each new incremental behavior forces the robot designer to retune all of the existing behaviors again to ensure that the new interactions with the freshly introduced behavior are all stable.

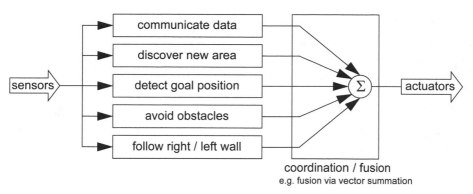

Figure 5.7
An architecture for behavior-based navigation.

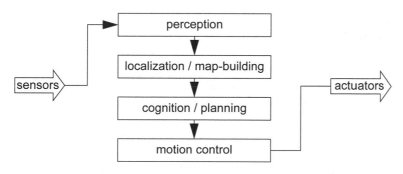

Figure 5.8
An architecture for map-based (or model-based) navigation.

In contrast to the behavior-based approach, the map-based approach includes both *localization* and *cognition* modules (see figure 5.8). In map-based navigation, the robot explicitly attempts to localize by collecting sensor data, then updating some belief about its position with respect to a map of the environment. The key advantages of the map-based approach for navigation are as follows:

- The explicit, map-based concept of position makes the system's belief about position transparently available to the human operators.

- The existence of the map itself represents a medium for communication between human and robot: the human can simply give the robot a new map if the robot goes to a new environment.

• The map, if created by the robot, can be used by humans as well, achieving two uses.

The map-based approach will require more up-front development effort to create a navigating mobile robot. The hope is that the development effort results in an architecture that can successfully map and navigate a variety of environments, thereby amortizing the up-front design cost over time.

Of course the key risk of the map-based approach is that an internal representation, rather than the real world itself, is being constructed and *trusted* by the robot. If that model diverges from reality (i.e., if the map is wrong), then the robot's behavior may be undesirable, even if the raw sensor values of the robot are only transiently incorrect.

In the remainder of this chapter, we focus on a discussion of map-based approaches and, specifically, the localization component of these techniques. These approaches are particularly appropriate for study given their significant recent successes in enabling mobile robots to navigate a variety of environments, from academic research buildings, to factory floors, and to museums around the world.

5.4   Belief Representation

The fundamental issue that differentiates various map-based localization systems is the issue of *representation*. There are two specific concepts that the robot must represent, and each has its own unique possible solutions. The robot must have a representation (a model) of the environment, or a map. What aspects of the environment are contained in this map? At what level of fidelity does the map represent the environment? These are the design questions for *map representation*.

The robot must also have a representation of its belief regarding its position on the map. Does the robot identify a single unique position as its current position, or does it describe its position in terms of a set of possible positions? If multiple possible positions are expressed in a single belief, how are those multiple positions ranked, if at all? These are the design questions for *belief representation*.

Decisions along these two design axes can result in varying levels of architectural complexity, computational complexity, and overall localization accuracy. We begin by discussing belief representation. The first major branch in a taxonomy of belief representation systems differentiates between single-hypothesis and multiple-hypothesis belief systems. The former covers solutions in which the robot postulates its unique position, whereas the latter enables a mobile robot to describe the degree to which it is uncertain about its position. A sampling of different belief and map representations is shown in figure 5.9.

5.4.1 Single-hypothesis belief
The single-hypothesis belief representation is the most direct possible postulation of mobile robot position. Given some environmental map, the robot's belief about position is

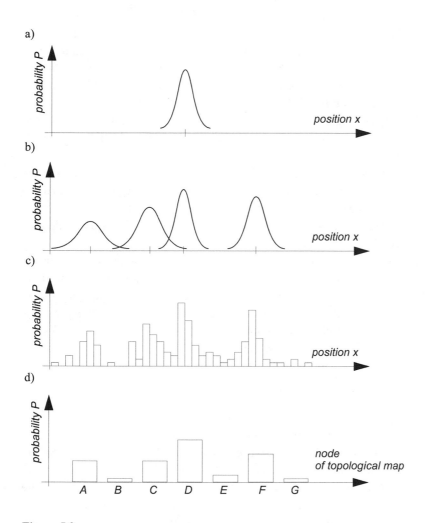

Figure 5.9
Belief representation regarding the robot position (1D) in continuous and discretized (tessellated) maps. (a) Continuous map with single-hypothesis belief, e.g., single Gaussian centered at a single continuous value. (b) Continuous map with multiple-hypothesis belief, e.g;. multiple Gaussians centered at multiple continuous values. (c) Discretized (decomposed) grid map with probability values for all possible robot positions, e.g., Markov approach. (d) Discretized topological map with probability value for all possible nodes (topological robot positions), e.g., Markov approach.

expressed as a single unique point on the map. In figure 5.10, three examples of a single-hypothesis belief are shown using three different map representations of the same actual environment (figure 5.10a). In figure 5.10b, a single point is geometrically annotated as the robot's position in a continuous 2D geometric map. In figure 5.10c, the map is a discrete, tessellated one, and the position is noted at the same level of fidelity as the map cell size. In figure 5.10d, the map is not geometric at all but abstract and topological. In this case, the single hypothesis of position involves identifying a single node i in the topological graph as the robot's position.

The principal advantage of the single-hypothesis representation of position stems from the fact that, given a unique belief, there is no position ambiguity. The unambiguous nature of this representation facilitates decision-making at the robot's cognitive level (e.g., path planning). The robot can simply assume that its belief is correct, and can then select its future actions based on its unique position.

Just as decision making is facilitated by a single-position hypothesis, so updating the robot's belief regarding position is also facilitated, since the single position must be updated by definition to a new, single position. The challenge with this position update approach, which ultimately is the principal disadvantage of single-hypothesis representation, is that robot motion often induces uncertainty due to effector and sensor noise. Therefore, forcing the position update process always to generate a *single* hypothesis of position is challenging and, often, impossible.

5.4.2 Multiple-hypothesis belief

In the case of multiple-hypothesis beliefs regarding position, the robot tracks not just a single possible position but also a possibly infinite set of positions.

In one simple example originating in the work of Jean-Claude Latombe [32, 188], the robot's position is described in terms of a convex polygon positioned in a 2D map of the environment. This multiple-hypothesis representation communicates the set of possible robot positions geometrically, with no preference ordering over the positions. Each point in the map is simply either contained by the polygon and, therefore, in the robot's belief set, or outside the polygon and thereby excluded. Mathematically, the position polygon serves to partition the space of possible robot positions. Such a polygonal representation of the multiple-hypothesis belief can apply to a continuous, geometric map of the environment [57] or, alternatively, to a tessellated, discrete approximation to the continuous environment.

It may be useful, however, to incorporate some ordering on the possible robot positions, capturing the fact that some robot positions are likelier than others. A strategy for representing a continuous multiple-hypothesis belief state along with a preference ordering over possible positions is to model the belief as a mathematical distribution. For example, [87] and [309] notate the robot's position belief using an $\{X, Y\}$ point in the 2D environment

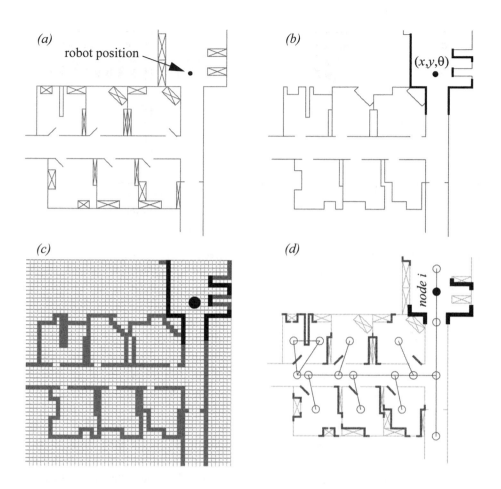

Figure 5.10
Three examples of single hypotheses of position using different map representations: (a) real map with walls, doors and furniture; (b) line-based map → around 100 lines with two parameters; (c) occupancy grid-based map → around 3000 grid cells size 50 × 50 cm; (d) topological map using line features (Z/S lines) and doors → around 50 features and 18 nodes.

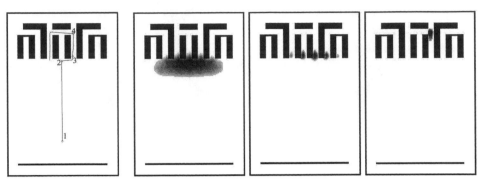

Path of the robot   *Belief states at positions 2, 3, and 4*

Figure 5.11
Example of multiple-hypothesis tracking (courtesy of W. Burgard [86]). The belief state that is largely distributed becomes very certain after moving to position 4. Note that darker coloring represents higher probability.

as the mean μ plus a standard deviation parameter σ, thereby defining a Gaussian distribution. The intended interpretation is that the distribution at each position represents the probability assigned to the robot being at that location. This representation is particularly amenable to mathematically defined tracking functions, such as the Kalman filter, that are designed to operate efficiently on Gaussian distributions.

An alternative is to represent the set of possible robot positions, not using a single Gaussian probability density function, but using discrete markers for each possible position. In this case, each possible robot position is individually noted along with a confidence or probability parameter (see figure 5.11). In the case of a highly tessellated map this can result in thousands or even tens of thousands of possible robot positions in a single-belief state.

The key advantage of the multiple-hypothesis representation is that the robot can explicitly maintain uncertainty regarding its position. If the robot only acquires partial information regarding position from its sensors and effectors, that information can conceptually be incorporated in an updated belief.

A more subtle advantage of this approach revolves around the robot's ability explicitly to measure its own degree of uncertainty regarding position. This advantage is the key to a class of localization and navigation solutions in which the robot not only reasons about reaching a particular goal but reasons about the future trajectory of its own belief state. For instance, a robot may choose paths that minimize its future position uncertainty. An example of this approach is [306], in which the robot plans a path from point A to point B that takes it near a series of landmarks in order to mitigate localization difficulties. This type of explicit reasoning about the effect that trajectories will have on the quality of localization requires a multiple-hypothesis representation.

One of the fundamental disadvantages of multiple-hypothesis approaches involves decision making. If the robot represents its position as a region or set of possible positions, then how shall it decide what to do next? Figure 5.11 provides an example. At position 3, the robot's belief state is distributed among five hallways separately. If the goal of the robot is to travel down one particular hallway, then given this belief state, what action should the robot choose?

The challenge occurs because some of the robot's possible positions imply a motion trajectory that is inconsistent with some of its other possible positions. One approach that we will see in the case studies that follow is to assume, for decision-making purposes, that the robot is physically at the most probable location in its belief state, then to choose a path based on that current position. But this approach demands that each possible position have an associated probability.

In general, the right approach to such decision-making problems would be to decide on trajectories that eliminate the ambiguity explicitly. But this leads us to the second major disadvantage of multiple-hypothesis approaches. In the most general case, they can be computationally very expensive. When one reasons in a 3D space of discrete possible positions, the number of possible belief states in the single-hypothesis case is limited to the number of possible positions in the 3D world. Consider this number to be N. When one moves to an arbitrary multiple-hypothesis representation, then the number of possible belief states is the power set of N, which is far larger: 2^N. Thus, explicit reasoning about the possible trajectory of the belief state over time quickly becomes computationally untenable as the size of the environment grows.

There are, however, specific forms of multiple-hypothesis representations that are somewhat more constrained, thereby avoiding the computational explosion while allowing a limited type of multiple-hypothesis belief. For example, if one assumes a Gaussian distribution of probability centered at a single position, then the problem of representation and tracking of belief becomes equivalent to Kalman filtering, a straightforward mathematical process described below. Alternatively, a highly tessellated map representation combined with a limit of ten possible positions in the belief state results in a discrete update cycle that is, at worst, only ten times more computationally expensive than a single-hypothesis belief update. And other ways to cope with the complexity problem, still being precise and computationally cheap, are hybrid metric-topological approaches [314, 317] or multi-Gaussian position estimation [57, 103, 157].

In conclusion, the most critical benefit of the multiple-hypothesis belief state is the ability to maintain a sense of position while explicitly annotating the robot's uncertainty about its own position. This powerful representation has enabled robots with limited sensory information to navigate robustly in an array of environments, as we shall see in the case studies that follow.

5.5   Map Representation

The problem of representing the environment in which the robot moves is a dual of the problem of representing the robot's possible position or positions. Decisions made regarding the environmental representation can have impact on the choices available for robot position representation. Often the fidelity of the position representation is bounded by the fidelity of the map.

Three fundamental relationships must be understood when choosing a particular map representation:

1. The precision of the map must appropriately match the precision with which the robot needs to achieve its goals.

2. The precision of the map and the type of features represented must match the precision and data types returned by the robot's sensors.

3. The complexity of the map representation has direct impact on the computational complexity of reasoning about mapping, localization, and navigation.

In the following sections, we identify and discuss critical design choices in creating a map representation. Each such choice has great impact on the relationships listed earlier and on the resulting robot localization architecture. As we shall see, the choice of possible map representations is broad. Selecting an appropriate representation requires understanding all of the trade-offs inherent in that choice as well as understanding the specific context in which a particular mobile robot implementation must perform localization. In general, the environmental representation and model can be roughly classified as presented in chapter 4, section 4.4.

5.5.1 Continuous representations

A continuous-valued map is one method for *exact* decomposition of the environment. The position of environmental features can be annotated precisely in continuous space. Mobile robot implementations to date use continuous maps only in 2D representations, as further dimensionality can result in computational explosion.

A common approach is to combine the exactness of a continuous representation with the compactness of the *closed-world assumption*. This means that one assumes that the representation will specify all environmental objects in the map, and that any area in the map that is devoid of objects has no objects in the corresponding portion of the environment. Thus, the total storage needed in the map is proportional to the density of objects in the environment, and a sparse environment can be represented by a low-memory map.

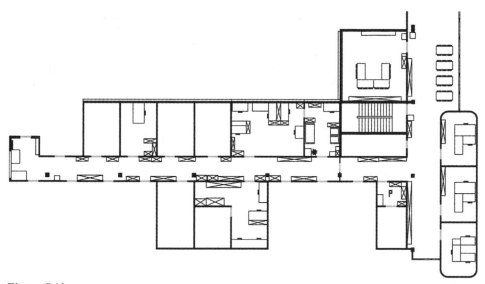

Figure 5.12
A continuous representation using polygons as environmental obstacles.

One example of such a representation, shown in figure 5.12, is a 2D representation in which polygons represent all obstacles in a continuous-valued coordinate space. This is similar to the method used by Latombe [32, 187] and others to represent environments for mobile robot path-planning techniques.

In the case of [32, 187], most of the experiments are in fact simulations run exclusively within the computer's memory. Therefore, no real effort would have been expended to attempt to use sets of polygons to describe a real-world environment, such as a park or office building.

In other work in which real environments must be captured by the maps, one sees a trend toward selectivity and abstraction. The human mapmaker tends to capture on the map, for localization purposes, only objects that can be detected by the robot's sensors and, furthermore, only a subset of the features of real-world objects.

It should be immediately apparent that geometric maps can capably represent the physical locations of objects without referring to their texture, color, elasticity, or any other such secondary features that do not relate directly to position and space. In addition to this level of simplification, a mobile robot map can further reduce memory usage by capturing only aspects of object geometry that are immediately relevant to localization. For example, all objects may be approximated using very simple convex polygons, sacrificing map felicity for the sake of computational speed.

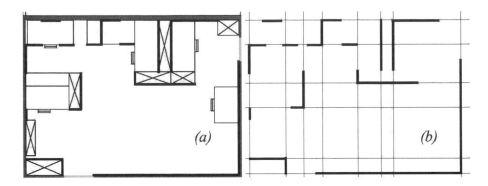

Figure 5.13
Example of a continuous-valued line representation of EPFL. (a) Real map. (b) Representation with
a set of infinite lines.

One excellent example involves line extraction. Many indoor mobile robots rely upon
laser rangefinding devices to recover distance readings to nearby objects. Such robots can
automatically extract best-fit lines from the dense range data provided by thousands of
points of laser strikes. Given such a line extraction sensor, an appropriate continuous mapping approach is to populate the map with a set of infinite lines. The continuous nature of
the map guarantees that lines can be positioned at arbitrary positions in the plane and at
arbitrary angles. The abstraction of real environmental objects such as walls and intersections captures only the information in the map representation that matches the type of information recovered by the mobile robot's rangefinding sensor.

Figure 5.13 shows a map of an indoor environment at EPFL using a continuous line representation. Note that the only environmental features captured by the map are straight
lines, such as those found at corners and along walls. This represents not only a sampling
of the real world of richer features but also a simplification, for an actual wall may have
texture and relief that is not captured by the mapped line.

The impact of continuous map representations on position representation is primarily
positive. In the case of single-hypothesis position representation, that position may be specified as any continuous-valued point in the coordinate space, and therefore extremely high
accuracy is possible. In the case of multiple-hypothesis position representation, the continuous map enables two types of multiple position representation.

In one case, the possible robot position may be depicted as a geometric shape in the
hyperplane, such that the robot is known to be within the bounds of that shape. This is
shown in figure 5.33, in which the position of the robot is depicted by an oval bounding
area.

Yet, the continuous representation does not disallow representation of position in the form of a discrete set of possible positions. For instance, in [119] the robot position belief state is captured by sampling nine continuous-valued positions from within a region near the robot's best-known position. This algorithm captures, within a continuous space, a discrete sampling of possible robot positions.

In summary, the key advantage of a continuous map representation is the potential for high accuracy and expressiveness with respect to the environmental configuration as well as the robot position within that environment. The danger of a continuous representation is that the map may be computationally costly. But this danger can be tempered by employing abstraction and capturing only the most relevant environmental features. Together with the use of the *closed-world assumption*, these techniques can enable a continuous-valued map to be no more costly, and sometimes even less costly, than a standard discrete representation.

5.5.2 Decomposition strategies

In the previous section, we discussed one method of simplification, in which the continuous map representation contains a set of infinite lines that approximate real-world environmental lines based on a 2D slice of the world. Basically this transformation from the real world to the map representation is a filter that removes all nonstraight data and furthermore extends line segment data into infinite lines that require fewer parameters.

A more dramatic form of simplification is *abstraction*: a general decomposition and selection of environmental features. In this section, we explore decomposition as applied in its more extreme forms to the question of map representation.

Why would one radically decompose the real environment during the design of a map representation? The immediate disadvantage of decomposition and abstraction is the loss of fidelity between the map and the real world. Both qualitatively, in terms of overall structure, and quantitatively, in terms of geometric precision, a highly abstract map does not compare favorably to a high-fidelity map.

Despite this disadvantage, decomposition and abstraction may be useful if the abstraction can be planned carefully so as to capture the relevant, *useful* features of the world while discarding all other features. The advantage of this approach is that the map representation can potentially be minimized. Furthermore, if the decomposition is hierarchical, such as in a pyramid of recursive abstraction, then reasoning and planning with respect to the map representation may be computationally far superior to planning in a fully detailed world model.

A standard, lossless form of *opportunistic decomposition* is termed *exact cell decomposition*. This method, introduced by Latombe [32], achieves decomposition by selecting boundaries between discrete cells based on geometric criticality.

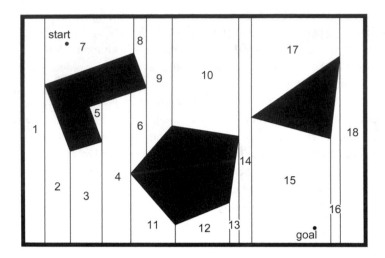

Figure 5.14
Example of exact cell decomposition.

Figure 5.14 depicts an exact decomposition of a planar workspace populated by polygonal obstacles. The map representation tessellates the space into areas of free space. The representation can be extremely compact because each such area is actually stored as a single node, resulting in a total of only eighteen nodes in this example.

The underlying assumption behind this decomposition is that the particular position of a robot within each area of free space does not matter. What matters is the robot's ability to traverse from each area of free space to the adjacent areas. Therefore, as with other representations we will see, the resulting graph captures the adjacency of map locales. If indeed the assumptions are valid and the robot does not care about its precise position within a single area, then this can be an effective representation that nonetheless captures the connectivity of the environment.

Such an exact decomposition is not always appropriate. Exact decomposition is a function of the particular environment obstacles and free space. If this information is expensive to collect or even unknown, then such an approach is not feasible.

An alternative is *fixed decomposition*, in which the world is tessellated, transforming the continuous real environment into a discrete approximation for the map. Such a transformation is demonstrated in figure 5.15, which depicts what happens to obstacle-filled and free areas during this transformation. The key disadvantage of this approach stems from its *inexact* nature. It is possible for narrow passageways to be lost during such a transformation, as shown in figure 5.15. Formally, this means that fixed decomposition is sound but not complete. Yet another approach is adaptive cell decomposition, as presented in figure 5.16.

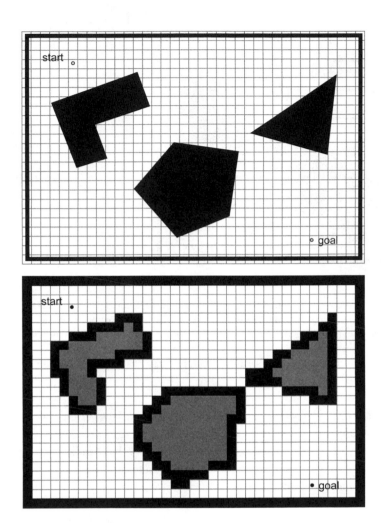

Figure 5.15
Fixed decomposition of the same space (narrow passage disappears).

The concept of fixed decomposition is extremely popular in mobile robotics; it is perhaps the single most common map representation technique currently utilized. One very popular version of fixed decomposition is known as the *occupancy grid* representation [233]. In an occupancy grid, the environment is represented by a discrete grid, where each cell is either filled (part of an obstacle) or empty (part of free space). This method is of particular value when a robot is equipped with range-based sensors because the range values

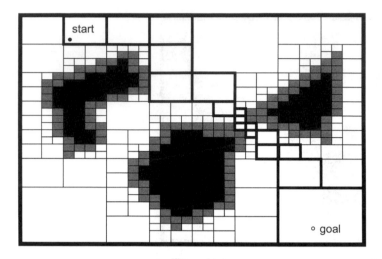

Figure 5.16
Example of adaptive (approximate variable-cell) decomposition of an environment [32]. The rectangle, bounding the free space, is decomposed into four identical rectangles. If the interior of a rectangle lies completely in free space or in the configuration space obstacle, it is not decomposed further. Otherwise, it is recursively decomposed into four rectangles until some predefined resolution is attained. The white cells lie outside the obstacles, the black are inside, and the gray are part of both regions.

of each sensor, combined with the absolute position of the robot, can be used directly to update the filled or empty value of each cell.

In the occupancy grid, each cell may have a counter, whereby the value 0 indicates that the cell has not been "hit" by any ranging measurements and, therefore, it is likely free space. As the number of ranging strikes increases, the cell's value is incremented and, above a certain threshold, the cell is deemed to be an obstacle. The values of cells are commonly discounted when a ranging strike travels *through* the cell, striking a further cell. By also discounting the values of cells over time, both hysteresis and the possibility of transient obstacles can be represented using this occupancy grid approach. Figure 5.17 depicts an occupancy grid representation in which the darkness of each cell is proportional to the value of its counter. One commercial robot that uses a standard occupancy grid for mapping and navigation is the Cye robot [342].

There remain two main disadvantages of the occupancy grid approach. First, the size of the map in robot memory grows with the size of the environment, and if a small cell size is used, this size can quickly become untenable. This occupancy grid approach is not compatible with the *closed-world assumption*, which enabled continuous representations to have potentially very small memory requirements in large, sparse environments. In contrast, the

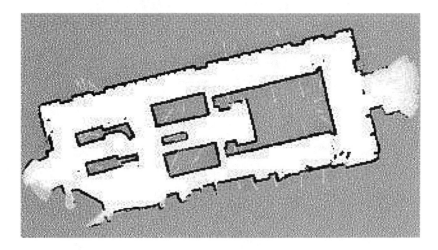

Figure 5.17
Example of an occupancy grid map representation. Courtesy of S. Thrun [314].

occupancy grid must have memory set aside for every cell in the matrix. Furthermore, any fixed decomposition method such as this imposes a geometric grid on the world *a priori*, regardless of the environmental details. This can be inappropriate in cases where geometry is not the most salient feature of the environment.

For these reasons, an alternative, called *topological* decomposition, has been the subject of some exploration in mobile robotics. Topological approaches avoid direct measurement of geometric environmental qualities, instead concentrating on characteristics of the environment that are most relevant to the robot for localization.

Formally, a topological representation is a graph that specifies two things: *nodes* and the *connectivity* between those nodes. Insofar as a topological representation is intended for the use of a mobile robot, nodes are used to denote areas in the world and arcs are used to denote adjacency of pairs of nodes. When an arc connects two nodes, then the robot can traverse from one node to the other without requiring traversal of any other intermediary node.

Adjacency is clearly at the heart of the topological approach, just as adjacency in a cell decomposition representation maps to geometric adjacency in the real world. However, the topological approach diverges in that the nodes are not of fixed size or even specifications of free space. Instead, nodes document an area based on any sensor discriminant such that the robot can recognize entry and exit of the node.

Figure 5.18 depicts a topological representation of a set of hallways and offices in an indoor environment. In this case, the robot is assumed to have an intersection detector, perhaps using sonar and vision to find intersections between halls and between halls and

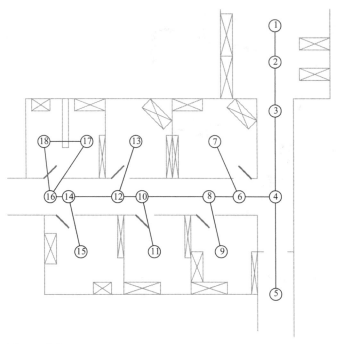

Figure 5.18
A topological representation of an indoor office area.

rooms. Note that nodes capture geometric space, and arcs in this representation simply represent connectivity.

Another example of topological representation is the work of Simhon and Dudek [290], in which the goal is to create a mobile robot that can capture the most interesting aspects of an area for human consumption. The nodes in their representation are visually striking locales rather than route intersections.

In order to navigate using a topological map robustly, a robot must satisfy two constraints. First, it must have a means for detecting its current position in terms of the nodes of the topological graph. Second, it must have a means for traveling between nodes using robot motion. The node sizes and particular dimensions must be optimized to match the sensory discrimination of the mobile robot hardware. This ability to "tune" the representation to the robot's particular sensors can be an important advantage of the topological approach. However, as the map representation drifts further away from true geometry, the expressiveness of the representation for accurately and precisely describing a robot position is lost. Therein lies the compromise between the discrete cell-based map representations and the topological representations. Interestingly, the continuous map representation has

Figure 5.19
An artificial landmark used by Chips during autonomous docking.

the potential to be both compact, like a topological representation, and precise, as with all direct geometric representations.

Yet, a chief motivation of the topological approach is that the environment may contain important nongeometric features—features that have no ranging relevance but are useful for localization. In chapter 4 we described such whole-image vision-based features.

In contrast to these whole-image feature extractors, often spatially localized landmarks are artificially placed in an environment to impose a particular visual-topological connectivity upon the environment. In effect, the artificial landmark can impose artificial structure. Examples of working systems operating with this landmark-based strategy have also demonstrated success. Latombe's landmark-based navigation research [188] has been implemented on real-world indoor mobile robots that employ paper landmarks attached to the ceiling as the locally observable features. Chips, the museum robot, is another robot that uses man-made landmarks to obviate the localization problem. In this case, a bright pink square serves as a landmark with dimensions and color signature that would be hard to accidentally reproduce in a museum environment [251]. One such museum landmark is shown in figure 5.19.

In summary, range is clearly not the only measurable and useful environmental value for a mobile robot. This is particularly true with the advent of color vision, as well as laser rangefinding, which provides reflectance information in addition to range information. Choosing a map representation for a particular mobile robot requires, first, understanding the sensors available on the mobile robot, and, second, understanding the mobile robot's functional requirements (e.g., required goal precision and accuracy).

5.5.3 State of the art: Current challenges in map representation

The previous sections describe major design decisions in regard to map representation choices. There are, however, fundamental real-world features that mobile robot map representations do not yet represent well. These continue to be the subject of open research, and several such challenges are described here.

The real world is dynamic. As mobile robots come to inhabit the same spaces as humans, they will encounter moving people, cars, strollers, and the transient obstacles placed and moved by humans as they go about their activities. This is particularly true when one considers the home environment with which domestic robots will someday need to contend.

The map representations described earlier do not, in general, have explicit facilities for identifying and distinguishing between permanent obstacles (e.g., walls, doorways, etc.) and transient obstacles (e.g., humans, shipping packages, etc.). The current state of the art in terms of mobile robot sensors is partly to blame for this shortcoming. Although vision research is rapidly advancing, robust sensors that discriminate between moving animals and static structures *from a moving reference frame* are not yet available. Furthermore, estimating the motion vector of transient objects remains a research problem.

Usually, the assumption behind the preceding map representations is that all objects on the map are effectively static. Partial success can be achieved by discounting mapped objects over time. For example, occupancy grid techniques can be more robust to dynamic settings by introducing temporal discounting, effectively treating transient obstacles as noise. The more challenging process of map creation is particularly fragile to environmental dynamics; most mapping techniques generally require that the environment be free of moving objects during the mapping process. One exception to this limitation involves topological representations. Because precise geometry is not important, transient objects have little effect on the mapping or localization process, subject to the critical constraint that the transient objects must not change the topological connectivity of the environment. Still, neither the occupancy grid representation nor a topological approach is actively recognizing and representing transient objects as distinct from both sensor error and permanent map features.

As vision sensing provides more robust and more informative content regarding the transience and motion details of objects in the world, mobile roboticists will in time propose representations that make use of that information. A classic example involves occlusion by human crowds. Museum tour guide robots generally suffer from an extreme amount of occlusion. If the robot's sensing suite is located along the robot's body, then the robot is effectively blind when a group of human visitors completely surround the robot. This is because its map contains only environmental features that are, at that point, fully hidden from the robot's sensors by the wall of people. In the best case, the robot should recognize its occlusion and make no effort to localize using these invalid sensor readings. In the worst case, the robot will localize with the fully occluded data, and will update its location incorrectly. A vision sensor that can discriminate the local conditions of the robot (e.g,. we are surrounded by people) can help eliminate this error mode.

A second open challenge in mobile robot localization involves the traversal of open spaces. Existing localization techniques generally depend on local measures such as range, thereby demanding environments that are somewhat densely filled with objects that the sensors can detect and measure. Wide-open spaces such as parking lots, fields of grass, and indoor atriums such as those found in convention centers pose a difficulty for such systems because of their relative sparseness. Indeed, when populated with humans, the challenge is exacerbated because any mapped objects are almost certain to be occluded from view by the people.

Once again, more recent technologies provide some hope of overcoming these limitations. Both vision and state-of-the-art laser rangefinding devices offer outdoor performance with ranges of up to a hundred meters and more. Of course, GPS performs even better. Such long-range sensing may be required for robots to localize using distant features.

This trend teases out a hidden assumption underlying most topological map representations. Usually, topological representations make assumptions regarding spatial locality: a node contains objects and features that are themselves within that node. The process of map creation thus involves making nodes that are, in their own self-contained way, recognizable by virtue of the objects contained within the node. Therefore, in an indoor environment, each room can be a separate node, and this is reasonable because each room will have a layout and a set of belongings that are unique to that room.

However, consider the outdoor world of a wide-open park. Where should a single node end and the next node begin? The answer is unclear because objects that are far away from the current node, or position, can yield information for the localization process. For example, the hump of a hill at the horizon, the position of a river in the valley, and the trajectory of the sun all are nonlocal features that have great bearing on one's ability to infer current position. The spatial locality assumption is violated and, instead, replaced by a visibility criterion: the node or cell may need a mechanism for representing objects that are measurable and visible from that cell. Once again, as sensors improve and, in this case, as outdoor locomotion mechanisms improve, there will be greater urgency to solve problems associated with localization in wide-open settings, with and without GPS-type global localization sensors.

We end this section with one final open challenge that represents one of the fundamental academic research questions of robotics: sensor fusion. A variety of measurement types are possible using off-the-shelf robot sensors, including heat, range, acoustic and light-based reflectivity, color, texture, friction, and so on. Sensor fusion is a research topic closely related to map representation. Just as a map must embody an environment in sufficient detail for a robot to perform localization and reasoning, sensor fusion demands a representation of the world that is sufficiently general and expressive that a variety of sensor types can have their data correlated appropriately, strengthening the resulting percepts well beyond that of any individual sensor's readings.

Perhaps the only general implementation of sensor fusion to date is that of neural network classifier. Using this technique, any number and any type of sensor values may be jointly combined in a network that will use whatever means necessary to optimize its classification accuracy. For the mobile robot that must use a human-readable internal map representation, no equally general sensor fusion scheme has yet been born. It is reasonable to expect that, when the sensor fusion problem is solved, integration of a large number of disparate sensor types may easily result in sufficient discriminatory power for robots to achieve real-world navigation, even in wide-open and dynamic circumstances such as a public square filled with people.

5.6 Probabilistic Map-Based Localization

5.6.1 Introduction

As stated earlier, multiple-hypothesis position representation is advantageous because the robot can explicitly track its own beliefs regarding its possible positions in the environment. Ideally, the robot's *belief state* will change, over time, as is consistent with its motor outputs and perceptual inputs. One geometric approach to multiple-hypothesis representation, mentioned earlier, involves identifying the possible positions of the robot by specifying a polygon in the environmental representation [187]. This method does not provide any indication of the relative chances between various possible robot positions.

Probabilistic map-based localization techniques differ from this because they explicitly identify probabilities with the possible robot positions, and for this reason these methods have been the focus of recent research. The reason probabilistic approaches to mobile robot localization have been developed is that the data coming from the robot sensors are affected by measurement errors, and therefore we can only compute the probability that the robot is in a given configuration. This new area of research is called *probabilistic robotics* [51]. The key idea in probabilistic robotics is to represent uncertainty using probability theory. Stating this in different words, instead of giving a single best estimate of the current robot configuration, probabilistic robotics represents the robot configuration as a probability distribution over the all possible robot poses. By doing so, ambiguity and degree of belief are represented using calculus of probability theory. The success of this theory applied to the mobile robot localization problem comes from the fact that probabilistic algorithms outperform alternative techniques in many real world applications.

In the following sections, we present two classes of probabilistic map-based localization. The first class, *Markov localization*, uses an explicitly specified probability distribution across all possible robot positions. The second method, *Kalman filter localization*, uses a Gaussian probability density representation of robot position. Unlike Markov localization, Kalman filter localization does not independently consider each possible pose in the robot's configuration space. Interestingly, the Kalman filter localization process results from the Markov localization axioms if the robot's position uncertainty is assumed to have a Gaussian form.

5.6.2 The robot localization problem

Before discussing each method in detail, we present the general robot localization problem and solution strategy. Consider a mobile robot moving in a known environment. As it starts to move, say from a precisely known location, it can keep track of its motion using odometry. Due to odometry uncertainty, after some movement the robot will become very uncertain about its position (see section 5.2.4). To keep position uncertainty from growing unbounded, the robot must localize itself in relation to its environment map. To localize, the robot might use its on-board exteroceptive sensors (e.g., ultrasonic, laser, vision sensors) to make observations of its environment. The information provided by the robot's odometry, plus the information provided by such exteroceptive observations, can be combined to enable the robot to localize as well as possible with respect to its map. The processes of updating based on proprioceptive sensor values and exteroceptive sensor values are often separated logically, leading to a general process for the robot position update that comprises two steps, which are called *prediction* (or *action*) *update* and *perception* (or *measurement*, or *correction*) *update*.

- During the *prediction* (or *action*) *update* the robot uses its proprioceptive sensors to estimate its configuration; for example, the robot estimates its motion using the encoders. In this phase, the uncertainty about the robot configuration increases due to the integration of the odometric error over time. In figure 5.20a, we illustrate this process for a robot moving in a one-dimensional environment.

- During the *perception* (or *measurement*, or *correction*) *update* the robot uses the information from its exteroceptive sensors to correct the position estimated during the prediction phase; for example, the robot uses a rangefinder to measure its current distance from a wall and corrects accordingly the position estimated during the prediction phase. During the perception phase, the uncertainty of the robot configuration shrinks (figure 5.20b).

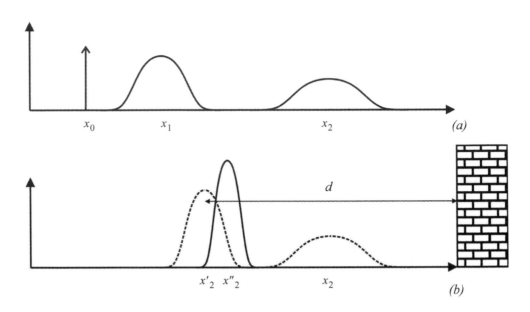

Figure 5.20 In probabilistic robotics, the *beliefs* about the robot configuration are represented as probability density functions. Note, in general the distributions can be any function. Not necessarily Gaussian. Only the Kalman filter assumes Gaussian distributions. (a) Prediction phase: the start position x_0 is assumed to be known, and therefore the probability density function is a *Dirac delta* function. As the robot starts moving, its uncertainty grows due to the odometric error, which accumulates over time. (b) Perception phase: the robot uses its exteroceptive sensor (e.g., a rangefinder) to measure the distance d from the right wall, and computes the position x'_2, which is in conflict with the current position x_2 estimated in the action phase. The perception update corrects the new position to x''_2 and as consequence its uncertainty shrinks (solid line).

In general, the prediction update contributes to increase the uncertainty to the robot's belief about position: encoders have error, and therefore motion is somewhat nondeterministic. By contrast, perception update generally refines the belief state (i.e., the uncertainty shrinks). Sensor measurements, when compared to the robot's environmental model, tend to provide clues regarding the robot's possible position.

In the next sections, we will describe two different methods of probabilistic map-based localization, which are the Markov localization and Kalman filter localization. In the case of Markov localization, the robot's belief state can be represented with any *arbitrary* probability density function. The prediction and perception phases update the probability of every possible robot pose.

In the case of Kalman filter localization, the robot's belief state is conversely represented using a single *Gaussian* probability density function (page 112), and thus it retains just a μ and σ parameterization of the robot's belief about position with respect to the map. Updating the parameters of the Gaussian distribution is all that is required. This fundamental difference in the representation of belief state leads to the following advantages and disadvantages of the two methods, as presented in [143]:

- Markov localization allows for localization starting from *any unknown* position and can thus recover from ambiguous situations because the robot can track multiple, completely disparate possible positions. However, to update the probability of all positions within the whole state space at any time requires a discrete representation of the space, such as a geometric grid or a topological graph (see section 5.5.2). The required memory and computational power can thus limit precision and map size.

- Kalman filter localization tracks the robot from an *initially known* position and is inherently both precise and efficient. In particular, Kalman filter localization can be used in continuous world representations. However, if the uncertainty of the robot becomes too large (e.g., due to a robot collision with an object) and thus not truly unimodal, the Kalman filter can fail to capture the multitude of possible robot positions and can become irrevocably lost.

In recent research projects, improvements are achieved or proposed by either only updating the state space of interest within the Markov approach [86] or by tracking multiple hypotheses with Kalman filters [57, 231], or by combining both methods to create a hybrid localization system [143, 317]. In the next sections we will review each approach in detail but first we will recall some concept of probability theory. For an in-depth study of probability theory, we refer the reader to [41].

5.6.3 Basic concepts of probability theory

Let X denote a random variable and x a specific value that X might assume. A typical example is dice rolling, where X can take on any value between 1 and 6. We denote with

$$p(X = x) \tag{5.16}$$

the probability that the random variable X has value x. For example, the result of a die roll is characterized by

$$p(X = 1) = p(X = 2) = p(X = 3) = p(X = 4) = p(X = 5) = p(X = 6) = \frac{1}{6}. \tag{5.17}$$

From now on, to simplify the notation we will omit explicit mention of the random variable, and will instead use the simple abbreviation $p(x)$.

In continuous spaces, random variables can take on a continuum of values and in this case we will talk about *probability density functions* (PDFs).

Both discrete and continuous probabilities integrate to one, therefore:

$$\sum_{x} p(x) = 1 \text{ (for discrete probabilities)}, \tag{5.18}$$

$$\int_{x} p(x)dx = 1 \text{ (for continuous probabilities)}. \tag{5.19}$$

Furthermore, probabilities are always non-negative, that is, $p(x) \geq 0$.

Gaussian distribution. As we have seen (page 112), a common probability density function is the Gaussian distribution

$$p(x) = \frac{1}{\sigma\sqrt{2\pi}} \exp\left(-\frac{(x-\mu)^2}{2\sigma^2}\right), \tag{5.20}$$

also called normal distribution, which is commonly abbreviated with

$$p(x) = N(x, \sigma^2) \tag{5.21}$$

where μ and σ^2 specify the mean and the variance of the random variable x. Note that in this case the random variable x is a scalar. However, when x is a k-dimensional vector, we will have a multivariate normal distribution characterized by a density function of the following form:

$$p(x) = \frac{1}{(2\pi)^{k/2} det(\Sigma)^{1/2}} \exp\left(-\frac{1}{2}(x-\mu)^T \Sigma^{-1}(x-\mu)\right), \tag{5.22}$$

where μ is the mean vector and Σ is a *positive semidefinite* and *symmetric* matrix called *covariance matrix*.

Joint distribution. The *joint distribution* of two random variables X and Y is given by $p(x, y)$, which describes the probability that the random variable X takes on the value x *and* that Y takes on the value y. If X and Y are *independent*, we have

$$p(x, y) = p(x)p(y) \tag{5.23}$$

Conditional probability. The *conditional probability* describes the probability that the random variable X takes on the value x *conditioned* on the knowledge that for sure Y takes on the value y. As an example, to compute the probability that the result of a die roll is 2 conditioned on the fact that we know with a probability of 100% that the result will be an even number (for example, we have a special die). Conditional probability is denoted with $p(x|y)$ and, if $p(y) > 0$, is defined as

$$p(x|y) = \frac{p(x, y)}{p(y)}.$$

$$(5.24)$$

If X and Y are independent, we have

$$p(x|y) = \frac{p(x)p(y)}{p(y)} = p(x).$$

$$(5.25)$$

In other words, if X and Y are independent, the knowledge of Y does not provide any useful information about the value of X.

Theorem of total probability. The *theorem of total probability* originates from the axioms of probability theory and is written as:

$$p(x) = \sum_y p(x|y)p(y) \text{ (for discrete probabilities),}$$

$$(5.26)$$

$$p(x) = \int_y p(x|y)p(y)dy \text{ (for continuous probabilities).}$$

$$(5.27)$$

As we will see in the next section, the theorem of total probability is used by both Markov and Kalman-filter localization algorithms during the prediction update.

Bayes rule. The *Bayes rule* relates the conditional probability $p(x|y)$ to its inverse $p(y|x)$. Under the condition that $p(y) > 0$, the Bayes rule is written as:

$$p(x|y) = \frac{p(y|x)p(x)}{p(y)}.$$

$$(5.28)$$

As we will see in the next section, the Bayes rule is used by both Markov and Kalman-filter localization algorithms during the measurement update.

Prior and posterior probability. A prior probability distribution of a random variable x is the probability distribution $p(x)$ that we have before incorporating the data y. For example, in mobile robotics the prior could be the probability distribution of the robot location over the whole space *prior* taking into account any sensor measurement. The probability $p(x|y)$, which is computed *after* incorporating the data is referred to as *posterior probability distribution*. As shown by equation (5.28), the Bayes rule provides a convenient way to compute the posterior probability $p(x|y)$ using the "inverse" conditional probability $p(y|x)$ and the prior probability $p(x)$. Using the preceding example from mobile robotics, if we want to know the probability $p(x|y)$ of the robot of occupying a specific position x after reading the sensor data y, we just need to multiply the conditional probability $p(y|x)$ of observing those measurements if the robot was at that position by the probability $p(x)$ of the robot to be there before reading the sensor. The result will then have to be divided by a certain normalization factor $p(y)$. The concept of prior and posterior probability and the benefit of the Bayes rule will be anyway clarified later on.

It is important to notice that the denominator of the Bayes rule, $p(y)$, does not depend on x. For this reason, in the Bayes rule the factor $p(y)^{-1}$ is usually written as a *normalization* factor, generically denoted η, which can be determined straightforwardly by remembering that the integral of a probability distribution is always 1. This way, the Bayes rule can be written as:

$$p(x|y) = \eta p(y|x)p(x).$$

(5.29)

5.6.4 Terminology

Path, input, and observations. Here we introduce the terminology that will be used throughout the next sections. The terminology and the notation are the same as introduced in [51]. Let t denote the time and x_t denote the robot location. For planar motion, $x_t = [x, y, \theta]^T$ is a three-dimensional vector consisting of the robot position and orientation. The robot *path* is given as

$$X_T = \{x_0, x_1, x_2, ..., x_T\},$$

(5.30)

where T could also be infinite.

Let u_t denote the proprioceptive sensor readings at time t. This can be, for instance, the reading of the robot's wheel encoders or IMU, or the control input given to the motors (e.g.

the speed).[23] If we assume that the robot receives exactly one data at each point in time, the sequence of proprioceptive data can be written as:

$$U_T = \{u_0, u_1, u_2, ..., u_T\} \tag{5.31}$$

In the absence of noise, we could obviously recover all the past robot locations X_T by integrating over U_T provided that the robot initial location x_0 is known. However, because of noise, the integrated path unavoidably drifts from the ground truth and therefore additional means are needed to reduce this drift. As we will see in the next sections, the exteroceptive sensor readings allow us to cancel the drift by keeping it bounded.

Exteroceptive sensors such as cameras, lasers, or ultrasonic rangefinders allow the robot to perceive the environment. The output of these sensors could be directly the 3D coordinates (in meters) of points, lines, or planes in the *sensor reference frame*. But they could also be simply the coordinates (in pixels) of point features (like corners) or lines (like doors). Whatever are the outputs returned by the sensor, they are typically referred to as *observations, measurement data,* or *exteroceptive sensor readings*. If we assume that the robot takes exactly one measurement at each point in time, the sequence of measurements is given as

$$Z_T = \{z_0, z_1, z_2, ..., z_T\} \ , \tag{5.32}$$

where it is important to remark that the coordinates of each observation are expressed in the *sensor reference frame* attached to the robot.

Finally, let M denote the *true* map of the environment and let us assume that the environment consists of two-dimensional point landmarks (or 2D infinite lines). In this case, the map is a vector of size $2n$, where n is the number of features in the world; therefore,

$$M = \{m_0, m_1, m_2, ..., m_{n-1}\} \ , \tag{5.33}$$

where m_i, $i = 0...n-1$, are vectors representing the 2D coordinates of the landmarks in the *world reference frame* (e.g., the 2D position of the point or the position and orientation of the line). The same discussion can be extended to the 3D case by incorporating the position of points, or location and orientation of lines and planes. Finally, we assume that the world is static, and therefore the environment map M is time-invariant.

---

23. To facilitate the language, in the remainder we will always refer to u_t as the *control input* but keep in mind that in general it can also represent the proprioceptive sensor readings.

Belief distributions. In section 5.4, we already introduced the concept of *belief represen-tation*. Here we review this concept in terms of probability. In general, a robot cannot mea-sure its *true* state (pose) directly, not even with GPS. What it can know is only the best estimate of its pose (for example $x_t = [3.0, 5.1, 180°]^T$) based on its sensors' data. There-fore, the knowledge the robot has about its state can only be inferred by the data. The best guess about the robot state is called *belief*. In probabilistic robotics, beliefs are represented through conditional probability distributions, and therefore they are posterior probabilities of the state variables conditioned on the available data. If we denote the *belief* over a state variable x_t by $bel(x_t)$, we can write

$$bel(x_t) = p(x_t | z_{1 \to t}, u_{1 \to t}),$$
(5.34)

where the posterior $p(x_t | z_{1 \to t}, u_{1 \to t})$ represents the probability of the robot of being at x_t given all its past observations $z_{1 \to t}$ and all its past control inputs $u_{1 \to t}$.

In Markov and Kalman localization, we will also often refer to the belief calculated *before* incorporating the new observation z_t just after the control input u_t. Such a posterior will be denoted as:

$$\overline{bel}(x_t) = p(x_t | z_{1 \to t-1}, u_{1 \to t}).$$
(5.35)

This probability distribution $\overline{bel}(x_t)$, which is computed just before including the new observation z_t, is often called *prediction* (or *action*) update, meaning that the current robot pose (belief) is only *predicted* on the basis of the motion control and the previous observa-tions. It is also called *action* because during this phase the robot physically moves. Con-versely, the calculation of $bel(x_t)$ from $\overline{bel}(x_t)$ is often called *correction* (or *perception*, or *measurement*) update because the robot pose is corrected after the observation.

5.6.5 The ingredients of probabilistic map-based localization
In order to solve the robot localization problem, the following information is required.

1. The initial probability distribution $bel(x_0)$. In the case the initial robot location is unknown, the initial belief $bel(x_0)$ is a uniform distribution over all poses. Conversely, if the location is perfectly known the initial belief is a *Dirac delta* function. As we will see, the Markov approach allows us to select any arbitrary initial distribution, while in the Kalman filter approach only a Gaussian distribution is allowed.

2. Map of the environment. The environment map $M = \{m_0, m_1, m_2, ..., m_n\}$ must be known. If the map is not known *a priori*, then the robot needs to build a map of the envi-ronment. Automatic map building will be covered in section 5.8.

3. Data. For localizing, the robot obviously needs to use data from its proprioceptive and exteroceptive sensors. We denote with z_t the current reading from the exteroceptive sensor. z_t is also called the observation. With u_t, we denote instead the reading from the proprioceptive sensor or the control input. For a differential drive robot, u_t could, for example, represent the encoder readings of the right and left wheel, and, therefore, we would write $u_t = [\Delta S_r, \Delta S_l]^T$.

4. The probabilistic motion model. The probabilistic motion model is derived from the kinematics of the robot. In the noise-free case, the robot current location x_t can be computed as a function f of the previous location x_{t-1} and the encoder readings u_t, that is:

$$x_t = f(x_{t-1}, u_t). \tag{5.36}$$

For example, for a differential drive robot f is simply the odometric-position-update formula (5.7).

To derive the probabilistic motion model, we need to model the error distributions over x_{t-1} and u_t and then use f to compute the error distribution over x_t. As we have seen in section 5.2.4, if we assume that both x_{t-1} and u_t are normally distributed, uncorrelated, and that f can be approximated by its first-order Taylor expansion, then the error distribution over x_t can be modeled as a multivariate Gaussian with mean value $f(x_{t-1}, u_t)$ and covariance matrix specified by the error propagation law (5.9).

5. The probabilistic measurement model. This is derived directly from the exteroceptive sensor model, e.g., the error model of the laser, the sonar, or the camera. As we have seen in chapter 4, laser and sonars provide range measurements, while a single camera provides bearing measurements. Because these measurements are always noisy, in order to characterize the sensor model we have to define the exact, noise-free measurement function. The measurement function h clearly depends on the environment map M and on the robot location x_t, therefore we can write:

$$z_t = h(x_t, M). \tag{5.37}$$

The measurement function h is typically a change of coordinates from the world frame to the sensor reference frame attached to the robot. For instance, in the example shown in figure 5.20 the robot uses a rangefinder to measure its distance d from the right wall. Here, d is the observation, and therefore $z_t = d$. The map M is represented by a single feature m (i.e., the wall), which we assume is at coordinate $m = 10$ (we assume one coordinate because we suppose that the robot moves in a one-dimensional environment). The measurement function in this simple example is therefore:

$$h(x_t, M) = 10 - x_t.$$ (5.38)

In the more general case, h is a change of coordinates from the world frame to the sensor reference frame attached to the robot.

To derive the probabilistic measurement model, we just need to add a noise term to the measurement function such that the probability distribution $p(z_t|x_t, M)$ peaks at the noise-free value $h(x_t, M)$. For example, if we assume Gaussian noise we can write

$$p(z_t|x_t, M) = N(h(x_t, M), R_t),$$ (5.39)

where, more generally, N denotes a multivariate normal distribution with mean value $h(x_t, M)$ and noise covariance matrix R_t.

5.6.6 Classification of localization problems

Before proceeding with Markov and Kalman filter localization, we have to understand the difference among three types of localization problems, which are: position tracking, global localization, and the kidnapped robot problem.

Position tracking. In position tracking, the robot current location is updated based on the knowledge of its previous position (tracking). This implies that the robot initial location is supposed to be known. Additionally, the uncertainty on the robot pose has to be small. If the uncertainty is too large, position tracking might fail to localize the robot. This concept will be investigated more in detail in Kalman filter localization, as in position tracking the robot belief is usually modeled with a unimodal distribution, such as the normal distribution.

Global localization. Global localization, conversely, assumes that the robot initial location is unknown. This means that the robot can be placed anywhere in the environment—without knowledge about it—and is able to localize globally within it. In global localization, the robot initial belief is usually a uniform distribution.

Kidnapped robot problem. The kidnapped robot problem tackles the case the robot gets kidnapped and moved to another location. The kidnapped robot problem is similar to the global localization problem only if the robot realizes having been kidnapped. The difficulty arises when the robot does not know that it has been moved to another location and it believes it knows where it is but in fact does not. The ability to recover from kidnapping is a necessary condition for the operation of any autonomous robot and even more for commercial robots.

5.6.7 Markov localization

Markov localization tracks the robot's belief state using an arbitrary probability density function to represent the robot's position (see also [87, 169, 249, 252]). In practice, all known Markov localization systems implement this generic belief representation by first tessellating the robot configuration space (x, y, θ) into a finite, discrete number of possible robot poses in the map. In actual applications, the number of possible poses can range from several hundred to millions of positions and orientations.

Markov localization addresses the *global localization problem*, the *position tracking problem*, and the *kidnapped robot problem*.

As we mentioned in section 5.6.2, the probabilistic robot localization process consists in the iteration of *prediction* and *measurement* updates. They compute the belief state that results when new information (e.g., encoder values and measurement data) is incorporated into a prior belief state with arbitrary probability density. As we will see, in both Markov and Kalman-filter localization the prediction update is based on the *theorem of total probability* (page 301), while the perception update is based on the *Bayes rule* (page 301).

In the next sections, we will explain separately these two steps for the case of Markov localization. We will first illustrate Markov localization in the continuous case and then in the discrete case (geometric grid-based). Finally, we will show an example of Markov localization using a topological map.

5.6.7.1 Prediction and measurement updates

Prediction (action) update. Let us recall that in this phase the robot estimates its current position (i.e. belief) based on the knowledge of the previous position (i.e., belief) and the odometric input. The *theorem of total probability* (page 301) is used to compute the robot's current belief $\overline{bel}(x_t)$ as a function of the previous belief $bel(x_{t-1})$ and the proprioceptive data (e.g., the encoder measurement or the control input) u_t :

$$\overline{bel}(x_t) = \int p(x_t | u_t, x_{t-1}) bel(x_{t-1}) dx_{t-1} \quad \text{(continuous case)}, \tag{5.40}$$

$$\overline{bel}(x_t) = \sum_{x_{t-1}} p(x_t | u_t, x_{t-1}) bel(x_{t-1}) \qquad \text{(discrete case)}. \tag{5.41}$$

As observed, the belief $\overline{bel}(x_t)$ that the robot assigns to the state x_t is obtained by the integral (or sum) of the product of two distributions: the prior assigned to x_{t-1}, and the probability that the control u_t induces a transition from x_{t-1} to x_t.

Let us try to clarify the reason of this integral (sum). In order to compute the probability of position x_t in the new belief state, one must integrate over all the possible ways in which the robot may have reached x_t according to the potential positions expressed in the former belief state. This is subtle but fundamentally important. The same location x_t can be reached from multiple source locations with the same encoder measurement u_t because the encoder measurement is uncertain.

Also observe that the integral (sum) in (5.40) and (5.41) must be computed over all possible robot positions x_t. This means that in real situations, where the number of cells used to represent the robot poses is several million, the computation (5.40) and (5.41) can become impractical and, hence, impede the real-time operation.

Finally, observe that (5.40) and (5.41) can be seen as a *convolution* (page 197, equation [4.102]) between the previous belief $bel(x_{t-1})$ and the probabilistic motion model $p(x_t|u_t, x_{t-1})$.[24] By thinking in terms of convolution, the reader should now have clear the reason why the prediction update causes the uncertainty of the robot location to grow (figure 5.20).

Perception (measurement) update. Let us recall that in this phase the robot corrects its previous position (i.e. its former belief) by opportunely combining it with the information from its exteroceptive sensors (figure 5.20). The *Bayes rule* (page 301) is used to compute the robot's new belief state $bel(x_t)$ as a function of its measurement data z_t and its former belief state $\overline{bel}(x_t)$:

$$bel(x_t) = \eta p(z_t|x_t, M)\overline{bel}(x_t), \tag{5.42}$$

where $p(z_t|x_t, M)$ is the probabilistic measurement model (page 305), that is, the probability of observing the measurement data z_t given the knowledge of the map M and the robot pose x_t. Therefore, the new belief state is simply the product between the probabilistic measurement model and the previous belief state. Observe that (5.42) does not update only one pose but all possible robot poses x_t.

The Markov localization algorithm. Figure 5.21 depicts the Markov localization algorithm in pseudo-algorithmic form.[25]

The critical challenge in Markov localization is the calculation of $p(z_t|x_t, M)$. The sensor model must calculate the probability of a specific perceptual measurement given the location of the robot and the map of the environment. Three key assumptions are used to construct this sensor model:

---

24. Note that (5.40) is not a real convolution because here the sign of none of the two functions is inverted.
25. Note that, since both Markov and Kalman localization make use of the Bayes rule, they are also called Bayes filters.

```
for all  x_t  do
```
$$\overline{bel}(x_t) = \int p(x_t | u_t, x_{t-1}) bel(x_{t-1}) dx_{t-1} \quad \text{(prediction update)}$$
$$bel(x_t) = \eta p(z_t | x_t, M) \overline{bel}(x_t) \qquad \text{(measurement update)}$$
```
endfor
return  bel(x_t)
```

Figure 5.21 The general algorithm for Markov localization.

1. If an object in the map is detected by, for instance, a range sensor, the measurement error can be described with a distribution that has a mean at the correct reading. The adopted distribution is usually a Gaussian.

2. There should always be a nonzero chance that a range sensor will read any measurement value, even if this measurement disagrees sharply with the environmental geometry. This means that the adopted distribution should always be nonzero in the range of all possible values returnable by the sensors. The peak should be centered at the correct sensor reading and the probability should be set to small values elsewhere. Again, a Gaussian distribution implicitly solves this problem.

3. In contrast to the generic error described in at the previous point, there is a specific failure mode in ranging sensors whereby the signal is absorbed or coherently reflected, causing the sensor's range measurement to be maximal. Therefore, there is a local peak in the probability density distribution at the maximal reading of a range sensor.

5.6.7.2 The Markov assumption

Equations (5.40) and (5.42) form the basis of Markov localization and incorporate the *Markov assumption*. Formally, this means that their output x_t is a function only of the robot's previous state x_{t-1} and its most recent actions (odometry) u_t and perception z_t. In a general, *non-Markovian* situation, the state of a system depends upon all of its history. After all, the values of a robot's sensors at time t do not really depend only on its position at time t. They depend to some degree on the trajectory of the robot over time, indeed, on the entire history of the robot. For example, the robot could have experienced a serious collision recently that has biased the sensor's behavior. Similarly, the position of the robot at time t does not really depend only on its position at time $t-1$ and its odometric measurements. Due to its history of motion, one wheel may have worn more than the other, causing a left-turning bias over time that affects its current position. Additionally, there might also be unmodeled dynamics of the environment such as moving people (which have effect on sensor measurements), inaccuracies in the probabilistic motion and measurement models, errors in the map used for a localizing robot, and software variables that influence multiple controls.

So the Markov assumption is, of course, not a valid assumption. However, the Markov assumption greatly simplifies tracking, reasoning, and planning, and so it is an approximation that continues to be extremely popular in mobile robotics. Indeed, Markov localization has been found to be surprisingly robust to such violations.

5.6.7.3 Illustration of Markov localization

In figure 5.22, we illustrate the working principle of the Markov localization in the continuous case. For simplicity, our environment is a one-dimensional hallway with three identical pillars.

In this example, we assume that at the beginning the robot does not know its initial location and has therefore to localize from scratch. Clearly, this is a global localization problem. According to the probabilistic framework described before, the robot initial belief $bel(x_0)$ is a uniform distribution over all locations as illustrated in figure 5.22a.

Now suppose that the robot uses its exteroceptive sensors and senses that it is next to a pillar. This is clearly the *perception update* of Markov localization. Then, according to the Bayes rule its belief $bel(x_0)$ has to be multiplied by $p(z_t|x_t, M)$ as stated in equation (5.42). How do we characterize $p(z_t|x_t, M)$? Because the three pillars are exactly identical, the robot does not know which one of the pillars is facing. Therefore, the probability $p(z_t|x_t, M)$ of observing a pillar is characterized by three peaks, each corresponding to one of the indistinguishable pillars in the environment. The upper plot in (b) visualizes $p(z_t|x_t, M)$. After this perception update the robot still does not know where it is. Indeed, it now has three, distinct hypotheses which are all equally plausible. Also notice that the probability in the regions not next to a pillar is nonzero. In probabilistic robot localization we can never be 100% sure that the robot is not somewhere. Therefore, it is important to maintain low-probability hypotheses. This is essential for achieving robustness, for example, if the robot gets lost or kidnapped. The lower plot in (b) visualizes the result of the multiplication. Because it results from a multiplication with a constant function, the result is still characterized by the three exactly identical peaks.

Now suppose that the robot moves to the right. We are now in the *action update* of Markov localization. Figure 5.22c shows the effect on the robot's belief. As a result of the *convolution* of the robot's previous belief with the motion model $p(x_t|u_t, x_{t-1})$, the new belief has been shifted in the direction of motion and also flattened out. The three peaks are now larger, which reflects the uncertainty that is introduced by the robot motion.

Figure 5.22d depicts the belief after observing another pillar. We are again in the *perception update*. Here, Markov localization algorithm multiples again the current belief with the perceptual probability $p(z_t|x_t, M)$. As observed, this time the result of the multiplication

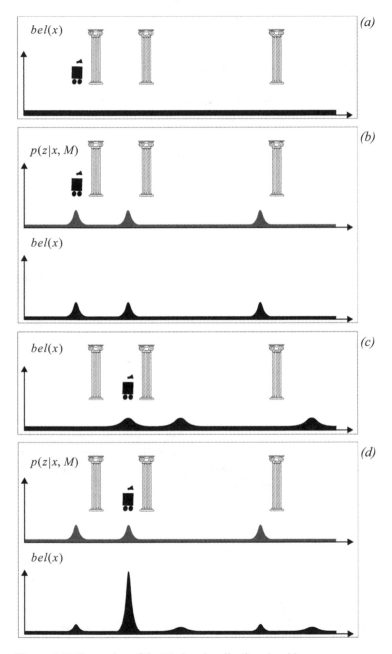

Figure 5.22 Illustration of the Markov localization algorithm.

is a single distinguishable peak near one of the pillars, and the robot is now quite confident of where it is.

5.6.7.4 Case study 1: Markov localization using a grid map

Markov localization is implemented in practice using a grid-space representation of the environment. Usually, a fixed decomposition is used, which consists in tessellating the state-space into fine-grained cells of uniform size (section 5.5.2).

For planar motion, the robot configuration is expressed by three parameters (x, y, θ). This means that also the space of all possible robot orientations must be discretized. The final state-space is therefore stored in the memory of the robot as a three-dimensional array (figure 5.24).

In this section, we illustrate the working principle of Markov localization using a grid map. For simplicity, we will assume again that the environment is one-dimensional. Considerations for the more general case of 2D environments will be done at the end of this section.

Let us tessellate our environment into ten equally spaced cells (figure 5.23). Suppose that the robot's initial belief $bel(x_0)$ is a uniform distribution from 0 to 3 as shown in figure 5.23a. Observe that all the elements were normalized so that their sum is 1.

Prediction update. Let us recall that in this phase, the robot moves and updates its belief using the motion model of the control input. Therefore, we need the probabilistic motion model (i.e. the odometric error model). Let us assume that the probabilistic motion model of the odometry $p(x_1|u_1, x_0)$ is the one represented in figure 5.23b. This model must be interpreted in this way: between time $t = 0$ and time $t = 1$, the robot may have moved either two or three units to the right. In this example, both movements have the same probability to occur. What will the robot belief be after this movement? The answer is again in the prediction-phase equation (5.41), that is, the final belief $\overline{bel}(x_1)$ is given by the theorem of total probability, which convolves the initial belief $bel(x_0)$ with the motion model $p(x_1|u_1, x_0)$. Using (5.41), we obtain:

$$\overline{bel}(x_1) = \sum_{x_0 = 0}^{3} p(x_1|u_1, x_0)bel(x_0).$$
(5.43)

Here, we would like to explain where this formula actually comes from. In order to compute the probability of position x_1 in the new belief state, one must sum over all the possible ways in which the robot may reach x_1 according to the potential positions expressed in the former belief state x_0 and the potential input expressed by u_1. Observe that because x_0 is limited between 0 and 3, the robot can only reach the states between 2 to 6, therefore:

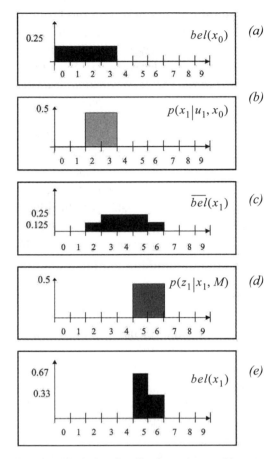

Figure 5.23 Markov localization using a grid-map.

$$p(x_1 = 2) = p(x_0 = 0)p(u_1 = 2) = 0.125, \tag{5.44}$$

$$p(x_1 = 3) = p(x_0 = 0)p(u_1 = 3) + p(x_0 = 1)p(u_1 = 2) = 0.25 \tag{5.45}$$

$$p(x_1 = 4) = p(x_0 = 1)p(u_1 = 3) + p(x_0 = 2)p(u_1 = 2) = 0.25 \tag{5.46}$$

$$p(x_1 = 5) = p(x_0 = 2)p(u_1 = 3) + p(x_0 = 3)p(u_1 = 2) = 0.25 \tag{5.47}$$

$$p(x_1 = 6) = p(x_0 = 3)p(u_1 = 3) = 0.125 \tag{5.48}$$

Expression (5.44) results from the fact that the state $x_1 = 2$ can only be reached with the combination $(x_0 = 0, u_1 = 2)$. Expression (5.45) comes from the fact that the state $x_1 = 3$ can only be reached with the combinations $(x_0 = 0, u_1 = 3)$ or $(x_0 = 1, u_1 = 2)$. The other expressions follow in a similar way. The reader can now verify that equations (5.44)–(5.48) are implementing nothing but the theorem of total probability (or convolution)[26] enunciated in (5.43). The result of the application of this theorem is shown in figure 5.23c.

Measurement update. Let us now assume that the robot uses its onboard rangefinder and measures the distance z from the origin. Assume that the statistical error model of the range sensor is the one shown in figure 5.23d. This plot tells us that the distance of the robot from the origin can be equally 5 or 6 units. What will the final robot belief be after this measurement? The answer is again in the measurement-update equation (5.42). The final belief $bel(x_1)$ is computed accordingly to the Bayes rule. It is the product between the robot current belief $\overline{bel}(x_1)$ and the measurement error model $p(z_1|x_1, M)$, where, in this case, the map M is simply the origin of the axes. Therefore:

$$bel(x_1) = \eta p(z_1|x_1, M)\overline{bel}(x_1).$$ (5.49)

The reader can verify that we need $\eta = 1/0.1875 \cong 5.33$ to make the final result $bel(x_1)$ normalized to one. The final belief is shown in figure 5.23e.

3D grid maps. As we said at the beginning of the section, in the more general planar motion case the grid-map is a three-dimensional array where each cell contains the probability of the robot to be in that cell (figure 5.24). In this case, the cell size must be chosen carefully. During each prediction and measurement steps, all the cells are updated. If the number of cells in the map is too large, the computation can become too heavy for real-time operations. The convolution in a 3D space is clearly the computationally most expensive step. As an example, consider a $30 \text{ m} \times 30 \text{ m}$ environment and a cell size of $0.1 \text{ m} \times 0.1 \text{ m} \times 1°$. In this case, the number of cells that need to be updated at each step would be $30 \times 30 \times 100 \times 360 = 32.4$ million cells!

One possible solution would then be to increase the cell size at the expense of localization accuracy. Another solution is to use an *adaptive* cell decomposition instead of a *fixed* cell decomposition as proposed by Burgard et al. [86, 87]. In this work, they overcame the problem of the huge state space by dynamically adapting the size of the cells according to the robot's certainty in its position, that is, smaller cells where the robot is more certain to

---

26. As mentioned earlier, notice that we are using improperly the term convolution. The only difference with convolution is that the theorem of total probability does not invert the sign of neither of the two argument functions.

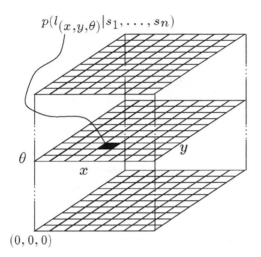

$p(l_{(x,y,\theta)}|s_1, \ldots, s_n)$

θ

x

y

$(0, 0, 0)$

Figure 5.24
The belief state representation 3D array used in Markov localization (courtesy of W. Burgard and S. Thrun).

be and bigger cells elsewhere. This way, they were able to localize a robot in a 30 m × 30 m with an error smaller than 4 cm using a number of cells varying only between 400 and 3600.

The resulting robot localization system of Burgard and his colleagues has been part of a navigation system that has demonstrated great success both at the University of Bonn (Germany) and at a public museum in Bonn. This is a challenging application because of the dynamic nature of the environment, as the robot's sensors are frequently subject to occlusion due to humans gathering around the robot. The robot ability to function well in this setting is a demonstration of the power of the Markov localization approach.

Reducing computational complexity: Randomized sampling. One class of techniques deserves mention because it can significantly reduce the computational overhead of techniques that employ fixed-cell decomposition representations. The basic idea, which we call *randomized sampling,* is known alternatively as *particle filter* algorithms, *condensation algorithms*, and *Monte Carlo* algorithms [129, 311].

Irrespective of the specific technique, the basic algorithm is the same in all these cases. Instead of representing *every* possible robot position by representing the complete and correct belief state, an approximate belief state is constructed by representing only a *subset* of the complete set of possible locations that should be considered.

For example, consider a robot with a complete belief state of 10,000 possible locations at time *t*. Instead of tracking and updating all 10,000 possible locations based on a new sensor measurement, the robot can select only 10% of the stored locations and update only those locations. By weighting this sampling process with the probability values of the locations, one can bias the system to generate more samples at local peaks in the probability density function. So the resulting 1000 locations will be concentrated primarily at the highest probability locations. This biasing is desirable, but only to a point.

We also wish to ensure that *some* less likely locations are tracked, as otherwise, if the robot does indeed receive unlikely sensor measurements, it will fail to localize. This *randomization* of the sampling process can be performed by adding additional samples from a flat distribution, for example. Further enhancements of these randomized methods enable the number of statistical samples to be varied on the fly, based, for instance, on the ongoing localization confidence of the system. This further reduces the number of samples required on average while guaranteeing that a large number of samples will be used when necessary [129].

These sampling techniques have resulted in robots that function indistinguishably as compared to their full belief state set ancestors, yet use computationally a fraction of the resources. Of course, such sampling has a penalty: completeness. The probabilistically complete nature of Markov localization is violated by these sampling approaches because the robot is failing to update *all* the nonzero probability locations, and thus there is a danger that the robot, due to an unlikely but correct sensor reading, could become truly lost. Of course, recovery from a lost state is feasible just as with all Markov localization techniques.

5.6.7.5 Case study 2: Markov localization using a topological map

A straightforward application of Markov localization is possible when the robot's environment representation already provides an appropriate decomposition. This is the case when the environmental representation is purely topological.

Consider a contest in which each robot is to receive a topological description of the environment. The description would include only the connectivity of hallways and rooms, with no mention of geometric distance. In addition, this supplied *map* would be imperfect, containing several false arcs (e.g., a closed door). Such was the case for the 1994 American Association for Artificial Intelligence (AAAI) National Robot Contest, at which each robot's mission was to use the supplied map and its own sensors to navigate from a chosen starting position to a target room.

Dervish, the winner of this contest, employed probabilistic Markov localization and used a multiple-hypothesis belief state over a topological environmental representation. We now describe Dervish as an example of a robot with a discrete, topological representation and a probabilistic localization algorithm.

Dervish, shown in figure 5.25, includes a sonar arrangement custom-designed for the 1994 AAAI National Robot Contest. The environment in this contest consisted of a rectilinear indoor office space filled with real office furniture as obstacles. Traditional sonars

Figure 5.25
Dervish exploring its environment.

were arranged radially around the robot in a ring. Robots with such sensor configurations are subject to both tripping over short objects below the ring and to decapitation by tall objects (such as ledges, shelves, and tables) that are above the ring.

Dervish's answer to this challenge was to arrange one pair of sonars diagonally upward to detect ledges and other overhangs. In addition, the diagonal sonar pair also proved to ably detect tables, enabling the robot to avoid wandering underneath tall tables. The remaining sonars were clustered in sets of sonars, such that each individual transducer in the set would be at a slightly varied angle to minimize specularity. Finally, two sonars near the robot's base were positioned to detect low obstacles, such as paper cups, on the floor.

We have already noted that the representation provided by the contest organizers was purely topological, noting the connectivity of hallways and rooms in the office environment. Thus, it would be appropriate to design Dervish's perceptual system to detect matching perceptual events: the detection and passage of connections between hallways and offices.

This *abstract* perceptual system was implemented by viewing the trajectory of sonar strikes to the left and right sides of Dervish over time. Interestingly, this perceptual system would use time alone and no concept of encoder value to trigger perceptual events. Thus, for instance, when the robot detects a 7–17 cm indentation in the width of the hallway for

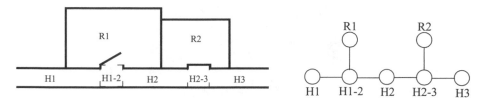

Figure 5.26
A geometric office environment (left) and its topological analog (right).

more than 1 second continuously, a *closed door* sensory event is triggered. If the sonar strikes jump well beyond 17 cm for more than 1 second, an *open door* sensory event triggers.

To reduce coherent reflection sensor noise (see section 4.1.9) associated with Dervish's sonars, the robot would track its angle relative to the hallway center line and completely suppress sensor events when its angle to the hallway exceeded 9 degrees. Interestingly, this would result in a conservative perceptual system that frequently misses features, particularly when the hallway is crowded with obstacles that Dervish must negotiate. Once again, the conservative nature of the perceptual system, and in particular its tendency to issue false negatives, would point to a probabilistic solution to the localization problem so that a complete trajectory of perceptual inputs could be considered.

Dervish's environmental representation was a discrete topological map, identical in abstraction and information to the map provided by the contest organizers. Figure 5.26 depicts a geometric representation of a typical office environment overlaid with the topological map for the same office environment. Recall that for a topological representation the key decision involves assignment of nodes and connectivity between nodes (see section 5.5.2). As shown on the left in figure 5.26, Dervish uses a topology in which node boundaries are marked primarily by doorways (and hallways and foyers). The topological graph shown on the right depicts the information captured in the example shown.

Note that in this particular topological model arcs are zero-length while nodes have spatial expansiveness and together cover the entire space. This particular topological representation is particularly apt for Dervish, given its task of navigating through hallways into a specific room and its perceptual capability of recognizing discontinuities in hallway walls.

In order to represent a specific belief state, Dervish associated with each topological node n a probability that the robot is at a physical position within the boundaries of n : $p(x_t = n)$. As will become clear, the probabilistic update used by Dervish was approximate, therefore technically one should refer to the resulting values as *likelihoods* rather than probabilities.

Table 5.1
Dervish's certainty matrix.

	Wall	Closed door	Open door	Open hallway	Foyer
Nothing detected	0.70	0.40	0.05	0.001	0.30
Closed door detected	0.30	0.60	0	0	0.05
Open door detected	0	0	0.90	0.10	0.15
Open hallway detected	0	0	0.001	0.90	0.50

The perception update process for Dervish functions precisely as in equation (5.42). Perceptual events are generated asynchronously, each time the feature extractor is able to recognize a large scale feature (e.g., doorway, intersection) based on recent ultrasonic values. Each perceptual event consists of a percept-pair (a feature on one side of the robot or two features on both sides).

Given a specific percept-pair z, equation (5.42) enables the likelihood of each possible position n to be updated using the formula:

$$p(n|z) = \eta p(z|n)p(n).$$

(5.50)

The value of $p(n)$ is already available from the current belief state of Dervish, and so the challenge lies in computing $p(z|n)$. The key simplification for Dervish is based upon the realization that, because the feature extraction system only extracts four total features and because a node contains (on a single side) one of five total features, every possible combination of node type and extracted feature can be represented in a 4×5 table.

Dervish's *certainty matrix* (shown in table 5.1) is just this lookup table. Dervish makes the simplifying assumption that the performance of the feature detector (i.e., the probability that it is correct) is only a function of the feature extracted and the actual feature in the node. With this assumption in hand, we can populate the *certainty matrix* with confidence estimates for each possible pairing of perception and node type. For each of the five world features that the robot can encounter (wall, closed door, open door, open hallway, and foyer) this matrix assigns a likelihood for each of the three one-sided percepts that the sensory system can issue. In addition, this matrix assigns a likelihood that the sensory system will fail to issue a perceptual event altogether (*nothing detected*).

For example, using the specific values in table 5.1, if Dervish is next to an open hallway, the likelihood of mistakenly recognizing it as an open door is 0.10. This means that for any node n that is of type *open hallway* and for the sensor value $z = open\,door$, $p(z|n) = 0.10$. Together with a specific topological map, the certainty matrix enables straightforward computation of $p(z|n)$ during the perception update process.

For Dervish's particular sensory suite and for any specific environment it intends to navigate, humans generate a specific certainty matrix that loosely represents its perceptual confidence, along with a global measure for the probability that any given door will be closed versus opened in the real world.

Recall that Dervish has no encoders and that perceptual events are triggered asynchronously by the feature extraction processes. Therefore, Dervish has no prediction update step as depicted by equation (5.41). When the robot does detect a perceptual event, multiple perception update steps will need to be performed to update the likelihood of every possible robot position given Dervish's former belief state. This is because there is a chance that the robot has traveled *multiple* topological nodes since its previous perceptual event (i.e., false-negative errors). Formally, the perception update formula for Dervish is in reality a combination of the general form of prediction update and measurement update. The likelihood of position n given perceptual event i is calculated as in equation (5.41):

$$p(n_t|z_t) = \sum p(n_t|n'_{t-i}, z_t)p(n'_{t-i}).$$ (5.51)

The value of $p(n'_{t-i})$ denotes the likelihood of Dervish being at position n' as represented by Dervish's former belief state. The temporal subscript $t - i$ is used in lieu of $t - 1$ because for each possible position n' the discrete topological distance from n' to n can vary depending on the specific topological map. The calculation of $p(n_t|n'_{t-i}, z_t)$ is performed by multiplying the probability of generating perceptual event z at position n by the probability of having failed to generate perceptual events at all nodes between n' and n:

$$p(n_t|n'_{t-i}, z_t) = p(z_t, n_t) \cdot p(\emptyset, n_{t-1}) \cdot p(\emptyset, n_{t-2}) \cdot \ldots \cdot p(\emptyset, n_{t-i+1}).$$ (5.52)

For example (figure 5.27), suppose that the robot has only two nonzero nodes in its belief state, {1-2, 2-3}, with likelihoods associated with each possible position: $p(1-2) = 1.0$ and $p(2-3) = 0.2$. For simplicity assume the robot is facing east with certainty. Note that the likelihoods for nodes 1–2 and 2–3 do not sum to 1.0. These values are not formal probabilities, and so computational effort is minimized in Dervish by avoiding normalization altogether. Now suppose that a perceptual event is generated: the robot detects an open hallway on its left and an open door on its right simultaneously.

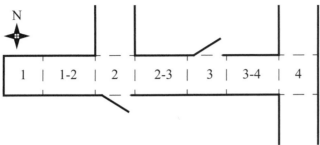

Figure 5.27
A realistic indoor topological environment.

State 2–3 will progress potentially to states 3, 3–4, and 4. But states 3 and 3–4 can be eliminated because the likelihood of detecting an open door when there is only wall is zero. The likelihood of reaching state 4 is the product of the initial likelihood for state 2–3, 0.2, the likelihood of not detecting anything at node 3, (a), and the likelihood of detecting a hallway on the left and a door on the right at node 4, (b). Note that we assume the likelihood of detecting nothing at node 3–4 is 1.0 (a simplifying approximation).

(a) occurs only if Dervish fails to detect the door on its left at node 3 (either closed or open), $[0.6 \cdot 0.4 + (1 - 0.6) \cdot 0.05]$, and correctly detects nothing on its right, 0.7.

(b) occurs if Dervish correctly identifies the open hallway on its left at node 4, 0.90, and mistakes the right hallway for an open door, 0.10.

The final formula, $0.2 \cdot [0.6 \cdot 0.4 + 0.4 \cdot 0.05] \cdot 0.7 \cdot [0.9 \cdot 0.1]$, yields a likelihood of 0.003 for state 4. This is a partial result for $p(4)$ following from the prior belief state node 2-3.

Turning to the other node in Dervish's prior belief state, 1–2 will potentially progress to states 2, 2–3, 3, 3–4, and 4. Again, states 2–3, 3, and 3–4 can all be eliminated since the likelihood of detecting an open door when a wall is present is zero. The likelihood of state 2 is the product of the prior likelihood for state 1–2, (1.0), the likelihood of detecting the door on the right as an open door, $[0.6 \cdot 0 + 0.4 \cdot 0.9]$, and the likelihood of correctly detecting an open hallway to the left, 0.9. The likelihood for being at state 2 is then $1.0 \cdot 0.4 \cdot 0.9 \cdot 0.9 = 0.3$. In addition, 1–2 progresses to state 4 with a certainty factor of $4.3 \cdot 10^{-6}$, which is added to the certainty factor above to bring the total for state 4 to 0.00328. Dervish would therefore track the new belief state to be {2, 4}, assigning a very high likelihood to position 2 and a low likelihood to position 4.

Empirically, Dervish's map representation and localization system have proved to be sufficient for navigation of four indoor office environments: the artificial office environment created explicitly for the 1994 National Conference on Artificial Intelligence; and the psychology, history, and computer science departments at Stanford University. All of these experiments were run while providing Dervish with no notion of the distance between adjacent nodes in its topological map. It is a demonstration of the power of probabilistic localization that, in spite of the tremendous lack of action and encoder information, the robot is able to navigate several real-world office buildings successfully.

One open question remains with respect to Dervish's localization system. Dervish was not just a localizer but also a navigator. As with all multiple hypothesis systems, one must ask the question, how does the robot decide how to move, given that it has multiple possible robot positions in its representation? The technique employed by Dervish is a common technique in the mobile robotics field: plan the robot's actions by assuming that the robot's actual position is its most likely node in the belief state. Generally, the most likely position is a good measure of the robot's actual world position. However, this technique has shortcomings when the highest and second highest most likely positions have similar values. In the case of Dervish, it nonetheless goes with the highest-likelihood position at all times, save at one critical juncture. The robot's goal is to enter a target room and remain there. Therefore, from the point of view of its goal, it is critical that Dervish finish navigating only when the robot has strong confidence in being at the correct final location. In this particular case, Dervish's execution module refuses to enter a room if the gap between the most likely position and the second likeliest position is below a preset threshold. In such a case, Dervish will actively plan a path that causes it to move farther down the hallway in an attempt to collect more sensor data and thereby increase the relative likelihood of one position in the multiple-hypothesis belief state.

Although computationally unattractive, one can go farther, imagining a planning system for robots such as Dervish for which one specifies a *goal belief state* rather than a goal position. The robot can then reason and plan in order to achieve a goal confidence level, thus explicitly taking into account not only robot position but also the measured likelihood of each position. An example of just such a procedure is the sensory uncertainty field of Latombe [306], in which the robot must find a trajectory that reaches its goal while maximizing its localization confidence on-line.

The major weakness of a purely topological decomposition of the environment is the resolution limitation imposed by such a granular representation. The position of the robot is usually limited to the resolution of a single node in such cases, and this may be undesirable for certain applications.

5.6.8 Kalman filter localization

5.6.8.1 Introduction
The Markov localization model can represent any arbitrary probability density function over the robot position. This approach is very general but, due to its generality, inefficient. Consider instead the key demands on a robot localization system. One can argue that it is not the exact replication of a probability density curve but the *sensor fusion* problem that is

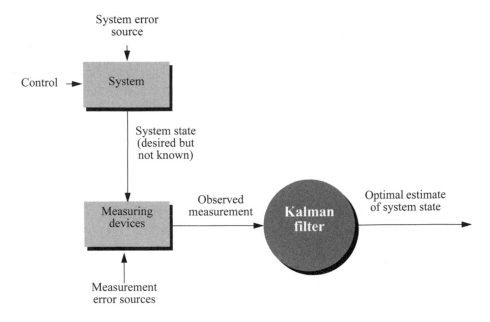

Figure 5.28
Typical Kalman filter application [209].

key to robust localization. Robots usually include a large number of heterogeneous sensors, each providing clues as to robot position and, critically, each suffering from its own failure modes. Optimal localization should take into account the information provided by all of these sensors. In this section we describe a powerful technique for achieving this sensor fusion, called the Kalman filter. This mechanism is in fact more efficient than Markov localization because of key simplifications when representing the probability density function of the robot's belief state and even its individual sensor readings, as described below. But the benefit of this simplification is a resulting *optimal recursive data-processing algorithm*. It incorporates all information, regardless of precision, to estimate the current value of the variable of interest (i.e., the robot's position). A general introduction to Kalman filters can be found in [209], and a more detailed treatment is presented in [3].

Figure 5.28 depicts the general scheme of Kalman filter estimation, where a system has a control signal and system error sources as inputs. A measuring device enables measuring some system states with errors. The Kalman filter is a mathematical mechanism for producing an optimal estimate of the system state based on the knowledge of the *system* and the *measuring device*, the description of the system noise and measurement errors and the uncertainty in the dynamics models. Thus the Kalman filter *fuses* sensor signals and system knowledge in an optimal way. Optimality depends on the criteria chosen to evaluate the performance and on the assumptions. Within the Kalman filter theory the system is assumed to be *linear* and with *white Gaussian* noise. For most mobile robot applications, the system is nonlinear. In this cases, the Kalman filter is usually applied after linearizing the system. The extension of Kalman filter to nonlinear systems is known as the *Extended Kalman Filter* (*EKF*), but its optimality cannot be guaranteed. As we have discussed earlier, the assumption of Gaussian error is invalid for our mobile robot applications, but nevertheless the results are extremely useful. In other engineering disciplines, the Gaussian error assumption has in some cases been shown to be quite accurate [209].

We begin with a section that illustrates the Kalman filter localization (section 5.6.8.2). Then, we introduce the Kalman filter theory (section 5.6.8.3), and present an application of that theory to the problem of mobile robot localization (5.6.8.4). Finally, section 5.6.8.5 presents a case study of a mobile robot that navigates indoor spaces by virtue of Kalman filter localization.

5.6.8.2 Illustration of Kalman filter localization

The *Kalman filter localization* algorithm, or *KF localization*, is a special case of Markov localization. Instead of using an arbitrary density function, the Kalman filter uses Gaussians to represent the robot belief $bel(x_t)$, the motion model, and the measurement model. Because a Gaussian is simply defined by its mean μ_t and covariance Σ_t, only these two parameters are updated during the prediction and measurement phase, resulting in a very efficient algorithm in comparison to Markov localization algorithm. However, the assumptions made by the Kalman filter limit the choice of the initial belief $bel(x_0)$ also to a Gaussian, which means that the robot initial location must be known with a certain approximation. Hence, the robot cannot recover its position if it gets lost. This is in contrast with the Markov localization. The Kalman filter therefore addresses the position-tracking problem but not the global localization or the kidnapped robot problem.

Figure 5.29 illustrates the Kalman filter localization algorithm using again our example of a mobile robot in a one-dimensional environment. As mentioned, the robot initial belief $bel(x_0)$ is represented by a Gaussian distribution. As shown in figure 5.29a, we assume that at the beginning the robot is near the first pillar. As the robot moves to the right (we are in the *action phase*), its uncertainty increases as a result of the convolution with the motion model (i.e., application of the theorem of total probability). The resulting belief is therefore a shifted Gaussian of increased width, figure 5.29b. Now, suppose that the robot uses its exteroceptive sensors (we are in the *perception phase*) and senses that it is near the second pillar. The posterior probability $p(z_t|x_t, M)$ of the observation is shown in figure 5.29c. This probability density is again a Gaussian. In order to compute the robot current belief, we must fuse this measurement probability with the robot's belief before the observation using the Bayes rule. The result of this fusion is again a Gaussian shown at the bottom of figure

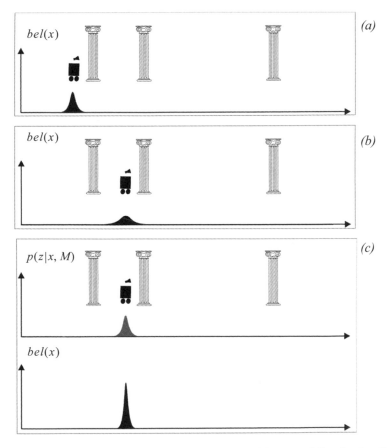

Figure 5.29 Application of the Kalman filter algorithm to mobile robot localization.

5.29c. Note that the variance of the resulting belief is smaller than the variances of both the measurement probability and the robot's previous belief. This result is obvious, since the fusion of two independent estimates should make the robot more certain than each individual estimate.

5.6.8.3 Introduction to Kalman filter theory

As said in the previous section, the Kalman filter theory is based on the assumptions that the system is linear and that overall the robot configuration, the odometric error model, and the measurement error model are affected by white Gaussian noise.

A Gaussian distribution is represented only by its first and second moments, which are the mean μ_t and the variance σ_t^2 (see equation 5.20). When the robot configuration is a vector (which is the case in practical applications) the distribution is a multivariate Gaussian represented by a mean vector μ_t and a covariance matrix Σ_t (see equation 5.22).

During the *prediction* and *measurement* updates only mean μ_t and covariance Σ_t are updated. Therefore, the Kalman filter is based on four equations: two for updating μ_t and Σ_t in the prediction update, and another two in the measurement update. In Kalman filtering the measurement update is also commonly called *correction update*.

As for Markov localization, the *prediction* and *measurement* update equations of Kalman filter are also based respectively on the *theorem of total probability* and on the *Bayes rule*. In this section, we will review these properties applied to special case of Gaussian distributions. For an in-depth study of these properties, we refer the reader to [41].

Applying the theorem of total probability. Let x_1, x_2 be two random variables that are independent and normally distributed:

$$x_1 = N(\mu_1, \sigma_1^2) \tag{5.53}$$

$$x_2 = N(\mu_2, \sigma_2^2). \tag{5.54}$$

Let also y be a function of x_1 and x_2, that is,

$$y = f(x_1, x_2). \tag{5.55}$$

What will the distribution of y be? The answer is much simpler when f is a linear function of the inputs, that is, when

$$y = Ax_1 + Bx_2. \tag{5.56}$$

In this case, if the inputs are independent, it can be shown that y is also normally distributed with mean and variance given by the following expressions:

$$\langle y \rangle = A\mu_1 + B\mu_2 \tag{5.57}$$

$$\sigma_y^2 = A^2\sigma_1^2 + B^2\sigma_2^2 \tag{5.58}$$

If x_1 and x_2 are vectors with covariances Σ_1 and Σ_2 respectively, then

$$\langle y \rangle = A\mu_1 + B\mu_2. \tag{5.59}$$

$$\Sigma_y = A\Sigma_1 A^T + B\Sigma_2 B^T. \tag{5.60}$$

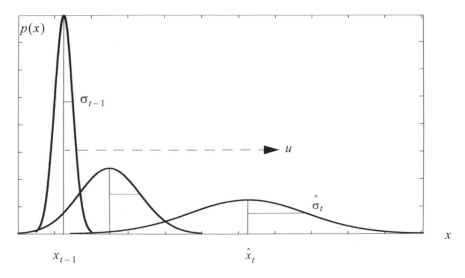

Figure 5.30
Propagation of probability density of a moving robot [209].

This result follows directly from the application of the total probability theorem. We can also look at this problem in terms of convolution, by reminding that the probability density function of the sum of two independent random variables is the convolution of each of their density functions [41]. It can also be shown that the convolution of two Gaussian random variables is another Gaussian [41].

In the case f is nonlinear, y is *not* normally distributed. However, it is a common practice to consider a first-order approximation by linearizing f about (μ_1, μ_2):

$$y \cong f(\mu_1, \mu_2) + F_{x_1}(x_1 - \mu_1) + F_{x_2}(x_2 - \mu_2) . \tag{5.61}$$

where F_{x_1} and F_{x_2} are the jacobians of f. This way, we obtain:

$$\langle y \rangle = f(\mu_1, \mu_2) . \tag{5.62}$$

$$\Sigma_y = F_{x_1} \Sigma_1 F_{x_1}^T + F_{x_2} \Sigma_2 F_{x_2}^T . \tag{5.63}$$

Equations (5.62) and (5.63) will be used in section 5.6.8.4 to implement the *prediction update* of Extended Kalman Filter (EKF)[27] localization. As we will show, in Kalman localization f is used to represent the odometric position update and its inputs are the robot previous position x_{t-1} and the control input u. In the simple case of a one-dimensional environment, the odometric position update is described by a simple sum, therefore, $f(x_{t-1}, u) = x_{t-1} + u$ and the update of the uncertainty over the time expressed in (5.63) is illustrated in figure 5.30. Observe that the uncertainty of the robot position after the application of (5.63) is larger.

Applying the Bayes rule. Let q denote the robot position, $p_1(q)$ the robot's belief resulting from the prediction update, and $p_2(q)$ the robot's belief resulting from some exteroceptive sensor measurement (for example, a rangefinder that returns the position of the robot directly in the global reference frame). The Bayes rule tells us how to compute the final distribution of the robot's belief $p(q)$ after the measurement has been taken. As usual, probability densities in Kalman filtering are assumed to be normally distributed; therefore

$$p_1(q) = N(\hat{q}_1, \sigma_1^2),\tag{5.64}$$

$$p_2(q) = N(\hat{q}_2, \sigma_1^2).\tag{5.65}$$

According to the Bayes rule, the final distribution $p(q)$, after the measurement, is proportional to the product $p_1(q) \cdot p_2(q)$ (figure 5.31). From the product of the two density functions (5.64) and (5.65), we obtain:

$$\frac{1}{\sigma_1\sqrt{2\pi}}\exp\left(-\frac{(q-\hat{q}_1)^2}{2\sigma_1^2}\right) \cdot \frac{1}{\sigma_2\sqrt{2\pi}}\exp\left(-\frac{(q-\hat{q}_2)^2}{2\sigma_2^2}\right) = \frac{1}{\sigma_1\sigma_2 2\pi}\exp\left(-\frac{(q-\hat{q}_1)^2}{2\sigma_1^2} - \frac{(q-\hat{q}_2)^2}{2\sigma_2^2}\right).$$

$$\tag{5.66}$$

As we can see, the argument of this exponential is quadratic in q, hence $p(q)$ is a Gaussian. We now need to determine its mean value $\hat{q}$ and variance σ that allow us to rewrite (5.66) in the form

$$\Omega\exp\left(-\frac{(q-\hat{q})^2}{2\sigma^2}\right).\tag{5.67}$$

---

27. The extended Kalman filter is the extension of the standard Kalman filter to nonlinear systems, by considering the first order approximation of the state transition function f and the observation model h.

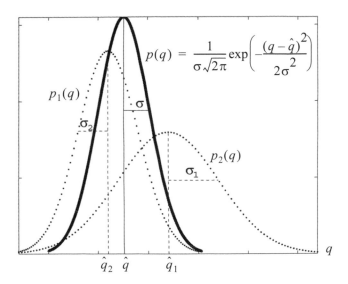

$$p(q) = \frac{1}{\sigma\sqrt{2\pi}} \exp\left(-\frac{(q-\hat{q})^2}{2\sigma^2}\right)$$

Figure 5.31
Fusing the probability density of two estimates [209]: the result of the product of two Gaussian function is another Gaussian. The result is then rescaled to make its area equal to 1.

By rearranging the exponential in (5.66), we get

$$\exp\left(-\frac{(q-\hat{q}_1)^2}{2\sigma_1^2} - \frac{(q-\hat{q}_2)^2}{2\sigma_2^2}\right) =$$

$$= \exp\left(-\frac{1}{2}\left(\frac{q^2(\sigma_1^2+\sigma_2^2) - 2q(\hat{q}_1\sigma_2^2+\hat{q}_2\sigma_1^2) + (\hat{q}_1^2\sigma_2^2+\hat{q}_2^2\sigma_1^2)}{\sigma_1^2\sigma_2^2}\right)\right) =$$

$$= \exp\left(-\frac{1}{2}\left(\frac{q^2 - \dfrac{2q(\hat{q}_1\sigma_2^2+\hat{q}_2\sigma_1^2)}{\sigma_1^2+\sigma_2^2} + \dfrac{\hat{q}_1^2\sigma_2^2+\hat{q}_2^2\sigma_1^2}{\sigma_1^2+\sigma_2^2}}{\dfrac{\sigma_1^2\sigma_2^2}{\sigma_1^2+\sigma_2^2}}\right)\right) =$$

$$= \exp\left(-\frac{1}{2}\frac{\left(q - \dfrac{\hat{q}_1\sigma_2^2+\hat{q}_2\sigma_1^2}{\sigma_1^2+\sigma_2^2}\right)^2}{\dfrac{\sigma_1^2\sigma_2^2}{\sigma_1^2+\sigma_2^2}}\right) \cdot \exp\left(-\frac{1}{2}\frac{\dfrac{\hat{q}_1^2\sigma_2^2+\hat{q}_2^2\sigma_1^2}{\sigma_1^2+\sigma_2^2} - \left(\dfrac{\hat{q}_1\sigma_2^2+\hat{q}_2\sigma_1^2}{\sigma_1^2+\sigma_2^2}\right)^2}{\dfrac{\sigma_1^2\sigma_2^2}{\sigma_1^2+\sigma_2^2}}\right) \cdot$$

$$\tag{5.68}$$

We can notice that the second term of this product depends only on $\hat{q}_1$ and $\hat{q}_2$ and, therefore, is constant. Hence, we can rewrite (5.68) as

$$\Omega \exp\left[-\frac{1}{2}\frac{\left(q - \frac{\hat{q}_1\sigma_2^2 + \hat{q}_2\sigma_1^2}{\sigma_1^2 + \sigma_2^2}\right)^2}{\frac{\sigma_1^2\sigma_2^2}{\sigma_1^2 + \sigma_2^2}}\right] = \Omega \exp\left(-\frac{(q-\hat{q})^2}{2\sigma^2}\right), \quad (5.69)$$

where

$$\hat{q} = \frac{\hat{q}_1\sigma_2^2 + \hat{q}_2\sigma_1^2}{\sigma_1^2 + \sigma_2^2}, \text{ or alternatively } \hat{q} = \frac{\frac{1}{\sigma_1^2}\hat{q}_1 + \frac{1}{\sigma_2^2}\hat{q}_2}{\frac{1}{\sigma_1^2} + \frac{1}{\sigma_2^2}}, \quad (5.70)$$

and

$$\sigma^2 = \frac{\sigma_1^2\sigma_2^2}{\sigma_1^2 + \sigma_2^2}, \text{ or alternatively } \frac{1}{\sigma^2} = \frac{1}{\sigma_1^2} + \frac{1}{\sigma_2^2} = \frac{\sigma_1^2 + \sigma_2^2}{\sigma_1^2\sigma_2^2}. \quad (5.71)$$

Notice that (5.70) and (5.71) can also be written as

$$\hat{q} = \hat{q}_1 + \frac{\sigma_1^2}{\sigma_1^2 + \sigma_2^2}(\hat{q}_2 - \hat{q}_1), \quad (5.72)$$

$$\sigma^2 = \sigma_1^2 - \frac{\sigma_1^4}{\sigma_1^2 + \sigma_2^2}. \quad (5.73)$$

These last two expressions will be valuable in the Kalman filter implementation. In Kalman filtering the factor $\sigma_1^2/(\sigma_1^2 + \sigma_2^2)$ is commonly called *Kalman gain*.

From equation (5.73) we can clearly see that the resulting variance σ^2 is smaller than both σ_1^2 and σ_2^2. Thus, the uncertainty of the position estimate has been decreased by combining the two measurements, that is, the previous robot's belief and the measurement from the exteroceptive sensor. Thus, even poor measurements will only increase the precision of an estimate. This is a result that we expect based on information theory. The solid probability density curve in figure 5.31 represents the result of the fusion operated by the Kalman filter.

Note that equations (5.72) and (5.73) are only valid for the one-dimensional case. For n-dimensional vectors, the final mean $\hat{q}$ and covariance $\hat{P}$ after fusion can be written respectively as:

$$\hat{q} = q_1 + P(P+R)^{-1}(q_2-q_1) \tag{5.74}$$

$$\hat{P} = P - P(P+R)^{-1}P, \tag{5.75}$$

where P and R are the covariances of q_1 and q_2 respectively.

Equations (5.74) and (5.75) will be used in section 5.6.8.4 to implement the *measurement update* of EKF localization. In Kalman filtering these equations are usually written as

$$\hat{q} = q_1 + K(q_2-q_1), \tag{5.76}$$

$$\hat{P} = P - K \cdot \Sigma_{IN} \cdot K^T, \tag{5.77}$$

where $K = P(P+R)^{-1}$ is the *Kalman gain*, (q_2-q_1) is the *innovation*, and $\Sigma_{IN} = (P+R)$ is the *innovation covariance*.

5.6.8.4 Application to mobile robots: Kalman filter localization

The Kalman filter is an optimal and efficient sensor fusion technique. Application of the Kalman filter to localization requires posing the robot localization problem as a sensor fusion problem. Recall that the basic probabilistic update of robot belief state can be segmented into two phases, *prediction update* and *measurement update*.

The key difference between the Kalman filter approach and our earlier Markov localization approach lies in the measurement update process. In Markov localization, the entire perception, that is, the robot's set of instantaneous sensor measurements, is used to update each possible robot position in the belief state individually. By contrast, the measurement update using a Kalman filter is a multistep process. The robot's total sensory input is treated not as a monolithic whole but as a set of extracted features that each relate to objects in the environment. Given a set of possible features, the Kalman filter is used to fuse the distance estimate from each feature to a matching object in the map. Instead of carrying out this matching process for many possible robot locations individually as in the Markov approach, the Kalman filter accomplishes the same probabilistic update by treating the whole, unimodal, and Gaussian belief state at once.

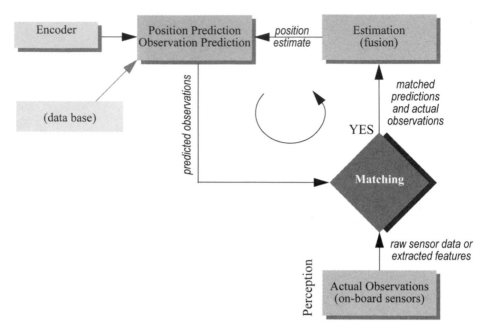

Figure 5.32
Schematic for Kalman filter mobile robot localization (see [35]).

Figure 5.32 depicts the particular schematic for Kalman filter localization. The first step is the *prediction update*, the straightforward application of a Gaussian error motion model to the robot's measured encoder travel. The *measurement update*, as just mentioned, is composed of multiple steps that are here summarized:

1. In the *observation step*, the robot collects actual sensor data and extracts appropriate features (e.g., lines, doors, or even the value of a specific sensor).

2. At the same time, based on its predicted position in the map, the robot generates a *measurement prediction* that consists in the features that the robot expects to observe from the position where it thinks it is (e.g. the position estimated in the prediction step).

3. In the *matching step* the robot computes the best matching between the features extracted during observation and the expected features selected during the measurement prediction.

4. Finally, the Kalman filter fuses the information provided by all of these matches to update the robot belief state in the *estimation step*.

In the following sections these steps are described in greater detail. The presentation is based on the work of Leonard and Durrant-Whyte [35, pages 61–65] and on that of Thrun, Burgard, and Fox [51].

Prediction update: Applying the theorem of total probability. The robot's position $\hat{x}_t$ at timestep t is predicted based on its old location at timestep $t-1$ and its movement due to the control input u_t:

$$\hat{x}_t = f(x_{t-1}, u_t). \tag{5.78}$$

For a differential-drive robot, $f(x_{t-1}, u_t)$ is given by (5.7), which describes the odometric position estimation.

Knowing the plant and the error model, we can also compute the variance P_{t-1} associated with this prediction using the equation derived from the total probability theorem applied to Gaussian distributions (see equation [5.63]):

$$\hat{P}_t = F_x \cdot P_{t-1} \cdot F_x^T + F_u \cdot Q_t \cdot F_u^T, \tag{5.79}$$

where P_{t-1} is the covariance of the previous robot state x_{t-1} and Q_t is the covariance of the noise associated to the motion model. This equation should not surprise the reader. This is in fact nothing but the application of the error propagation law (section 4.1.3.2).

Equations (5.78) and (5.79) are the two key equations of the prediction update in EKF localization. They allow us to predict the robot's position and its uncertainty after a movement specified by the control input u_t.

Again, note that because the belief state is assumed to be Gaussian, we are just updating two values: the mean value and the covariance of the distribution. Conversely, notice that in Markov localization *all* the robot's possible states (i.e., all the cells) are updated!

Measurement update. As we said before, this phase consists of four steps:

1. Observation. The first step is to obtain sensor measurements z_t from the robot at time t. In general, the observation z_t consists of a set of n single observations z_t^i ($i=0...n$) extracted from the sensor. Formally, each single observation can represent an extracted feature like a point landmark, a line, or even a single, raw sensor value.

The parameters of the features are usually specified in the sensor frame and therefore in a local reference frame of the robot. However, for matching we need to represent the observations and measurement predictions in the same frame $\{S\}$. In our presentation we will transform the measurement predictions from the global world coordinate frame $\{W\}$ to the sensor frame $\{S\}$. This transformation is specified by the function h, as discussed in section 5.6.5 when we talked about the *probabilistic measurement model*.

2. Measurement prediction. We use the robot predicted position $\hat{x}_t$ and the map M to generate multiple predicted feature observations $\hat{z}_t^j$.[28] The predicted observations are what the robot *expects* to see if it was at that particular position. Assume, for example, that, based only on the motion estimated by the odometry, the robot expects to be in front of a door. Assume now that the robot checks this hypothesis using its sensors and detects that it is actually facing a wall. Then, in this case the *door* is the predicted observation $\hat{z}_t$, while the *wall* is the actual observation z_t.

In order to compute the predicted observation, the robot must transform all the features m^j in the map M into the *local* sensor coordinate frame. If we define the transformation for the feature j through the function h^j, then we can write:

$$\hat{z}_t^j = h^j(\hat{x}_t, m^j), \tag{5.80}$$

which obviously depends on the position of each feature in the map (represented by m^j) and the current robot position $\hat{x}_t$.

3. Matching. At this point we have a set of actual observations, positioned in the sensor space, and we also have a set of predicted features, also positioned in the sensor space. The matching step has the purpose of identifying all of the single observations that match specific predicted features well enough to be used during the estimation process. In other words, we will, for a subset of the observations and a subset of the predicted features, find pairings that intuitively say "this observation is the robot's measurement of this predicted feature based on the map."

Formally, the goal of the matching procedure is to produce an assignment from the observation z_t^i to the predicted observations $\hat{z}_t^j$. For each measurement prediction for which a corresponding observation is found, we calculate the innovation v_t^{ij}. The *innovation* is a measure of the difference between the predicted and observed measurements:

$$v_t^{ij} = [z_t^i - \hat{z}_t^j] = [z_t^i - h^j(\hat{x}_t, m^j)]. \tag{5.81}$$

The *innovation covariance* $\Sigma_{IN_t}^{ij}$ can be found by applying the error propagation law (section 4.1.3.2, equation [4.15]):

---

28. Note that we use index j because observed and predicted features are not yet matched, that is, the observed feature i might not correspond with the feature j in the map.

$$\Sigma^{ij}_{IN_t} = H^j \cdot \hat{P}_t \cdot H^{j^T} + R^i_t,$$ (5.82)

where H^j is the jacobian of h^j and R^i_t represents the covariance (noise) of the actual observation z^i_t.

To determine the validity of the correspondence between measurement prediction and observation, a *validation gate g* has to be specified. A possible choice for the validation gate is the Mahalanobis distance:

$$v^{ij^T}_t \cdot (\Sigma^{ij}_{IN_t})^{-1} \cdot v^{ij}_t \le g^2 .$$ (5.83)

However, depending on the application, the sensors, and the environment models, more sophisticated validation gates might be employed.

The validation equation is used to test the observation z^i_t for membership in the validation gate for each predicted measurement. When a single observation falls in the validation gate, we get a successful match. If one observation falls in multiple validation gates, the best matching candidate is selected or multiple hypotheses are tracked. Observations that do not fall in the validation gate are simply ignored for localization. Such observations could have resulted from objects not in the map, such as new objects (e.g., someone places a large box in the hallway) or transient objects (e.g., humans standing next to the robot may form a line feature). One approach is to take advantage of such unmatched observations to populate the robot's map.

4. Estimation: Applying the Bayes rule. In this step, we compute the best estimate x_t of the robot's position based on the position prediction $\hat{x}_t$ and all the observations z^i_t at time t. To do this position update, we first stack the validated observations z^i_t into a single vector to form z_t and designate the composite innovation v_t. Then, we stack the measurement Jacobians H^j for each validated measurement together to form the composite Jacobian H and the measurement error (noise) vector $R_t = diag[R^i_t]$. From these, we can then compute the composite innovation covariance Σ_{IN_t} using equation (5.82). Finally, by using the results from the application of the Bayes rule to Gaussian distributions, equations (5.74) and (5.75), we can update the robot's position estimate x_t and its associated covariance P_t as

$$x_t = \hat{x}_t + K_t v_t,$$ (5.84)

$$P_t = \hat{P}_t - K_t \cdot \Sigma_{IN_t} \cdot K^T_t,$$ (5.85)

where

$$K_t = \hat{P}_t \cdot H_t^T \cdot (\Sigma_{IN_t})^{-1} \tag{5.86}$$

is the Kalman gain.

As an exercise, the reader can verify that when the h is an identity (5.84) and (5.85) reduce exactly to equations (5.74) and (5.75). Indeed, by imposing H equal to the identity matrix, equation (5.84) simplifies to

$$x_t = \hat{x}_t + \hat{P}_t(\hat{P}_t + R_t)^{-1}(z_t - \hat{x}_t) \tag{5.87}$$

$$P_t = \hat{P}_t - \hat{P}_t(\hat{P}_t + R_t)^{-1}\hat{P}_t, \tag{5.88}$$

which correspond respectively to (5.74) and (5.75).

Equation (5.84) says that the best estimate x_t of the robot state at time t is equal to the best prediction of the value $\hat{x}_t$ before the new measurement z_t is taken, plus a correction term of an optimal weighting value K_t times the difference between z_t and the best prediction $\hat{z}_t$ at time t.

The new, fused estimate of the robot position is again subject to a Gaussian probability density curve. Its mean and covariance are simply functions of two inputs, mean and covariance. Thus the Kalman filter provides both a compact, simplified representation of uncertainty and an extremely efficient technique for combining heterogeneous estimates to yield a new estimate for our robot's position.

In the next section, we will implement a Kalman filter localization algorithm for a differential drive robot.

5.6.8.5 Case study: Kalman filter localization with line feature extraction

The Pygmalion robot at the EPFL is a differential-drive robot that uses a laser rangefinder as its primary sensor [59, 60]. In contrast to Dervish, the environmental representation of Pygmalion is continuous and abstract: the map consists of a set of infinite lines describing the environment. Pygmalion's belief state is, of course, represented as a Gaussian distribution since this robot uses the Kalman filter localization algorithm. The value of its mean position x_t is represented to a high level of precision, enabling Pygmalion to localize with very high precision when desired. We next present details for Pygmalion's implementation of the Kalman filter localization steps. For simplicity we assume that the sensor frame $\{S\}$ is equal to the robot frame $\{R\}$. If not specified, all the vectors are represented in the world coordinate system $\{W\}$.

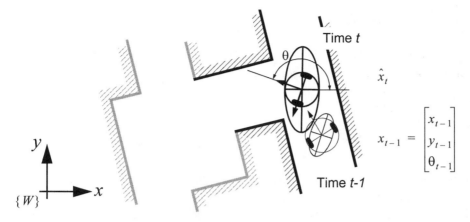

Figure 5.33
Prediction of the robot's position (thick) based on its former position (thin) and the executed movement. The ellipses drawn around the robot positions represent the uncertainties in the x, y direction (e.g., 3σ). The uncertainty of the orientation θ is not represented in the picture.

1. Robot position prediction. Suppose that at time $t-1$ the robot best position estimate is $x_{t-1} = [x_{t-1}, y_{t-1}, \theta_{t-1}]^T$. The control input u_t drives the robot to the position $\hat{x}_t$ (figure 5.33).

The robot position prediction $\hat{x}_t$ at time t can be computed from the previous estimate x_{t-1} and the odometric integration of the movement. For the differential drive that Pygmalion has, we can use the odometry model developed in section 5.2.4:

$$\hat{x}_t = f(x_{t-1}, u_t) = \begin{bmatrix} x_{t-1} \\ y_{t-1} \\ \theta_{t-1} \end{bmatrix} + \begin{bmatrix} \dfrac{\Delta s_r + \Delta s_l}{2} \cos\left(\theta_{t-1} + \dfrac{\Delta s_r - \Delta s_l}{2b}\right) \\ \dfrac{\Delta s_r + \Delta s_l}{2} \sin\left(\theta_{t-1} + \dfrac{\Delta s_r - \Delta s_l}{2b}\right) \\ \dfrac{\Delta s_r - \Delta s_l}{b} \end{bmatrix}, \tag{5.89}$$

where Δs_l, Δs_r characterize the displacement of the left and right wheel. Therefore, the control input is exactly $u_t = [\Delta s_l, \Delta s_r]^T$.

The updated covariance matrix is

$$\hat{P}_t = F_x \cdot P_{t-1} \cdot F_x^T + F_u \cdot Q_t \cdot F_u^T, \qquad (5.90)$$

where P_{t-1} is the covariance of the previous robot state x_{t-1} and Q_t is the covariance of the noise associated to the motion model (see equation [5.8]), that is,

$$Q_t = \begin{bmatrix} k_r |\Delta s_r| & 0 \\ 0 & k_l |\Delta s_l| \end{bmatrix}. \qquad (5.91)$$

2. Observation. For line-based localization, each single observation (i.e., a line feature) is extracted from the raw laser rangefinder data and consists of the two line parameters α_t^i, r_t^i (figure 4.88), because for a rotating laser rangefinder a representation in the polar coordinate frame is more appropriate:

$$z_t^i = \begin{bmatrix} \alpha_t^i \\ r_t^i \end{bmatrix}. \qquad (5.92)$$

After acquiring the raw data at time t, lines and their uncertainties are extracted (figure 5.34a–b. This leads to n observed lines with $2n$ line parameters (figure 5.34c) and a covariance matrix R_t^i for each line that can be calculated from the uncertainties of all the measurement points contributing to each line as developed for line extraction in section 4.7.1:

$$R_t^i = \begin{bmatrix} \sigma_{\alpha\alpha} & \sigma_{\alpha r} \\ \sigma_{r\alpha} & \sigma_{rr} \end{bmatrix}_t^i. \qquad (5.93)$$

3. Measurement prediction. Based on the stored map and the predicted robot position $\hat{x}_t$, the measurement predictions $\hat{z}_t^j$ of the expected features are generated (figure 5.35).[29] These features are stored in the map M and specified in the world coordinate system $\{W\}$. In order to compute the predicted observation, the robot must transform all the line features m^j in the map M into its local robot coordinate frame $\{R\}$. According to figure 5.35, the transformation is given by

---

29. To reduce the required calculation power, there is often an additional step that first selects the possible features, in this case lines, from the whole set of features in the map.

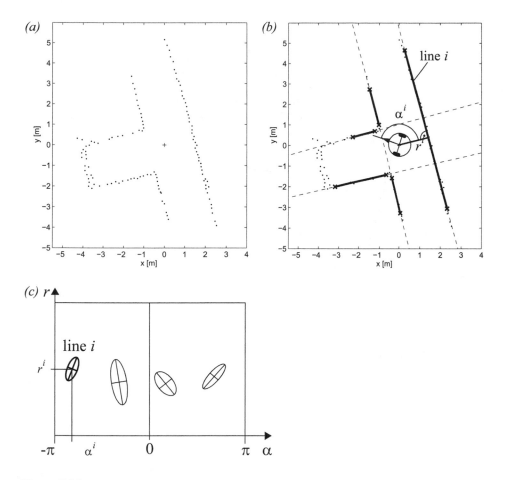

Figure 5.34
Observation: From the raw data (a) acquired by the laser scanner at time t, lines are extracted (b). The line parameters α^i and r^i and its uncertainties can be represented in the model space (c).

$$
\hat{z}_t^j = \begin{bmatrix} \hat{\alpha}_t^j \\ \hat{r}_t^j \end{bmatrix} = h^j(\hat{x}_t, m^j) = \begin{bmatrix} {}^{\{W\}}\alpha_t^j - \hat{\theta}_t \\ {}^{\{W\}}r_t^j - (\hat{x}_t \cos({}^{\{W\}}\alpha_t^j) + \hat{y}_t \sin({}^{\{W\}}\alpha_t^j)) \end{bmatrix}, \tag{5.94}
$$

and its Jacobian H^j by

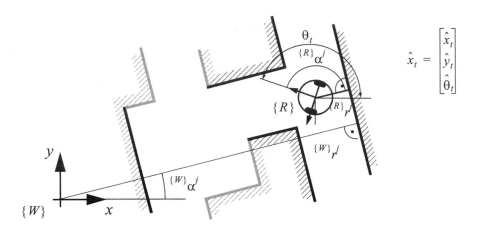

Figure 5.35
Representation of the target position in the world coordinate frame $\{W\}$ and robot coordinate frame $\{R\}$.

$$H^j = \begin{bmatrix} \dfrac{\partial \alpha_t^j}{\partial \hat{x}} & \dfrac{\partial \alpha_t^j}{\partial \hat{y}} & \dfrac{\partial \alpha_t^j}{\partial \hat{\theta}} \\[2ex] \dfrac{\partial r_t^j}{\partial \hat{x}} & \dfrac{\partial r_t^j}{\partial \hat{y}} & \dfrac{\partial r_t^j}{\partial \hat{\theta}} \end{bmatrix} = \begin{bmatrix} 0 & 0 & -1 \\[1ex] -\cos(^{\{W\}}\alpha_t^j) & -\sin(^{\{W\}}\alpha_t^j) & 0 \end{bmatrix}, \tag{5.95}$$

where we used

$$m^j = \begin{bmatrix} ^{\{W\}}\alpha_t^j \\[1ex] r_t^j \end{bmatrix}. \tag{5.96}$$

The measurement prediction results in predicted lines represented in the robot coordinate frame (figure 5.36). They are uncertain, because the prediction of robot position is uncertain.

4. Matching. For matching, we must find correspondence (or a pairing) between predicted and observed features (figure 5.37). In our case we take the Mahalanobis distance

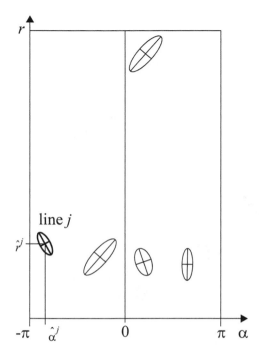

Figure 5.36
Measurement predictions: Based on the and the estimated robot position the targets (visible lines) are predicted. They are represented in the model space similar to the observations.

$$v_t^{ijT} \cdot (\Sigma_{IN_t}^{ij})^{-1} \cdot v_t^{ij} \le g^2 \tag{5.97}$$

with

$$v_t^{ij} = [z_t^i - \hat{z}_t^j] = [z_t^i - h^j(\hat{x}_t, m^j)] =$$

$$= \begin{bmatrix} \alpha_t^i \\ r_t^i \end{bmatrix} - \begin{bmatrix} {}^{\{W\}}\alpha_t^j - \hat{\theta}_t \\ {}^{\{W\}}r_t^j - (\hat{x}_t \cos({}^{\{W\}}\alpha_t^j) + \hat{y}_t \sin({}^{\{W\}}\alpha_t^j)) \end{bmatrix}, \tag{5.98}$$

$$\Sigma_{IN_t}^{ij} = H^j \cdot \hat{P}_t \cdot H^{jT} + R_t^i \tag{5.99}$$

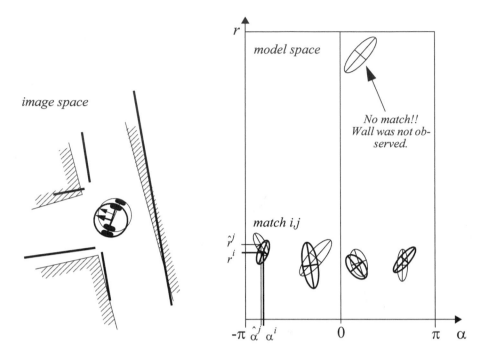

Figure 5.37
Matching: The observations (thick) and the predicted observation (thin) are matched and the innovation and its uncertainties are calculated.

to enable finding the best matches while eliminating all other remaining observed and predicted unmatched features.

5. Estimation. Applying the Kalman filter results in a final pose estimate corresponding to the weighted sum of (figure 5.38);

- the pose estimates of each matched pairing of observed and predicted features;

- the robot position estimation based on odometry and observation positions.

5.7   Other Examples of Localization Systems

Markov localization and Kalman filter localization have been two extremely popular strategies for research mobile robot systems navigating indoor environments. They have strong formal bases and therefore well-defined behavior. But there are other probabilistic localiza-

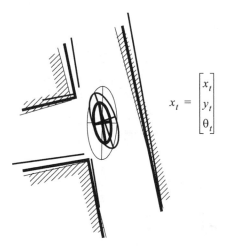

$$x_t = \begin{bmatrix} x_t \\ y_t \\ \theta_t \end{bmatrix}$$

Figure 5.38
Kalman filter estimation of the new robot position: By fusing the prediction of robot position (thin) with the innovation gained by the measurements (thick) we get the updated estimate x_t of the robot position (very thick).

tion techniques that have been used with varying degrees of success on commercial and research mobile robot platforms. Some techniques that deserve mention are Unscented Kalman Filter (UKF) localization, grid localization, and Monte Carlo localization. UKF is similar to EKF in that it also assumes Gaussian distributions but relies on a different way to linearize the motion and measurements models, which is called the *unscented transform*. Conversely, grid localization and Monte Carlo localization are not limited to unimodal distributions. While grid localization uses the so-called histogram filter to represent the robot belief, Monte Carlo localization uses particle filters. The latter is probably the most popular localization algorithm (this was already introduced on page 315). Because a description of these techniques goes beyond the scope of this book, we refer the reader to [51] for such information.

There are, however, several categories of localization techniques that deserve mention. Not surprisingly, many implementations of these techniques in commercial robotics employ modifications of the robot's environment, something that the Markov localization and Kalman filter localization communities eschew. In the following sections, we briefly identify the general strategy incorporated by each category and reference example systems, including, as appropriate, those that modify the environment and those that function without environmental modification.

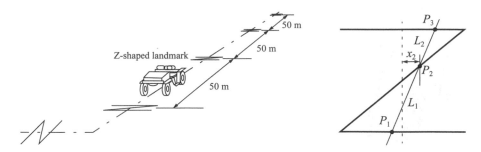

Figure 5.39
Z-shaped landmarks on the ground. Komatsu Ltd., Japan [6, pages 179-180].

5.7.1 Landmark-based navigation

Landmarks are generally defined as passive objects in the environment that provide a high degree of localization accuracy when they are within the robot's field of view. Mobile robots that make use of landmarks for localization generally use artificial markers that have been placed by the robot's designers to make localization easy.

The control system for a landmark-based navigator consists of two discrete phases. When a landmark is in view, the robot localizes frequently and accurately, using action update and perception update to track its position without cumulative error. But when the robot is in a no-landmark "zone," then only action update occurs, and the robot accumulates position uncertainty until the next landmark enters the robot's field of view.

The robot is thus effectively *dead-reckoning* from landmark zone to landmark zone. This in turn means the robot must consult its map carefully, ensuring that each motion between landmarks is sufficiently short, given its motion model, that it will be able to localize successfully upon reaching the next landmark.

Figure 5.39 shows one instantiation of landmark-based localization. The particular shape of the landmarks enables reliable and accurate pose estimation by the robot, which must travel using *dead reckoning* between the landmarks.

One key advantage of the landmark-based navigation approach is that a strong formal theory has been developed for this general system architecture [187]. In this work, the authors have shown precise assumptions and conditions which, when satisfied, guarantee that the robot will always be able to localize successfully. This work also led to a real-world demonstration of landmark-based localization. Standard sheets of paper were placed on the ceiling of the Robotics Laboratory at Stanford University, each with a unique checkerboard pattern. A Nomadics 200 mobile robot was fitted with a monochrome CCD camera aimed vertically up at the ceiling. By recognizing the paper landmarks, which were placed approximately 2 m apart, the robot was able to localize to within several centimeters, then move, using dead reckoning, to another landmark zone.

The primary disadvantage of landmark-based navigation is that in general it requires significant environmental modification. Landmarks are local, and therefore a large number are usually required to cover a large factory area or research laboratory. For example, the Robotics Laboratory at Stanford made use of approximately thirty discrete landmarks, all affixed individually to the ceiling.

5.7.2 Globally unique localization

The landmark-based navigation approach makes a strong general assumption: when the landmark is in the robot's field of view, localization is essentially perfect. One way to reach the Holy Grail of mobile robotic localization is effectively to enable such an assumption to be valid no matter *where* the robot is located. It would be revolutionary if a look at the robot's sensors immediately identified its particular location, uniquely and repeatedly.

Such a strategy for localization is surely aggressive, but the question of whether it can be done is primarily a question of sensor technology and sensing software. Clearly, such a localization system would need to use a sensor that collects a very large amount of information. Since vision does indeed collect far more information than previous sensors, it has been used as the sensor of choice in research toward globally unique localization. If humans were able to look at an individual picture and identify the robot's location in a well-known environment, then one could argue that the information for globally unique localization does exist within the picture; it must simply be teased out. As described in section 4.6, an important milestone toward this direction has been achieved with "bag of features" approaches, where the current image is first converted into a "bag" of distinctive local features (section 4.5) which are then used to find the most similar images in a dataset of million of pictures in less than one second. This approach successfully demonstrated robust localization on a more than 1,000 km trajectory using purely images collected from vehicle-mounted camera [108,109].

If one would like to use laser scans instead of camera images, then the angular histogram depicted in figure 4.95 of the previous chapter is another example in which the robot's laser sensor values are transformed into an identifier of location. In this case, the identifier is a histogram instead of a bag of features. However, due to the limited information content of laser scans, it is likely that two *places* in the robot's environment may have angular histograms that are too similar to be differentiated successfully. Therefore, image-based localization should be preferred for large-scale environments, since images provide better globally unique localization than laser-based strategies.

The key advantage of globally unique localization is that, when these systems function correctly, they greatly simplify robot navigation. The robot can move to any point and will always be assured of localizing by collecting a sensor scan.

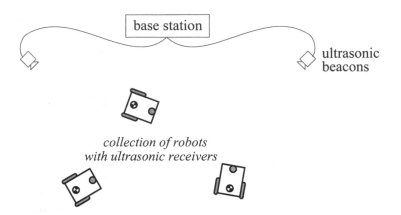

Figure 5.40
Active ultrasonic beacons.

But the main disadvantage of globally unique localization is that it is likely that this method will *never* offer a complete solution to the localization problem. There will always be cases where local sensory information is truly ambiguous, and globally unique localization using only current sensor information is therefore unlikely to succeed (e.g., in a forest!). Humans often have excellent *local positioning systems*, particularly in nonrepeating and well-known environments such as their homes. However, there are a number of environments in which such immediate localization is challenging even for humans: consider hedge mazes and large new office buildings with repeating halls that are identical.

5.7.3 Positioning beacon systems

One of the most reliable solutions to the localization problem is to design and deploy an active beacon system specifically for the target environment. This is the preferred technique used by both industry and military applications as a way of ensuring the highest possible reliability of localization. The GPS system can be considered as just such a system (see section 4.1.8.1).

Figure 5.40 depicts one such beacon arrangement for a collection of robots. Just as with GPS, by designing a system whereby the robots localize passively while the beacons are active, any number of robots can simultaneously take advantage of a single beacon system. As with most beacon systems, the design depicted depends foremost upon geometric principles to effect localization. In this case the robots must know the positions of the two active ultrasonic beacons in the global coordinate frame in order to localize themselves to the global coordinate frame.

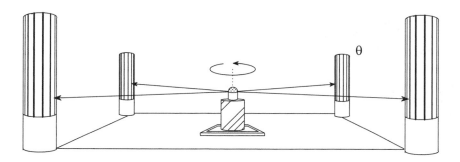

Figure 5.41
Passive optical beacons.

A popular type of beacon system in industrial robotic applications is depicted in figure 5.41. In this case, beacons are retroreflective markers that can be easily detected by a mobile robot based on their reflection of energy back to the robot. Given known positions for the optical retroreflectors, a mobile robot can identify its position whenever it has three such beacons in sight simultaneously. Of course, a robot with encoders can localize over time as well, and it does not need to measure its angle to all three beacons at the same instant.

The advantage of such beacon-based systems is usually extremely high engineered reliability. By the same token, significant engineering usually surrounds the installation of such a system in a specific commercial setting. Therefore, moving the robot to a different factory floor will be both time-consuming and expensive. Usually, even changing the routes used by the robot will require serious reengineering.

5.7.4 Route-based localization

Even more reliable than beacon-based systems are route-based localization strategies. In this case, the route of the robot is explicitly marked so that it can determine its position, not relative to some global coordinate frame but relative to the specific path it is allowed to travel. There are many techniques for marking such a route and the subsequent intersections. In all cases, one is effectively creating a railway system, except that the railway system is somewhat more flexible and certainly more human-friendly than a physical rail. For example, high ultraviolet-reflective, optically transparent paint can mark the route such that only the robot, using a specialized sensor, easily detects it. Alternatively, a guidewire buried underneath the hall can be detected using inductive coils located on the robot chassis.

In all such cases, the robot localization problem is effectively trivialized by forcing the robot to always follow a prescribed path. To be fair, there are new industrial *unmanned*

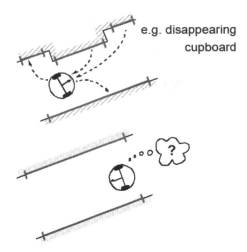

Figure 5.42 An autonomous robot should be able to track changes in the environments for localizing.

guided vehicles that do deviate briefly from their route in order to avoid obstacles. Nevertheless, the cost of this extreme reliability is obvious: the robot is much more inflexible given such localization means, and therefore any change to the robot's behavior requires significant engineering and time.

5.8 Autonomous Map Building

5.8.1 Introduction

All the localization strategies that we have discussed so far require the existence of a map of the environment. The map is normally built by hand. This means that, for accurate localization, the position of the landmarks (e.g., walls, artificial beacons, etc.) that the robot uses for self-localizing must be accurately measured and included in the map. Unfortunately, this approach can be hard, costly, and very time-consuming when the size of the environment is large or when the environment changes due to artificial modifications or dynamic objects. Assume, for instance, a domestic robot that is supposed to work in home environments. In this case the robot should be able to detect changes in the map due to the re-arrangement of the furniture (figure 5.42). Another drawback of handmade maps is that the look of the map can be different depending on the different perception of who makes the map.

The alternative to handmade map building is therefore "automatic map building." Indeed, a robot that localizes successfully has the right sensors for detecting the environment, and so the robot ought to build its own map. This ambition goes to the heart of autonomous mobile robotics. In prose, we can express our eventual goal as follows: *starting from an arbitrary initial point, a mobile robot should be able to explore autonomously the environment with its on-board sensors, gain knowledge about it, interpret the scene, build an appropriate map, and localize itself relative to this map.*

The recent advances in both robotics and computer vision have made this goal somewhat achieved. An important subgoal has been the invention of techniques for place recognition and for autonomous creation and modification of an environmental map. Of course a mobile robot's sensors have only a limited range, and so the robot must physically explore its environment to build such a map. So, the robot must not only create a map, but it must also do this while moving and localizing to explore the environment. In the robotics community, this is often called the Simultaneous Localization and Mapping (SLAM) problem. The relevance of the SLAM problem for the robotics community owes to the fact that the solution to this problem would make a robot truly autonomous.

After a short introduction to the simultaneous localization and mapping problem (section 5.8.2), we will review three major algorithms from which a large number of published methods have been derived. First, we will review the traditional approach, which is based on the extended Kalman filter (section 5.8.4). As an application of one of these methods, we will present the Visual-SLAM algorithm (section 5.8.5), which uses a single camera as only sensor. Second, we will review the Graph-SLAM algorithm (section 5.8.7), which is born from the intuition that the SLAM problem can be interpreted as a sparse graph of constraints. Third, we will review the particle-filter SLAM (section 5.8.8). Finally, we will discuss open problems in SLAM (section 5.8.9).

It is important to point out that none of these method is the favorite solution to the SLAM problem. The choice of the right method will depend on the number and type of features in the environments, the resolution of the desired map, the computational time, and so on.

For an in-depth study of SLAM algorithms, we refer the reader to [51]. Up-to-date references and on-line software can instead be found in these tutorials [62, 120, 313]. In the following sections, we will keep the same notation as in the section on localization, which is also the same as in [51].

5.8.2 SLAM: The simultaneous localization and mapping problem

As we have seen in section 5.6.2, localization is the problem of estimating the robot position (and therefore its path) given a known map of the environment. Conversely, mapping is the construction of the map of the environment knowing the true path of the robot. The aim of SLAM is to recover both the robot path and the environment map using only the data gathered by its proprioceptive and exteroceptive sensors. These data are typically the robot displacement estimated from the odometry and features (e.g., corners, lines, planes) extracted from laser, ultrasonic, or camera images.

SLAM is difficult because both the estimated path and the extracted features are corrupted by noise. The problem is illustrated in figure 5.43. Let us assume that the robot

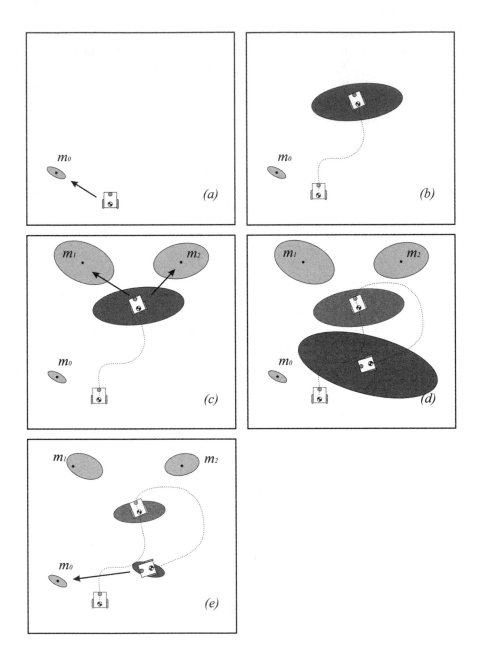

Figure 5.43 Illustration of the SLAM problem.

uncertainty at its initial location is zero. From this position, the robot observes a feature which is mapped with an uncertainty related to the exteroceptive sensor error model (a). As the robot moves, its pose uncertainty increases under the effect of the errors introduced by the odometry (b). At this point, the robot observes two features and maps them with an uncertainty that results from the combination of the measurement error with the robot pose uncertainty (c). From this, we can notice that the map becomes correlated with the robot position estimate. Similarly, if the robot updates its position based on an observation of an imprecisely known feature in the map, the resulting position estimate becomes correlated with the feature location estimate. In order to reduce its uncertainty, the robot must observe features whose location is relatively well known. These features can, for instance, be landmarks that the robot has already observed before. In this case, the observation is called *loop closure detection*. When a loop closure is detected, the robot pose uncertainty shrinks. At the same time, the map is updated and the uncertainty of other observed features and all previous robot poses also reduce (figure 5.43e).

The general problem of map-building is thus an example of the chicken-and-egg problem. For localization the robot needs to know where the features are, whereas for map-building the robot needs to know where it is on the map.

5.8.3 Mathematical definition of SLAM

As we already did for map-based localization, we also describe SLAM in probabilistic terminology. The terminology used in the section is the same introduced in sections 5.6.4 and 5.6.5 for probabilistic map-based localization. The reader may take a moment to review those sections before proceeding.

Let us recall that we define the robot pose at time t by x_t. The robot path is given as

$$X_T = \{x_0, x_1, x_2, ..., x_T\}, \tag{5.100}$$

where T might also be infinite. In SLAM, the robot initial location x_0 is assumed to be known, while the others locations are not.

Let u_t denote the robot motion between time $t-1$ and time t. Let us recall that these data can be the proprioceptive sensor readings (e.g., from the robot's wheel encoders) or the control inputs given to the motors. The sequence of the robot relative motions can then be written as:

$$U_T = \{u_0, u_1, u_2, ..., u_T\}. \tag{5.101}$$

Let M denote the *true* map of the environment

$$M = \{m_0, m_1, m_2, ..., m_{n-1}\}, \tag{5.102}$$

then m_i, $i = 0...n-1$, are vectors representing the positions of the landmarks that, again, might be points, lines, planes, or any sort of high-level feature (e.g., doors). Observe that, for simplicity, we assume that the map is static.

Finally, if we assume that the robot takes one measurement at each time, we can denote by

$$Z_T = \{z_0, z_1, z_2, ..., z_T\} \tag{5.103}$$

the sequence of landmark observations in the sensor reference frame attached to the robot. For example, if the robot is equipped with an on-board camera, the observation z_i can be a vector representing the coordinates of a corner or those of a line in the image. If, instead, the robot is equipped with a laser rangefinder, such a vector can represent the position of a corner or a line in the laser sensor frame.

According to this terminology, we can now define SLAM as the problem of recovering a model of the map M and the robot path X_T from the odometry U_T and observations Z_T. In the literature, we distinguish between the *full* SLAM problem and the *online* SLAM problem. The full SLAM problem consists in estimating the joint posterior probability over X_T and M from the data, that is

$$p(X_T, M|Z_T, U_T) . \tag{5.104}$$

The online SLAM problem, conversely, consists in estimating the joint posterior over x_t and M from the data, that is

$$p(x_t, M|Z_T, U_T) . \tag{5.105}$$

Therefore, the full SLAM problem tries to recover the entire robot path X_T, while the online SLAM problem tries to estimates only the current robot pose x_t.

In order to solve the SLAM problem, we need to know the probabilistic motion model and probabilistic measurement model. These models have been introduced in section 5.6.5. In particular, let us recall that

$$p(x_t|x_{t-1}, u_t) \tag{5.106}$$

represents the probability that the robot pose is x_t given the robot previous pose x_{t-1} and proprioceptive data (or control input) u_t. Similarly, let us recall that

$$p(z_t|x_t, M) \tag{5.107}$$

is the probability of measuring z_t given the known map M and assuming that the robot takes the observation at location x_t.

We encourage the reader to take a moment to review these concepts in section 5.6.5.

In the next sections, we will describe the three main paradigms developed over the last two decades to solve the SLAM problem, which are EKF SLAM, graph-based SLAM, and particle filter SLAM. From these paradigms, many other algorithms have been derived. For an in-depth study of these algorithms, we refer the reader to [51].

5.8.4 Extended Kalman Filter (EKF) SLAM

In this section, we will see the application of the EKF to the online SLAM problem. EKF-based SLAM is historically the first formulation proposed and was introduced in several papers [100, 294, 295, 228, 229].

The EKF SLAM proceeds exactly like the standard EKF that we have seen for robot localization (section 5.6.8), with the only difference that it uses an extended state vector y_t which comprises both the robot pose x_t and the position of all the features m_i in the map, that is:

$$y_t = [x_t, m_0, ..., m_{n-1}]^T.$$
(5.108)

In our localization example based on line features (section 5.6.8.5), the dimension of y_t would be $3+2n$, since we need three variables to represent the robot pose (x, y, θ) and $2n$ variables for the n line-landmarks having vector components (α^i, r^i). Therefore, the state vector would be written as

$$y_t = [x_t, y_t, \theta_t, \alpha_0, r_0, \alpha_1, r_1, ..., \alpha_{n-1}, r_{n-1}]^T.$$
(5.109)

As the robot moves and takes measurements, the state vector and covariance matrix are updated using the standard equations of the extended Kalman filter. Clearly, the state vector in EKF SLAM is much larger that the state vector in EKF localization where only the robot pose was being updated. This makes EKF SLAM computationally much more expensive.

Notice that, because of its formulation, maps in EKF SLAM are supposed to be feature-based (i.e., points, lines, planes). As new features are observed, they are added to the state vector. Thus, the noise covariance matrix grows quadratically, with size $(3 + 2n) \times (3 + 2n)$. For computational reasons, the size of the map is therefore usually limited to less than a thousand features. However, numerous approaches have been developed to cope with a larger number of features, which decompose the map into smaller submaps, for which covariances are updated separately [63].

As we mentioned, the implementation of the EKF SLAM is nothing but the straightforward application of the EKF equations to the online SLAM problem, that is, equations

(5.78)–(5.79) and (5.84)–(5.85). In order to do this, we need to specify the functions that characterize the prediction and measurement model. If we use again our line-based localization example of section 5.6.8.5, the measurement model is then the same as in equation (5.94). The prediction model, conversely, has to take into account that the motion will only update the robot pose according to (5.89), while the features will remain unchanged. Therefore, we can write the prediction model of the EKF SLAM as

$$
\hat{y}_t = y_t +
\begin{bmatrix}
\dfrac{\Delta s_r + \Delta s_l}{2} \cos\left(\theta_{t-1} + \dfrac{\Delta s_r - \Delta s_l}{2b}\right) \\[2ex]
\dfrac{\Delta s_r + \Delta s_l}{2} \sin\left(\theta_{t-1} + \dfrac{\Delta s_r - \Delta s_l}{2b}\right) \\[2ex]
\dfrac{\Delta s_r - \Delta s_l}{b} \\[2ex]
0 \\
0 \\
\dots \\
0 \\
0
\end{bmatrix}.
\tag{5.110}
$$

At the start, when the robot takes the first measurements, the covariance matrix is populated by assuming that these (initial) features are uncorrelated, which implies that the off diagonal elements are set to zero. However, when the robot starts moving and takes new measurements, both the robot pose and features start becoming correlated. Accordingly, the covariance matrix becomes *nonsparse*.[30] The existence of this correlation can be explained by recalling that the uncertainty of the features in the map depends on the uncertainty associated to the robot pose. But it also depends on the uncertainty of other features that have been used to update the robot pose. This means that when a new feature is observed, this contributes to correct not only the estimate of the robot pose but also that of the other features as well. The more observations are made, the more the correlations between the features will grow. Therefore, the correlations between the features—and so the fact that the covariance matrix is nonsparse—are of significant important in SLAM [105]: the bigger these correlations, the better the solution of the SLAM.

Figure 5.43 illustrates the working principle of EKF SLAM in a simple environment with three features. The robot initial location is assumed as the origin of the system reference frame and therefore the initial uncertainty of the robot pose is zero. From this position, the robot observes a feature and maps it with an uncertainty related to the sensor error

---

30. In numerical analysis, a sparse matrix is a matrix populated primarily with zeros.

model (a). As the robot moves, its pose uncertainty increases under the effect of the errors introduced by the odometry (b). At some point, the robot observes two features and maps them with an uncertainty which results from the combination of the measurement error with the robot pose uncertainty (c). From this, we can notice that the map becomes correlated with the robot position estimate. Now, the robot drives back toward its starting position and its pose uncertainty increases again (d). At this point, it reobserves the first feature, whose location is relatively well known compared to the other features. This makes the robot more sure about its current location and therefore its pose uncertainty shrinks (e). Notice that so far we only considered the online SLAM problem. Therefore, only the robot current position was being updated. The full SLAM problem, conversely, updates the entire robot path and thus all its previous poses. In this case, after reobserving the first feature, also the robot previous pose uncertainty will shrink and so will also the uncertainties associated to the other features. The position of these features is in fact correlated with the robot previous poses.

EKF SLAM has been successfully applied in many different domains, including airborne, underwater, outdoor, and indoor environments. Figure 5.44 shows results of a 6DoF SLAM using a 3D laser rangefinder. The robot starts at the center and makes three rounds. Figure 5.44a shows the resulting map using only odometry. As you can see, the map is inconsistent (the scans are not aligned) due to the accumulated odometry drift. In (b), the accumulated odometry error is drastically reduced by using scan matching and alignment techniques.[31] Finally, in (c), the accumulated drift and the offset error are no longer present after application of EKF SLAM. Notice that in this particular application, horizontal and vertical planes have been used as features.

The basic formulation of EKF SLAM assumes that the position of the features is fully measurable from a single robot location. This is because most SLAM applications have been realized using rangefinders (i.e., lasers, sonars, or stereo cameras) that provide both range and bearing information about the features. However, there are situations where either the range [190] or the bearing information (the angle) is available. The latter occurs, for example, when using a single camera. As seen in section 4.2.3, a calibrated camera is a bearing sensor (figure 4.31). In this case, the SLAM problem is usually called *monocular Visual SLAM* or *bearing-only SLAM* [110, 227]. In this case, the standard EKF can still be applied, as we will see in the next section.

---

31. One of the most popular techniques for aligning two different laser scans is the Iterative Closest Point (ICP) algorithm [72]. This, however, works well only if the relative motion between the two scans is known with good approximation (for instance from the odometry) otherwise other global optimization techniques are required [97,204].

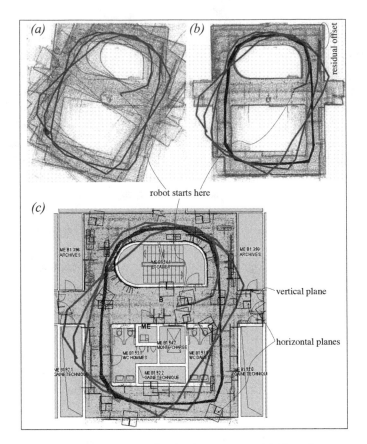

Figure 5.44 EKF SLAM using 3D laser scanner. (a) The robot starts at the center and makes three rounds. (a) Aligned 3D scans using odometry only leading to an inconsistent map. (b) Aligned 3D scans after using scan matching. The accumulated odometry error can be drastically reduced, but a small residual error remains (see offset).(c) The result of the EKF SLAM. The map built has been superimposed on a building plan for visual comparison. Notice that the offset is no longer present. Image courtesy of J. Weingarten [331].

5.8.5 Visual SLAM with a single camera

The term Visual SLAM (V-SLAM) was coined in 2003 by Davison [110, 112], who presented the first real-time EKF SLAM system with a single hand-held camera. No odometry, range finder, or GPS was used but just a single perspective camera. V-SLAM can be seen as a multiview Structure-from-Motion (SfM) (which we introduced in section 4.2.6). Indeed, both attempt to recover simultaneously both the camera motion and the structure (feature positions) of the environment from a single camera by tracking interest points in the images. The main difference between V-SLAM and SfM is in that V-SLAM takes into account the feature uncertainty using a probabilistic framework. Another difference is that V-SLAM needs to process the images chronologically while SfM works also for unordered datasets. The original V-SLAM implementation by Davison used an extended Kalman filter. As we mentioned at the end of section 5.8.4, V-SLAM is also called bearing-only SLAM, to emphasize the fact that it uses only angle observations. This, again, is in contrast to laser-based or ultrasound-based SLAM, which need instead both angle and range information. Because of this, bearing-only SLAM is more challenging than range-bearing SLAM. In laser-based SLAM, the position of the features in the robot frame can be estimated from a single robot position. In V-SLAM, conversely, we need to move the camera to recover the position of the features, as we know from structure-from-motion.

The first problem in monocular V-SLAM is the estimation of the position of the features at the time the system starts. Using a rangefinder, this is obviously not a problem, but for V-SLAM this is not possible. To overcome this problem, in his original implementation Davison used a planar pattern of known geometry where the relative position of at least four boundary corners is known. From four corners of known position, the 6DoF camera pose with respect to these points can be determined uniquely.[32] As long as the camera is moved in front of the pattern, the camera pose can be estimated from single images. As the camera starts moving away from the pattern, new features must be triangulated and added to the map. At this point, the EKF V-SLAM process starts.

The implementation of the EKF V-SLAM is again the vanilla EKF applied to the SLAM problem. To implement it, we need to know the motion and measurement update functions.

As we did for the standard EKF SLAM, the state vector y contains both the camera pose and the feature position but this time also the camera velocity. Also, observe that Davison chose to parametrize the camera orientation with *quaternions* in order to avoid singularities, and thus the camera orientation is represented with four variables. The dimension of the state vector in the EKF V-SLAM is therefore $13 + 3n$; in fact, we need three parameters for the position r, four for the orientation-quaternion q, three for the translational velocity v, another three for the angular velocity ω, and $3n$ for the feature positions m_i. Also observe that in his implementation, the observed features are not lines but image points and therefore the feature position is represented by three cartesian coordinates. The vector state at time t can therefore be written as

$$y_t = [x_t, m_0, m_1, ..., m_{n-1}]^T, \tag{5.111}$$

where

---

32. The problem of determining the camera position and orientation from a set of 2D-3D point correspondences is known as *camera pose estimation* [29].

$$x_t = [r_t, q_t, v_t, \omega_t]^T. \tag{5.112}$$

Prediction step. Notice that in V-SLAM we do not use odometry to predict the next camera position. To overcome this problem, Davison proposed the use of a constant velocity model. This means that between consecutive frames the velocity is assumed to be constant, and therefore the position of the camera at time t is computed by integrating the motion starting at time t-1, assuming that the initial velocity is the one estimated at time t-1. By keeping this in mind, we can actually write the motion prediction function f as:

$$\hat{x}_t = f(x_{t-1}, u_t) = \begin{bmatrix} r + (v + V)\Delta t \\ q \times q((\omega + \Omega)\Delta t) \\ v + V \\ \omega + \Omega \end{bmatrix}, \tag{5.113}$$

where the unknown intentions (in terms of velocity and acceleration) of the carrier of the camera are taken into account in the constant velocity model by V and Ω, which are computed as:

$$V = a\Delta t \text{ and } \Omega = \alpha\Delta t, \tag{5.114}$$

where a and α are the unknown translational and angular accelerations that are modeled as zero mean Gaussian distributions. The prediction update equation of the EKF can therefore be written as

$$\begin{bmatrix} \hat{x}_t \\ \hat{\alpha}_t^0 \\ \hat{r}_t^0 \\ \dots \\ \hat{\alpha}_t^{n-1} \\ \hat{r}_t^{n-1} \end{bmatrix} = \begin{bmatrix} f(x_{t-1}, u_t) \\ \alpha_{t-1}^0 \\ r_{t-1}^0 \\ \dots \\ \alpha_{t-1}^{n-1} \\ r_{t-1}^{n-1} \end{bmatrix}, \tag{5.115}$$

where $(\hat{\alpha}_t^i, \hat{r}_t^i)$ and $(\alpha_{t-1}^i, r_{t-1}^i)$ denote the positions of the i-th feature at times t and t-1 respectively.

Measurement update. In the measurement update, the camera pose is corrected based on the reobservation of features. In addition, new features are initialized and added to the map.

Figure 5.45 (Left) Feature image patches. (Right) Search regions predicted from the previous frame using a constant velocity motion model. Image courtesy of Andrew Davison [110].

In V-SLAM, the features are interest points (figure 5.45) extracted using one of the interest point detectors described in section 4.5. Therefore, the features are expressed in image pixel coordinates.

In this phase, we also have to define the measurement function h. This function is used to compute the predicted observations, that is, to predict where the features are going to appear after the motion update. To determine h, we need to take into account the transformation from the world coordinate frame to the local camera frame and, in addition, the perspective transformation from the camera frame onto the image plane (figure 4.32, equation [4.44]). Therefore, function h is given exactly by equation (4.44). Finally, after computing the uncertainty of each predicted observations (which is drawn as an ellipse in figure 5.45), we can update the state vector and its covariance using the standard EKF measurement update equations. The main steps of Davison's V-SLAM are illustrated in figure 5.46.

5.8.6 Discussion on EKF SLAM

As we mentioned earlier, EKF SLAM is nothing but the application of the vanilla extended Kalman filter with a joint state composed of the robot pose and the feature locations. At every iteration both the state and the joint covariance matrix are updated, which means that the computation grows quadratically with the number of features. In order to overcome these limitations, efficient real-time implementations of EKF SLAM have been proposed in the last years, which can cope with thousand of features. The main idea is to decompose the map into smaller submaps for which covariances are updated separately.

Another problem of EKF SLAM is in the linearization made by the extended Kalman filter, which is reflected by the use of the Jacobians in the motion and measurement updates. Unfortunately, both the motion and the measurement model are typically nonlin-

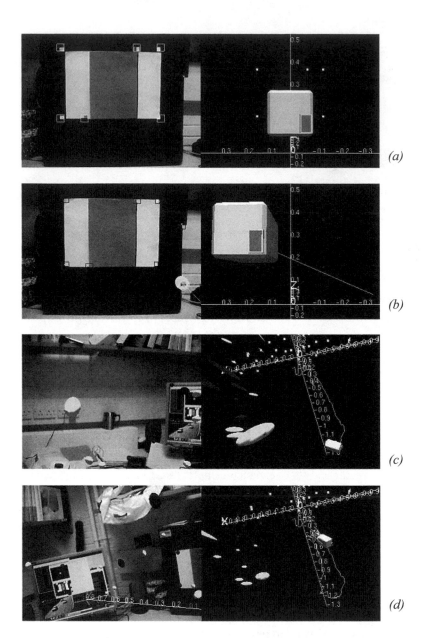

Figure 5.46 (a) The camera starts moving with six known features on a pattern. (b) Nearby unknown features are initialized and added to the map. (c) As the camera moves, the uncertainty of the estimated features in the map increases. (d) As the camera revisits some of the features seen at the beginning, its uncertainty shrinks. Image courtesy of Andrew Davison [110,112].

ear, therefore their linearization can sometimes lead to inconsistency or divergence of the solution.

Another issue in EKF SLAM is its sensitiveness to incorrect data associations of the features, which happens when the robot incorrectly matches feature m_i with features m_j. This problem becomes even more important at the loop closure, that is, when the robot returns to reobserve features after a long traverse. Incorrect data association can occur frequently with 2D laser rangefinders due to the difficulty of identifying distinctive features in their point clouds. This task is, however, facilitated with cameras thanks to the huge availability of feature detectors (section 4.5). Some of these detectors, like SIFT (section 4.5.5.1), have recently demonstrated very successful results in loop closure detection over very long traverses (1000 km) [109] by employing the "bag of features" approach (see section 4.6 on location recognition).

We have also seen that the correlations between features is of significant importance in SLAM. The more observations are made, the more the correlations between features grow, the better the solution of the SLAM. Eliminating or ignoring the correlations between features (like it was done at the beginning of the research in EKF SLAM) is exactly contrary to the nature of the SLAM problem. As the robot moves and observes some features, these become more and more correlated. In the limit, they become fully correlated, that is, given the exact position of any feature, the location of any other feature can be determined with absolute precision.[33]

Regarding the convergence of the map, as the robot moves making observations, the determinant of the map covariance matrix and of all covariance submatrices converges monotonically toward zero. This means that the error in the relative position between the features decreases to the point where the map is known with absolute certainty or, alternatively, it reaches a lower bound that depends on the error introduced when the first observation was taken.

5.8.7 Graph-based SLAM

Graph-based SLAM was introduced for the first time in [200], which influenced many other implementations. Most of the graph-based SLAM techniques attempt to solve the full SLAM problem, but several approaches can also be found in the literature to solve the online SLAM problem.

Graph-based SLAM is born from the intuition that the SLAM problem can be interpreted as a sparse graph of nodes and constraints between nodes. The nodes of the graph are the robot locations $x_0, x_1, ..., x_T$ and the n features in the map $m_0, m_1, ..., m_{n-1}$. The constraints are the relative position between consecutive robot poses x_{t-1}, x_t (given by the

---

33. Note, this is only possible in principle. In real scenarios, there is always some uncertainty left (e.g., measurement uncertainty).

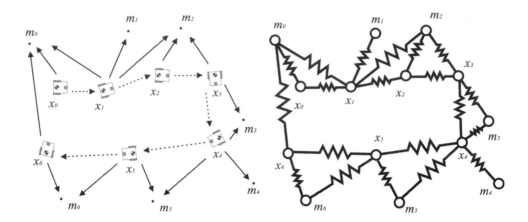

Figure 5.47 Evocative illustration of the graph construction. Constraints between nodes in the graph are represented as "soft" constraints (like springs). The solution of the SLAM problem can then be computed as the configuration with minimal energy.

odometry input u_t) and the relative position between the robot locations and the features observed from those locations.

The key property to remember about graph-based SLAM is that the constraints are not to be thought as rigid constraints but as soft constraints (figure 5.47). It is by relaxing these constraints that we can compute the solution to the full SLAM problem, that is, the best estimate of the robot path and the environment map. In other words, graph-based SLAM represents robot locations and features as the nodes of an elastic net. The SLAM solution can then be found by computing the state of minimal energy of this net [139]. Common optimization techniques to find the solution are based on gradient descent, and similar ones. A very efficient minimization procedure, along with open source code, was proposed in [141].

There is a significant advantage of graph-based SLAM techniques over EKF SLAM. As we have seen, in EKF SLAM the amount of computation and memory requirement to update and store the covariance matrix grows quadratically in the number of features. Conversely, in graph-based SLAM the update time of the graph is constant and the required memory is linear in the number of features. However, the final graph optimization can become computationally costly if the robot path is long. Nevertheless, graph-based SLAM algorithms have shown impressing and very successful results with even hundred million features [79, 116, 117, 173, 315]. However, these algorithms attempt to optimize over the entire robot path and were therefore implemented to work offline. Some of the online implementations used submap approaches.

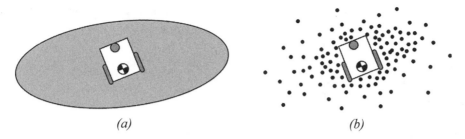

Figure 5.48 The standard EKF SLAM represent the probability distribution of the robot location is a parametric form, which is a two-dimensional Gaussian (a). Conversely, particle filter SLAM represent this the probability distribution as a set of particles drawn randomly from the parametric distribution (b). For the specific example of a Gaussian distribution, (b) the density of particles is higher toward the center of the Gaussian and decreases with the distance.

5.8.8 Particle filter SLAM

This particular solution to the SLAM problem is based on the randomized sampling of the belief distribution that we introduced already on page 315. The term *particle* filter is born from the fact that it represents the robot belief distribution not in a parametric form (like a Gaussian) but rather as a set of samples (i.e., *particles*) drawn randomly from this distribution. This concept is pictorially illustrated in figure 5.48. The power of this representation is in its ability to model any sort of distribution (e.g., non-Gaussian) and also nonlinear transformations.

Particle filters find their origin in Monte Carlo methods [215], but a step that makes them practically applicable to the SLAM problem is based on the work of Rao and Blackwell [75, 263], from which these filters inherited the name of *Rao-Blackwellized* particle filters. Finally, the Rao-Blackwellized particle filter was applied for the first time to the SLAM problem by Murphy and Russel [239] and found a very efficient implementation in the work of Montemerlo et al. [231], who also coined the name of FastSLAM.

Now we will give a general overview of the particle filter SLAM. For a detailed explanation of the solution to this problem, we refer the reader to the original paper on Fast-SLAM [231].

At every time step, the particle filter maintains always the same number K of particles (e.g., $K = 1000$). Each particle contains an estimate of the robot path $X_t^{[k]}$ and estimates of the position of each feature in the map, which are represented as two-dimensional Gaussians with mean values $\mu_{t,i}^{[k]}$ and covariance matrices $\Sigma_{t,i}^{[k]}$. Therefore, a particle is characterized by

$$X_t^{[k]}; \ (\mu_{t,0}^{[k]}, \Sigma_{t,0}^{[k]}); \ (\mu_{t,1}^{[k]}, \Sigma_{t,1}^{[k]}); \ \dots ; \ (\mu_{t,1n-1}^{[k]}, \Sigma_{t,n-1}^{[k]}) \tag{5.116}$$

where k denotes the index of the particle and n the number of features in the map. Note that in particle filter SLAM, the mean and covariance of each feature are updated using distinct Kalman filters, one for each feature in the map.

When the robot moves, the motion model specified by the odometry reading u_t is applied to each particle $x_{t-1}^{[k]}$ to generate the new location $x_t^{[k]}$.

When the robot makes an observation z_t, we compute for each particle the so-called *importance factor* $w_t^{[k]}$, which is determined as the probability of observing z_t given the particle $x_t^{[k]}$ and all previous observations $z_{0 \to t-1}$, that is,

$$w_t^{[k]} = p(z_t | x_t^{[k]}, z_{0 \to t-1}). \tag{5.117}$$

Notice that computing the importance factor for each particle is like sampling the probability distribution $p(z_t | x_t, z_{0 \to t-1})$.

The final step in particle filter SLAM is called *resampling*. This step replaces the current set of particles with another set according to the importance factor determined above. Finally, the mean and covariance of each feature are updated according to the standard EKF update rule.

Although this description of the algorithm may appear rather complex, the FastSLAM algorithm can be readily implemented and is one of the easiest-to-implement SLAM algorithms. Furthermore, FastSLAM has the big advantage over EKF SLAM that its complexity grows logarithmically in the number of features (and thus not quadratically as in EKF SLAM). This is mainly because instead of using a single covariance matrix for both the robot pose and the map (as in EKF SLAM), it uses separate Kalman filters, one for each feature. A very efficient implementation of FastSLAM is FastSLAM 2.0, which was proposed by the same authors in [232].

Finally, another great advantage over EKF SLAM is that due to the use of randomized sampling it does not require the linearization of the motion model and can also represent non-Gaussian distributions.

5.8.9 Open challenges in SLAM

One of the first assumptions we made is that the map is time-invariant, that is, static. However, real-world environments present moving objects such as vehicles, animals, and humans. A good SLAM algorithm should then be robust against dynamic objects. One way to tackle this problem is by treating them as outliers. However, the ability to recognize these objects, or even to predict where they are moving to, would improve the efficiency and the quality of the final map.

Another current topic of research is multiple robot mapping [312], that is, how to combine the individual readings of a team of multiple robots exploring the environment.

Another issue in SLAM is its sensitiveness to incorrect data associations. This problem is particularly important at the loop closure, that is, when the robot returns to previously visited locations after a long traverse. 2D laser rangefinders are more prone than cameras to incorrect data associations due to the difficulty of identifying unique and distinctive features in laser point clouds. The task of recognizing loop closures is, however, facilitated with cameras. Cameras provide much richer information than lasers. Furthermore, the development of distinctive feature detectors such as SIFT and SURF (section 4.5.5), which are also robust under large changes in the camera viewpoint and scale, has allowed researchers to cope with very challenging and large-scale environments (even 1000 km [109] without GPS). This is made possible by the use of the "bag of features" approach, which we described in section 4.6.

As we have seen in section 5.8.5, visual SLAM is a very recent active field of research that is fascinating more and more researchers around the world. Although laser scanners are still the most used sensors for SLAM, cameras are more appealing because they are cheaper and provide much richer information. Furthermore, they are lighter than laser, which enables the use on-board micro lightweight helicopters [76]. However, monocular cameras have the disadvantage that they provide only bearing information rather than depth; therefore, the solution to the SLAM will always be up to a scale. The absolute scale can, however, be recovered using some prior information such as the knowledge of the size of one element in the scene (a window, a table, etc.) or using other sensors such as GPS, odometry, or IMU. A recent solution even demonstrated the ability to recover the absolute scale by exploiting nonholonomic constraints of wheeled vehicles [277]. Stereo cameras conversely provide measurements directly in the absolute scale, but their resolution degrades with the measured distance.

5.8.10 Open source SLAM software and other resources

Here is a list of some of open source SLAM software and datasets available online.

1. http://www.openslam.org contains one of the most comprehensive lists of SLAM software currently available. There you can find up-to-date resources for both C/C++ and Matlab. Furthermore, here you also can to upload your own SLAM algorithm.

2. http://www.doc.ic.ac.uk/~ajd/software.html contains both C/C++ and Matlab implementations of Davison's real-time monocular visual SLAM.

3. http://www.robots.ox.ac.uk/~gk/PTAM/ is an alternative real-time monocular visual SLAM algorithm known as PTAM and implemented by Klein and Murray [167].

4. http://webdiis.unizar.es/~neira/software/slam/slamsim.htm is a Matlab EKF SLAM simulator

5. http://www.rawseeds.org/home provides a large collection of benchmarked datasets for SLAM. Several sensors were used to acquire the data, among them laser rangefinder, multiple cameras, IMUs, and GPS. Furthermore, these datasets come with a ground truth that can be used to evaluate the performance of your SLAM algorithm.

6. Additional software, datasets, and lectures about SLAM can be found on the websites of the past SLAM summer schools: google SLAM summer school.

5.9   Problems

1. Consider a differential-drive robot that uses wheel encoders only. The wheels are a distance d apart, and each wheel has radius r. Suppose this robot uses only its encoders to attempt to describe a square, with sides of length $1000r$, returning to the origin. For each of range error, turn error, and drift error, supposing an error rate of 10%, compute the worst-case effect of each type of error on the different between the final actual robot position and the original position in both position and orientation.

2. Consider the environment of figure 5.6. Your robot begins in the top left room and has the goal of stopping in the large room at position B. Design a sequence for a behavior-based robot to navigate successfully to B. Behaviors available are:
     LWF: left wall follow
     RWF: right wall follow
     HF: go down centerline of a hallway
     Turn X: turn X degrees left/right
     Move X: move X centimeters forward/backward
     EnterD: center and enter through a doorway
   Termination conditions available are:
     DoorL: doorway on left
     DoorR: doorway on right
     HallwayI: hallway intersection

3. Consider exact cell decomposition. What is the worst-case and best-case number of nodes that may be created when using this method as a function of the number of convex polygons and the number of sides of each polygon?

4. Consider the case of figure 5.27 and the method of 5.6.7.5. Suppose an initial belief state: {1,1–2,2–3}, with the robot facing east with certainty and with uncertainty {0.4, 0.4, 0.2} respectively. Two perceptual events occur. First: {door on left; door on right}. Second: {nothing on left; hall on right}. Complete the resulting belief update and describe the belief state. There is no need to normalize the results.

5. Challenge Question.

Implement a simple EKF Visual SLAM for the case of a omnidirectional-drive robot moving in a two-dimensional environment. Assume a constant velocity model. For simplicity, you may also assume that the robot is constrained to move along a line, which means Visual SLAM in a one-dimensional environment.

6 Planning and Navigation

6.1 Introduction

This book has focused on the elements of a mobile robot that are critical to robust mobility: the kinematics of locomotion, sensors for determining the robot's environmental context, and techniques for localizing with respect to its map. We now turn our attention to the robot's cognitive level. Cognition generally represents the purposeful decision making and execution that a system utilizes to achieve its highest-order goals.

In the case of a mobile robot, the specific aspect of cognition directly linked to robust mobility is *navigation competence*. Given partial knowledge about its environment and a goal position or series of positions, navigation encompasses the ability of the robot to act based on its knowledge and sensor values so as to reach its goal positions as efficiently and as reliably as possible. The focus of this chapter is how the tools of the previous chapters can be combined to solve this navigation problem.

Within the mobile robotics research community, a great many approaches have been proposed for solving the navigation problem. As we sample from this research background, it will become clear that in fact there are strong similarities between all of these approaches, even though they appear, on the surface, quite disparate. The key difference between various navigation architectures is the manner in which they decompose the problem into smaller subunits. In sections 6.3, 6.4, and 6.5, we describe the most popular of these architectures, contrasting their relative strengths and weaknesses.

First, however, in section 6.2 we discuss two additional key competences required for mobile robot navigation. Given a map and a goal location, *path planning* involves identifying a trajectory that will cause the robot to reach the goal location when executed. Path planning is a strategic problem-solving competence, since the robot must decide what to do over the long term to achieve its goals.

The second competence is equally important but occupies the opposite, tactical extreme. Given real-time sensor readings, *obstacle avoidance* means modulating the trajectory of the robot in order to avoid collisions. A great variety of approaches have demonstrated competent obstacle avoidance, and we survey a number of these approaches as well.

6.2   Competences for Navigation: Planning and Reacting

In the artificial intelligence community, planning and reacting are often viewed as contrary approaches or even opposites. When applied to physical systems such as mobile robots, however, planning and reacting have a strong complementarity, each being critical to the other's success. The navigation challenge for a robot involves executing a course of action (or plan) to reach its goal position. During execution, the robot must react to unforeseen events (e.g., obstacles) in such a way as to still reach the goal. Without reacting, the planning effort will not pay off because the robot will never physically reach its goal. Without planning, the reacting effort cannot guide the overall robot behavior to reach a distant goal—again, the robot will never reach its goal.

An information-theoretic formulation of the navigation problem will make this complementarity clear. Suppose that a robot R at time i has a map M_i and an initial belief state b_i. The robot's goal is to reach a position p while satisfying some temporal constraints: $loc_g(R) = p$; $(g \leq n)$. Thus, the robot must be at location p at or before timestep n.

Although the goal of the robot is distinctly physical, the robot can only really sense its belief state, not its physical location, and therefore we map the goal of reaching location p to reaching a belief state b_g, corresponding to the belief that $loc_g(R) = p$. With this formulation, a plan q is nothing more than one or more trajectories from b_i to b_g. In other words, plan q will cause the robot's belief state to transition from b_i to b_g if the plan is executed from a world state consistent with both b_i and M_i.

Of course, the problem is that the latter condition may not be met. It is entirely possible that the robot's position is not quite consistent with b_i, and it is even likelier that M_i is either incomplete or incorrect. Furthermore, the real-world environment is dynamic. Even if M_i is correct as a single snapshot in time, the planner's model regarding how M changes over time is usually imperfect.

In order to reach its goal nonetheless, the robot must incorporate new information gained during plan execution. As time marches forward, the environment changes and the robot's sensors gather new information. This is precisely where reacting becomes relevant. In the best of cases, reacting will modulate robot behavior locally in order to correct the planned-upon trajectory so that the robot still reaches the goal. At times, unanticipated new information will require changes to the robot's strategic plans, and so ideally the planner also incorporates new information as that new information is received.

Taken to the limit, the planner would incorporate every new piece of information in real time, instantly producing a new plan that in fact reacts to the new information appropriately. This extreme, at which point the concept of planning and the concept of reacting merge, is called *integrated planning and execution* and is discussed in section 6.5.4.3.

Completeness. A useful concept throughout this discussion of robot architecture involves whether particular design decisions sacrifice the system's ability to achieve a desired goal whenever a solution exists. This concept is termed *completeness*. More formally, the robot system is *complete* if and only if, for all possible problems (i.e., initial belief states, maps, and goals), when there exists a trajectory to the goal belief state, the system will achieve the goal belief state (see [40] for further details). Thus when a system is incomplete, then there is at least one example problem for which, although there is a solution, the system fails to generate a solution. As you may expect, achieving completeness is an ambitious goal. Often, completeness is sacrificed for computational complexity at the level of representation or reasoning. Analytically, it is important to understand how completeness is compromised by each particular system.

In the following sections, we describe key aspects of planning and reacting as they apply to mobile robot path planning and obstacle avoidance and describe how representational decisions impact the potential completeness of the overall system. For greater detail, refer to [32, 44, chapter 25].

6.3 Path Planning

Even before the advent of affordable mobile robots, the field of path planning was heavily studied because of its applications in the area of industrial manipulator robotics. Interestingly, the path-planning problem for a manipulator with, for instance, six degrees of freedom is far more complex than that of a differential-drive robot operating in a flat environment. Therefore, although we can take inspiration from the techniques invented for manipulation, the path-planning algorithms used by mobile robots tend to be simpler approximations owing to the greatly reduced degrees of freedom. Furthermore, industrial robots often operate at the fastest possible speed because of the economic impact of high throughput on a factory line. So, the dynamics and not just the kinematics of their motions are significant, further complicating path planning and execution. In contrast, a number of mobile robots operate at such low speeds that dynamics are rarely considered during path planning, further simplifying the mobile robot instantiation of the problem.

Configuration space. Path planning for manipulator robots and, indeed, even for most mobile robots, is formally done in a representation called *configuration space*. Suppose that a robot arm (e.g., SCARA robot) has k degrees of freedom. Every state or configuration of the robot can be described with k real values: q_1, ..., q_k. The k-values can be regarded as a point p in a k-dimensional space called the configuration space C of the robot. This description is convenient because it allows us to describe the complex 3D shape of the robot with a single k-dimensional point.

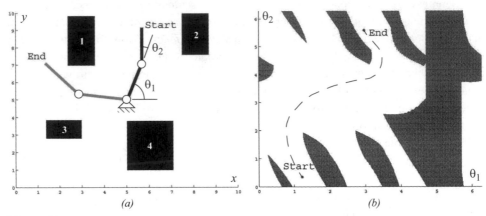

Figure 6.1
Physical space (a) and configuration space (b): (a) A two-link planar robot arm has to move from the configuration *start* to *end*. The motion is thereby constraint by the obstacles 1 to 4. (b) The corresponding configuration space shows the free space in joint coordinates (angle θ_1 and θ_2) and a path that achieves the goal.

Now consider the robot arm moving in an environment where the workspace (i.e., its physical space) contains known obstacles. The goal of path planning is to find a path in the physical space from the initial position of the arm to the goal position, avoiding all collisions with the obstacles. This is a difficult problem to visualize and solve in the physical space, particularly as k grows large. But in configuration space the problem is straightforward. If we define the *configuration space obstacle O* as the subspace of C where the robot arm bumps into something, we can compute the free space $F = C - O$ in which the robot can move safely.

Figure 6.1 shows a picture of the physical space and configuration space for a planar robot arm with two links. The robot's goal is to move its end effector from position *start* to *end*. The configuration space depicted is 2D because each of two joints can have any position from 0 to 2π. It is easy to see that the solution in C-space is a line from *start* to *end* that remains always within the free space of the robot arm.

For mobile robots operating on flat ground, we generally represent robot position with three variables (x, y, θ), as in chapter 3. But, as we have seen, most robots are nonholonomic, using differential-drive systems or Ackerman steered systems. For such robots, the nonholonomic constraints limit the robot's velocity $(\dot{x}, \dot{y}, \dot{\theta})$ in each configuration (x, y, θ). For details regarding the construction of the appropriate *free space* to solve such path-planning problems, see [32, p. 405].

In mobile robotics, the most common approach is to assume for path-planning purposes that the robot is in fact holonomic, simplifying the process tremendously. This is especially common for differential-drive robots because they can rotate in place, and so a holonomic path can be easily mimicked if the rotational position of the robot is not critical.

Furthermore, mobile roboticists will often plan under the further assumption that the robot is simply a *point*. Thus we can further reduce the configuration space for mobile robot path planning to a 2D representation with just x- and y-axes. The result of all this simplification is that the configuration space looks essentially identical to a 2D (i.e., flat) version of the physical space, with one important difference. Because we have reduced the robot to a point, we must inflate each obstacle by the size of the robot's radius to compensate. With this new, simplified configuration space in mind, we can now introduce common techniques for mobile robot path planning.

Path-planning overview. The robot's environment representation can range from a continuous geometric description to a decomposition-based geometric map or even a topological map, as described in section 5.5. The first step of any path-planning system is thus to transform this possibly continuous environmental model into a discrete map suitable for the chosen path-planning algorithm. Path planners differ in how they use this discrete decomposition. In this book, we describe two general strategies:

1. Graph search: a connectivity graph in free space is first constructed and then searched. The graph construction process is often performed offline.

2. Potential field planning: a mathematical function is imposed directly on the free space. The gradient of this function can then be followed to the goal.

6.3.1 Graph search
Graph search techniques have traditionally been strongly rooted in the field of mathematics. Nonetheless, in recent years much of the innovation has been devised in the robotics community. This may be largely attributed to the need for real-time capable algorithms, which can accommodate evolving maps and thus changing graphs. For most of these methods we distinguish two main steps: graph construction, where nodes are placed and connected via edges, and graph search, where the computation of an (optimal) solution is performed.

6.3.1.1 Graph construction
Starting from a representation of free and occupied space, several methods are known to decompose this representation into a graph that can then be searched using any of the algorithms described in section 6.3.1.2 and 6.3.1.3. The challenge lies in constructing a set of nodes and edges that enable the robot to go anywhere in its free space while limiting the total size of the graph.

First, we describe two road map approaches that achieve this result with dramatically different types of roads. In the case of the *visibility graph*, roads come as close as possible

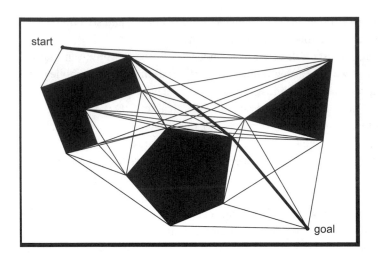

Figure 6.2
Visibility graph [32]. The nodes of the graph are the initial and goal points and the vertices of the configuration space obstacles (polygons). All nodes which are visible from each other are connected by straight-line segments, defining the road map. This means there are also edges along each polygon's sides.

to obstacles and resulting optimal paths are minimum-length solutions. In the case of the *Voronoi diagram*, roads stay as far away as possible from obstacles. We then detail cell decomposition methods where the idea is to discriminate between free and occupied geometric areas. Exact cell decomposition is a lossless decomposition, whereas approximate cell decomposition represents an approximation of the original map. A graph is then formed through a specified connectivity relation between cells. Finally, we describe the construction of lattice graphs, which are formed by shifting an underlying base set of edges over the free space. Lattice graphs are typically constructed by employing a mathematical model of the robot so that their edges become directly executable.

Visibility graph. The visibility graph for a polygonal configuration space C consists of edges joining all pairs of vertices that can see each other (including both the initial and goal positions as vertices as well). The unobstructed straight lines (roads) joining those vertices are obviously the shortest distances between them. The task of the path planner is then to find a (shortest) path from the initial position to the goal position along the roads defined by the visibility graph (figure 6.2).

Visibility graphs are moderately popular in mobile robotics, partly because their implementation is quite simple. Particularly when the environmental representation describes objects in the environment as polygons in either continuous or discrete space, the visibility graph can employ the obstacle polygon descriptions readily.

There are, however, two important caveats when employing visibility graph search. First, the size of the representation and the number of edges and nodes increase with the number of obstacle polygons. Therefore, the method is extremely fast and efficient in sparse environments, but it can be slow and inefficient compared to other techniques when used in densely populated environments.

The second caveat is a much more serious potential flaw: solution paths found by graph search tend to take the robot as close as possible to obstacles on the way to the goal. More formally, we can prove that shortest solutions on the visibility graph are *optimal* in terms of path length. This powerful result also means that all sense of safety, with respect to staying a reasonable distance from obstacles, is sacrificed for this optimality. The common solution is to grow obstacles by significantly more than the robot's radius, or, alternatively, to modify the solution path after path planning to distance the path from obstacles where possible. Of course such actions sacrifice the optimal-length results of visibility graph path planning.

Voronoi diagram. Contrasting with the visibility graph approach, a Voronoi diagram is a complete road map method that tends to *maximize* the distance between the robot and obstacles in the map. For each point in free space, its distance to the nearest obstacle is computed. If you plot that distance as the height coming out of the page, it increases as you move away from an obstacle (see figure 6.3). At points that are equidistant from two or more obstacles, such a distance plot has sharp ridges. The Voronoi diagram consists of the edges formed by these sharp ridge points. When the configuration space obstacles are polygons, the Voronoi diagram consists of straight line and parabolic segments only. Algorithms that find paths on the Voronoi road map are complete, just as are visibility graph methods, because the existence of a path in the free space implies the existence of one on the Voronoi diagram as well (i.e., both methods guarantee completeness). However, the solution paths on the Voronoi diagram are usually far from optimal in the sense of total path length.

The Voronoi diagram has an important weakness in the case of limited range localization sensors. Since its edges maximize the distance to obstacles, any short-range sensor on the robot will be in danger of failing to sense its surroundings. If such short-range sensors are used for localization, then the chosen path will be quite poor from a localization point of view. On the other hand, the visibility graph method can be designed to keep the robot as close as desired to objects in the map.

There is, however, an important subtle advantage that the Voronoi diagram method has over most other graphs: *executability*. Given a particular planned path via Voronoi diagram planning, a robot with range sensors, such as a laser rangefinder or ultrasonics, can follow a Voronoi edge in the physical world using simple control rules that match those used to

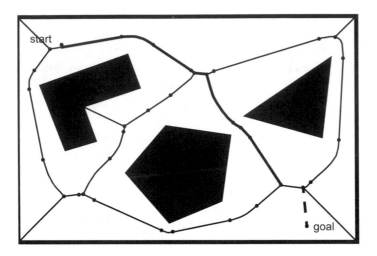

Figure 6.3
Voronoi diagram [32]. The Voronoi diagram consists of the lines constructed from all points that are equidistant from two or more obstacles. The initial q_{init} and goal q_{goal} configurations are mapped into the Voronoi diagram to q'_{init} and q'_{goal}, each by drawing the line along which its distance to the boundary of the obstacles increases the fastest. The points on the Voronoi diagram represent transitions from straight line segments (minimum distance between two lines) to parabolic segments (minimum distance between a line and a point).

create the Voronoi diagram: the robot maximizes the readings of local minima in its sensor values. This control system will naturally keep the robot on Voronoi edges, so that Voronoi motion can mitigate encoder inaccuracy. This interesting physical property of the Voronoi diagram has been used to conduct automatic mapping of an environment by finding and moving on unknown Voronoi edges, then constructing a consistent Voronoi map of the environment [103].

Exact cell decomposition. Figure 6.4 depicts exact cell decomposition, whereby the boundary of cells is based on geometric criticality. The resulting cells are each either completely free or completely occupied, and therefore path planning in the network is complete, like the road-map–based methods seen earlier. The basic abstraction behind such a decomposition is that the particular position of the robot within each cell of free space does not matter; what matters is rather the robot's ability to traverse from each free cell to adjacent free cells.

The key disadvantage of exact cell decomposition is that the number of cells and, therefore, the overall computational planning efficiency depends on the density and complexity of objects in the environment, just as with road-map–based systems. The key advantage is

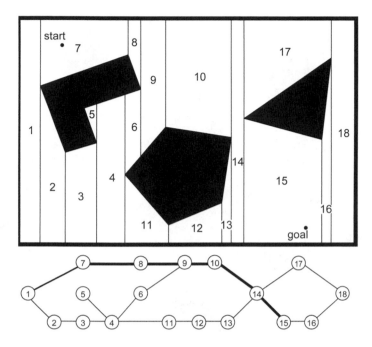

Figure 6.4
Example of exact cell decomposition. Cells are for example divided according to the horizontal coordinate of extremal obstacle points.

a result of this same correlation. In environments that are extremely sparse, the number of cells will be small, even if the geometric size of the environment is very large. Thus the representation will be efficient in the case of large, sparse environments. Practically speaking, due to complexities in implementation, the exact cell decomposition technique is used relatively rarely in mobile robot applications, although it remains a solid choice when a lossless representation is highly desirable—for instance, to preserve completeness fully.

Approximate cell decomposition. By contrast, approximate cell decomposition is one of the most popular graph construction techniques in mobile robotics. This is partly due to the popularity of grid environmental representations. These grid representations are themselves fixed grid-size decompositions and so they are identical to an approximate cell decomposition of the environment.

The most popular form of this, shown in figure 5.15, is the fixed-size cell decomposition. The cell size is not dependent on the particular objects in an environment, and so narrow passage ways can be lost due to the inexact nature of the tessellation. Practically

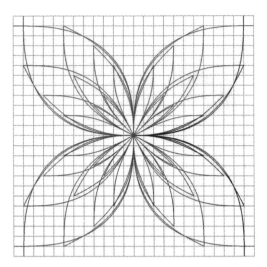

Figure 6.5 16-directional state lattice constructed for a planetary exploration rover. The state includes 2D position, heading, and curvature (x, y, θ, k). Note that straight segments are part of the lattice set but occluded by longer curved segments. The lattice is 2D-shift invariant and partially invariant to rotation. All successor edges of state $(0, 0, 0, 0)$ are depicted in black. Image courtesy of M. Pivtoraiko [260].

speaking, this is rarely a problem owing to the very small cell size used (e.g., 5 cm on each side).

Figure 5.16 illustrates a variable-size approximate cell decomposition method. The free space is externally bounded by a rectangle and internally bounded by three polygons. The rectangle is recursively decomposed into smaller rectangles. Each decomposition generates four identical new rectangles. At each level of resolution only the cells whose interiors lie entirely in the free space are used to construct the connectivity graph. Path planning in such adaptive representations can proceed in a hierarchical fashion. Starting with a coarse resolution, the resolution is reduced until either the path planner identifies a solution or a limit resolution is attained (e.g, k • size of robot).

The great benefit of approximate cell decomposition is the low computational complexity induced to path planning.

Lattice graph. Lattice structures have only recently been adapted to graph search. They are formed by first constructing a base set of edges (such as the one depicted in figure 6.5) and then repeating it over the whole configuration space to form a graph. As such, the approximate cell decomposition technique could be interpreted as a simple lattice: the neighborhood structure of each cell forms a cross, which is then repeated by 2D shifts of multiples of a single cell increment. The main benefit with respect to other graph construction methods lies in the design freedom in creating feasible edges, that is, edges that can be inherently executed by a robotic platform, however. To this end, Bicchi et al. [73] succeeded in applying an input discretization to a mathematical model of their robotic platform resulting in a configuration space lattice for certain simple kinematic vehicle models. Later, more broadly applicable methods have been devised in the configuration space directly: Pivtoraiko et al. [260] a priori fixed a problem specific configuration space discretization and dimensionality (e.g., in 2D position, orientation, curvature; figure 6.5). They then computed solutions to two point boundary problems between any two discrete states in the configuration space by also using a robot model. In a final step, the resulting large number of edges was pruned to a more manageable subset (the base lattice) by discarding edges which are similar to or can be decomposed into other edges already part of the subset.

Lattice graphs are typically precomputed for a given robotic platform and stored in memory. They thus belong to the class of approximate decomposition methods. Due to their inherent executability, edges along the solution path may be directly used as feed-forward commands to the controller.

Discussion. The fundamental cost of any fixed decomposition approach is memory. For a large environment, even when sparse, the grid must be represented in its entirety. Practically, because of the falling cost of RAM computer memory, this disadvantage has been mitigated in recent years.

In contrast to the exact decomposition methods, approximate approaches can sacrifice completeness but are mathematically less involved and thus easier to implement. In contrast to the fixed-size decompositions, variable-size decompositions will adapt to the complexity of the environment. Sparse environments will therefore contain appropriately fewer nodes and edges and consume dramatically less memory.

6.3.1.2 Deterministic graph search

Suppose now that our environment map has been converted into a connectivity graph using one of the graph generation methods presented earlier. Whatever map representation is chosen, the goal of path planning is to find the *best* path in the map's connectivity graph between the start and the goal, where *best* refers to the selected optimization criteria (e.g., the shortest path). In this section, we present several search algorithms that have become quite popular in mobile robotics. For an in-depth study on graph-search techniques we refer the reader to [44].

Discriminators. Due to the similarity between many graph search algorithms, we begin this section with an elaboration on their respective differences. To this end, it is beneficent to introduce the concepts of expected total cost $f(n)$, path cost $g(n)$, edge traversal cost $c(n, n')$, and heuristic cost $h(n)$, which are all functions of the node n (and an adjacent node

n'). In particular, we denote the accumulated cost from the start node to any given node n with $g(n)$. The cost from a node n to an adjacent node n' becomes $c(n, n')$, and the expected cost (heuristic cost) from a node n to the goal node is described with $h(n)$. The total expected cost from start to goal via state n can then be written as

$$f(n) = g(n) + \varepsilon \cdot h(n),$$ (6.1)

where ε is a parameter that assumes algorithm-dependent values.

In the special case that every individual edge in the graph assumes the same traversal cost (such as in an occupancy grid, introduced in section 5.5.2), optimal implementations may be developed in a simpler form and obtain faster execution speeds compared to the general instance. Examples of such algorithms include *depth-first* and *breadth-first* searches. On the other hand, Dijkstra's algorithm and variants allow for the computation of optimal paths in nonuniform cost maps as well. This comes at the cost of higher algorithmic complexity, however. In all of these implementations, $\varepsilon = 0$.

In the case of $\varepsilon \neq 0$, a heuristic function $h(n)$ is employed, which essentially incorporates additional information about the problem set and thus often allows for faster convergence of the search query. In this book, we restrict our attention to heuristics that are both consistent and underestimate the true cost. Most practical heuristics fulfill these requirements. For $\varepsilon = 1$, the optimal A* algorithm results, whereas for $\varepsilon > 1$ suboptimal or greedy A* variants are obtained.

Now that we have a general idea on the relation between some of the most popular graph search algorithms, we can proceed to introduce them in more detail.

Breadth-first search. This graph-search algorithm begins with the start node (denoted by A in figure 6.6) and explores all of its neighboring nodes. Then, for each of these nodes, it explores all their unexplored neighbors and so on. This process (that is, marking a node "active", exploring each of its neighbors and marking them "open", and finally marking the parent node "visited") is called node expansion. In breadth-first search, nodes are expanded in order of proximity to the start node with proximity defined as the shortest number of edge transitions. The algorithm proceeds until it reaches the goal node where it terminates. The computation of a solution is fast, since a reordering of nodes waiting for expansion is not necessary. They are already sorted in increasing order of proximity to the start node. Figure 6.6 illustrates the working principle of the breadth-first algorithm for a given graph.

It can be seen that the search always returns the path with the fewest number of edges between the start and goal node. If we assume that the cost of all individual edges in the graph is constant, then breadth-first search is also optimal in that it always returns the minimum-cost path. In this case, a node's reexpansion (as in the case of node G) can be easily circumvented by assigning a flag to a visited node. This addition does not affect solution

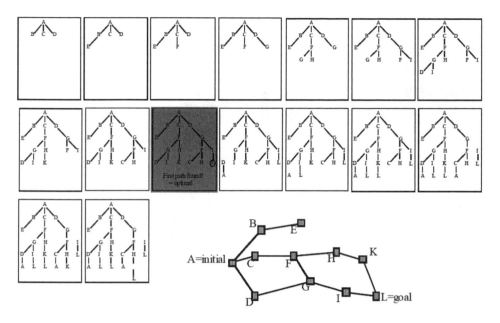

Figure 6.6 Working principle of breadth-first search.

optimality, since nodes are expanded in order of proximity to the start. However, if the graph has nonuniform costs associated with each edge, then breadth-first search is not guaranteed to be cost-optimal. Indeed, the path with the minimum number of edges does not necessarily coincide with the cheapest path, since there might be another path with more edges but lower total cost.

An example of breadth-first search algorithm in the context of robotics is the *wavefront expansion algorithm*, which is also known as *NF1* or *grassfire* [183]. This algorithm is an efficient and simple-to-implement technique for finding routes in fixed-size cell arrays. The algorithm employs wavefront expansion from the goal position outward, marking for each cell its L^1 (Manhattan) distance to the goal cell [154] (see figure 6.7). This process continues until the cell corresponding to the initial robot position is reached. At this point, the path planner can estimate the robot's distance to the goal position as well as recover a specific solution trajectory by simply linking together cells that are adjacent and always closer to the goal.

Given that the entire array can be in memory, each cell is only visited once when looking for the shortest discrete path from the initial position to the goal position. So, the search is linear in the number of cells only. Thus, complexity does not depend on the sparseness and

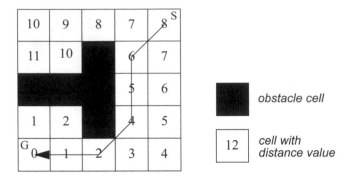

Figure 6.7
An example of the distance transform and the resulting path as it is generated by the NF1 function. S
denotes the start, G the goal. The neighbors of each cell i are defined as the four adjacent cells that
share an edge with i (4-neighborhood).

density of the environment, nor on the complexity of the objects' shapes in the environment.

Depth-first search. The working principle of depth-first search algorithm is shown in
figure 6.8. In contrast to breadth-first search, depth-first search expands each node up to the
deepest level of the graph (until the node has no more successors). As those nodes are
expanded, their branch is removed from the graph and the search backtracks by expanding
the next neighboring node of the start node until its deepest level and so on. An inconve-
nience of this algorithm is that it may revisit previously visited nodes or enter redundant
paths. However, these situations may be easily avoided through an efficient implementation. A significant advantage of depth-first over breadth-first is space complexity. In fact,
depth-first needs to store only a single path from the start node to the goal node along with
all the remaining unexpanded neighboring nodes for each node on the path. Once each node
has been expanded and all its children nodes have been explored, it can be removed from
memory.

Dijkstra's algorithm. Named after its inventor, E.W. Dijkstra, this algorithm is similar to
breadth-first search, except that edge costs may assume any positive value and the search
still guarantees solution optimality [114]. This introduces some additional complexity into
the algorithm for which we need to introduce the concept of the *heap*, a specialized tree-
based data structure. Its elements (which are comprise to-be-expanded graph nodes) are
ordered according to a key, which in our case amounts to the expected total path cost $f(n)$
at that given node n. Dijkstra's algorithm then expands nodes starting from the start similar

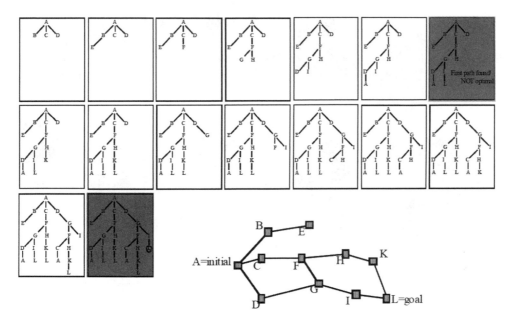

Figure 6.8 Working principle of depth-first search.

to breadth-first search, except that the neighbors of the expanded node are placed in the heap and reordered according to their $f(n)$ value, which corresponds to $g(n)$ since no heuristic is used. Subsequently, the cheapest state on the heap (the top element after reordering) is extracted and expanded. This process continues until the goal node is expanded, or no more nodes remain on the heap. A solution can then be backtracked from the goal to the start. Due to reorder operations on the heap, the time complexity rises from $O(n+m)$ in breadth-first search to $O(n\log(n)+m)$, with n the number of nodes, and m the number of edges.

In robotic applications, Dijkstra's search is typically computed from the robot's goal position. Consequently, not only the best path from the start node to the goal is computed, but also all lowest cost paths from any starting position in the graph to the goal node. The robot may thus localize and determine the best route toward the goal based on its current position. After moving some distance along this path, the process is repeated until the goal is reached, or the environment changes (which would require a recomputation of the solution). This technique allows the robot to reach the goal without replanning even in presence of localization and actuation noise.

A* algorithm. For consistent heuristics, the A* algorithm (pronounced "a star") [147] is similar to Dijkstra's algorithm. However, the inclusion of a heuristic function $h(n)$, which

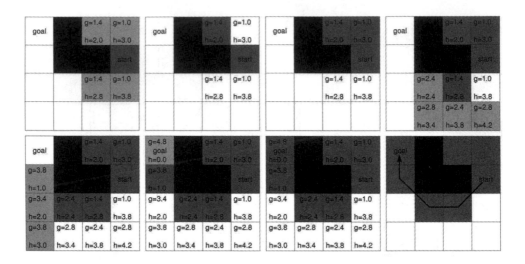

Figure 6.9 Working principle of the A* algorithm. Nodes are expanded in order of lowest $f(n) = g(n) + h(n)$ cost. $g(n)$ is indicated at the top left corner, $h(n)$ at the bottom right corner of each cell. The neighborhood of each cell is selected as the 8-neighborhood (all eight adjacent cells). Diagonal moves cost $\sqrt{2}$ times as much as horizontal and vertical moves. Obstacle cells are colored in black, expanded cells in dark gray, and cells put on the heap during this expansion step in light gray. Image courtesy of M. Rufli.

encodes additional knowledge about the graph, makes this algorithm especially efficient for single node to single node queries. In order to guarantee solution optimality, the heuristic is required to be an underestimating function of the cost to go. In robotics, A* is mainly employed on a grid, and the heuristic is then often chosen as the distance between any cell and the goal cell in absence of any obstacles. If such knowledge is available, it can be used to guide the search toward the goal node. Generally, this dramatically reduces the number of node expansions required to arrive at a solution compared to Dijkstra's algorithm.

A* search begins by expanding the start node and placing all of its neighbors on a heap. In contrast to Dijkstra's algorithm, the heap is ordered according to the smallest $f(n)$ value that includes the heuristic function $h(n)$. The lowest cost state is then extracted and expanded. This continues until the goal node is explored. The lowest cost solution can again be backtracked from the goal. For an example, see figure 6.9. The time complexity of A* largely depends on the chosen heuristic $h(n)$. On average, much better performance than with Dijkstra's algorithm can be expected, however.

Often it is not necessary to obtain an optimal solution, as long as there are guarantees on its suboptimality level. In such cases, a solution that costs at most ε times the (unknown) optimal solution may be obtained by setting $\varepsilon > 1$. The solution may then be improved as

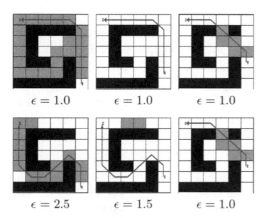

$\epsilon = 1.0$	$\epsilon = 1.0$	$\epsilon = 1.0$
$\epsilon = 2.5$	$\epsilon = 1.5$	$\epsilon = 1.0$

Figure 6.10 comparison of the number of expanded cells for D* (top) and Anytime D* (bottom, starting with a sub-optimality value of 2.5) in a planning and re-planning scenario. Note that an opening in the top wall is detected in the third frame, after the robot has moved upward twice. Obstacle cells are colored in black, cells expanded during a given time-step in gray. Image courtesy of M. Rufli.

search time allows, by reusing parts of the previous queries. This procedure results in the Anytime Replanning A* algorithm [191]. If the heuristic is accurate, far fewer states can be expected to be expanded than for optimal A*.

D* algorithm. The D* algorithm [304, 170] represents an incremental replanning version of A*, where the term incremental refers to the algorithm's reuse of previous search effort in subsequent search iterations. Let us illustrate this with an example (see figure 6.10): our robot is initially provided with a crude map of the environment (i.e., obtained from an aerial image). In this map, the navigation module plans an initial path by employing A*. After executing this path for a while, the robot observes some changes in the environment with its onboard sensors. Subsequent to updating the map, a new solution path needs to be computed. This is where D* comes into play. Instead of generating a new solution from scratch (as A* would do), only states affected by the added (or removed) obstacle cells are recomputed. Because changes to the map are most often observed locally (due to proprioceptive sensors), the planning problem is usually reversed; node expansion begins from the robot goal state. In this way, large parts of the previous solution remain valid for the new computation. Compared to A*, search time may decrease by a factor of one to two orders of magnitude. For more detail and a description on computing affected states, consult [170].

Analogous to A*, the D* algorithm has also been extended to an anytime version, called Anytime D* [192].

6.3.1.3 Randomized graph search

When encountering complex high-dimensional path planning problems (such as in manipulation tasks on robotic arms, or molecule folding and docking queries for drug placement, and so on) it becomes infeasible to solve them exhaustively within reasonable time limits. Reverting to heuristic search methods is often not possible due to the lack of an appropriate heuristic function and a reduction of the problem dimensionality frequently fails due to velocity and acceleration constraints imposed on the model, which should not be violated for security reasons. In such situations, randomized search becomes useful, since it forgoes solution optimality for faster solution computation.

Rapidly Exploring Random Trees (RRTs). RRTs typically grow a graph online during the search process and thus a priori only require an obstacle map but no graph decomposition. The algorithm begins with an initial tree (which might be empty) and then successively adds nodes, connected via edges, until a termination condition is triggered. Specifically, during each step a random configuration q_{rand} in the free space is selected. The tree node that is closest to q_{rand}, denoted as q_{near}, is then computed. Starting from q_{near}, an edge (with fixed length) is grown toward q_{rand} using an appropriate robot motion model. The configuration q_{new} at the end of this edge is then added to the tree, if the connecting edge is collision-free [185]

Typical extensions to the algorithm aim at speeding-up solution computation: bidirectional versions grow partial trees from both the start and goal configuration. Besides parallelization capability, faster convergence in nonconvex environments can be expected [186]. Another often employed modification biases the process of selecting a random free space configuration q_{rand}. The goal node is then selected instead of q_{rand} with a fixed nonzero probability, thus guiding tree growth toward the goal state. This process is especially efficient in sparse environments, but it may lead to a slowdown in presence of concave obstacles [33].

Even though the RRT algorithm and its extensions lack solution optimality guarantees and deterministic completeness, it can be proven that they are probabilistically complete. This signifies that if a solution exists, the algorithm will eventually find it as the number of nodes added to the tree grows toward infinity (see figure 6.11).

6.3.2 Potential field path planning

Potential field path planning creates a field, or gradient, across the robot's map that directs the robot to the goal position from multiple prior positions (see [32]). This approach was originally invented for robot manipulator path planning and is used often and under many variants in the mobile robotics community. The potential field method treats the robot as a point under the influence of an artificial potential field $U(q)$. The robot moves by follow-

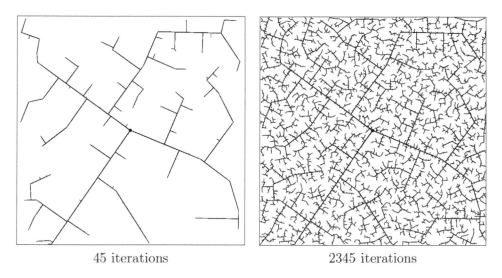

45 iterations   2345 iterations

Figure 6.11 The evolution of a RRT. Image courtesy of S.M. LaValle [33].

ing the field, just as a ball would roll downhill. The goal (a minimum in this space) acts as an attractive force on the robot, and the obstacles act as peaks, or repulsive forces. The superposition of all forces is applied to the robot, which, in most cases, is assumed to be a point in the configuration space (see figure 6.12). Such an artificial potential field smoothly guides the robot toward the goal while simultaneously avoiding known obstacles.

It is important to note, though, that this is more than just path planning. The resulting field is also a control law for the robot. Assuming the robot can localize its position with respect to the map and the potential field, it can always determine its next required action based on the field.

The basic idea behind all potential field approaches is that the robot is attracted toward the goal, while being repulsed by the obstacles that are known in advance. If new obstacles appear during robot motion, one could update the potential field in order to integrate this new information. In the simplest case, we assume that the robot is a point; thus the robot's orientation θ is neglected, and the resulting potential field is only 2D (x, y). If we assume a differentiable potential field function $U(q)$, we can find the related artificial force $F(q)$ acting at the position $q = (x, y)$.

$$F(q) = -\nabla U(q), \tag{6.2}$$

where $\nabla U(q)$ denotes the gradient vector of U at position q.

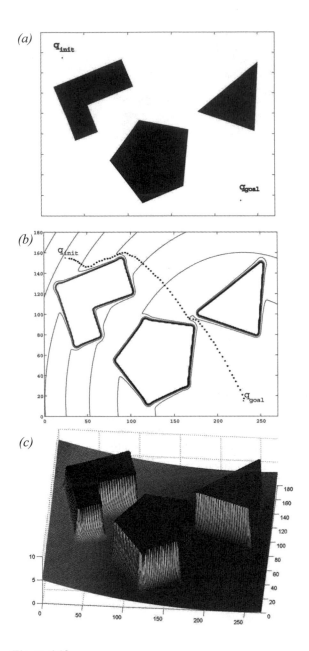

Figure 6.12
Typical potential field generated by the attracting goal and two obstacles (see [32]). (a) Configuration of the obstacles, start (top left) and goal (bottom right). (b) Equipotential plot and path generated by the field. (c) Resulting potential field generated by the goal attractor and obstacles.

$$\nabla U = \begin{bmatrix} \dfrac{\partial U}{\partial x} \\ \dfrac{\partial U}{\partial y} \end{bmatrix}. \tag{6.3}$$

The potential field acting on the robot is then computed as the sum of the attractive field of the goal and the repulsive fields of the obstacles:

$$U(q) = U_{att}(q) + U_{rep}(q). \tag{6.4}$$

Similarly, the forces can also be separated in a attracting and repulsing part:

$$\begin{aligned} F(q) &= F_{att}(q) - F_{rep}(q) \\ &= -\nabla U_{att}(q) - \nabla U_{rep}(q). \end{aligned} \tag{6.5}$$

Attractive potential. An attractive potential can, for example, be defined as a parabolic function.

$$U_{att}(q) = \frac{1}{2}k_{att} \cdot \rho^2_{goal}(q), \tag{6.6}$$

where k_{att} is a positive scaling factor and $\rho_{goal}(q)$ denotes the Euclidean distance $\|q - q_{goal}\|$. This attractive potential is differentiable, leading to the attractive force F_{att}

$$F_{att}(q) = -\nabla U_{att}(q) \tag{6.7}$$

$$= -k_{att} \cdot \rho_{goal}(q)\nabla\rho_{goal}(q) \tag{6.8}$$

$$= -k_{att} \cdot (q - q_{goal}) \tag{6.9}$$

that converges linearly toward 0 as the robot reaches the goal.

Repulsive potential. The idea behind the repulsive potential is to generate a force away from all known obstacles. This repulsive potential should be very strong when the robot is close to the object, but it should not influence its movement when the robot is far from the object. One example of such a repulsive field is

$$
U_{rep}(q) = \begin{cases} \dfrac{1}{2}k_{rep}\left(\dfrac{1}{\rho(q)} - \dfrac{1}{\rho_0}\right)^2 & \text{if } \rho(q) \le \rho_0 \\ 0 & \text{if } \rho(q) \ge \rho_0 , \end{cases}
\tag{6.10}
$$

where k_{rep} is again a scaling factor, $\rho(q)$ is the minimal distance from q to the object and ρ_0 the distance of influence of the object. The repulsive potential function U_{rep} is positive or zero and tends to infinity as q gets closer to the object.

If the object boundary is convex and piecewise differentiable, $\rho(q)$ is differentiable everywhere in the free configuration space. This leads to the repulsive force F_{rep} :

$$
F_{rep}(q) = -\nabla U_{rep}(q)
\tag{6.11}
$$

$$
= \begin{cases} k_{rep}\left(\dfrac{1}{\rho(q)} - \dfrac{1}{\rho_0}\right)\dfrac{1}{\rho^2(q)}\dfrac{q - q_{obstacle}}{\rho(q)} & \text{if } \rho(q) \le \rho_0 \\ 0 & \text{if } \rho(q) \ge \rho_0 . \end{cases}
$$

The resulting force $F(q) = F_{att}(q) + F_{rep}(q)$ acting on a point robot exposed to the attractive and repulsive forces moves the robot away from the obstacles and toward the goal (see figure 6.12). Under ideal conditions, by setting the robot's velocity vector proportional to the field force vector, the robot can be smoothly guided toward the goal, similar to a ball rolling around obstacles and down a hill.

However, there are some limitations with this approach. One is local minima that appear dependent on the obstacle shape and size. Another problem might appear if the objects are concave. This might lead to a situation for which several minimal distances $\rho(q)$ exist, resulting in oscillation between the two closest points to the object, which could obviously sacrifice completeness. For more detailed analyses of potential field characteristics, refer to [32].

The extended potential field method. Khatib and Chatila proposed the extended potential field approach [164]. Like all potential field methods, this approach makes use of attractive and repulsive forces that originate from an artificial potential field. However, two additions to the basic potential field are made: the *rotation potential field* and the *task potential field*.

The rotation potential field assumes that the repulsive force is a function of the distance from the obstacle and the orientation of the robot relative to the obstacle. This is done using

a) Classical Potential

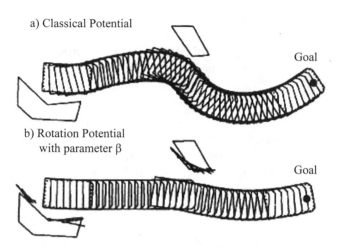

Goal

b) Rotation Potential
   with parameter β

Goal

Figure 6.13
Comparison between a classical potential field and an extended potential field. Image courtesy of
Raja Chatila [164].

a gain factor that reduces the repulsive force when an obstacle is parallel to the robot's
direction of travel, since such an object does not pose an immediate threat to the robot's
trajectory. The result is enhanced wall following, which was problematic for earlier implementations of potential fields methods.

The task potential field considers the present robot velocity, and from that it filters out
those obstacles that should not affect the near-term potential based on robot velocity. Again
a scaling is made, this time of all obstacle potentials when there are no obstacles in a sector
named Z in front of the robot. The sector Z is defined as the space that the robot will sweep
during its next movement. The result can be smoother trajectories through space. An example comparing a classical potential field and an extended potential field is depicted in figure
6.13.

Other extensions. A great variety of improvements to the artificial potential field method
have been proposed and implemented since the development of Khatib's original approach
in 1986 [163]. The most promising of these methods seems to be related to the harmonic
potential field which is a solution to the Laplace equation [126, 230]

$$\nabla^2 U(q) \equiv 0 \, , q \in \Omega \, , \tag{6.12}$$

where U again denotes the potential field as a function of the robot configuration q and Ω represents the workspace the robot operates in.

The main benefit of the harmonic potential field method with respect to earlier implementations is the complete absence of local minima inside the workspace. A unique solution may be generated through equation (6.12) and the specification of boundary conditions at start and goal locations and along the obstacle and workspace borders. In particular, the start location is pulled up to a high potential, whereas the goal position is pulled down to ground. For obstacle and workspace borders, we distinguish between two types of boundary conditions each resulting in a characteristic potential field. The Dirichlet condition requires that the potential is a known function along object boundaries (denoted with Γ)

$$U(q) = f(q) \, , \, q \in \Gamma \, . \tag{6.13}$$

For $f(q) = const$, obstacle boundaries become equipotential lines. The robot then follows a path perpendicular to objects in their close vicinity. Excessively long but safe paths tend to emerge.

On the other hand, the von Neumann boundary condition requires

$$\frac{\partial U(q)}{\partial q} = g(q) \, , \, q \in \Gamma \, . \tag{6.14}$$

where n is the normal vector to the obstacle boundary Γ. For $g(q) = 0$, robot motion parallel to object boundaries emerges. For all but the most elementary of obstacle geometries the Laplace equation needs to be solved numerically through a discretization of the workspace into cells. An iterative update rule (e.g the Gauss-Seidel method [155]) can then be applied until convergence. In several extensions, regional and directional constraints have been added to the harmonic potential field method to account for uneven terrain, nonholonomic and kinematic vehicle constraints [195], one-way roads [206], and external forces acting on the robot [207].

Potential fields are extremely easy to implement, much like the breadth-first search described on page 380. Thus, it has become a common tool in mobile robot applications in spite of its theoretical limitations.

This completes our brief summary of the path-planning techniques that are most popular in mobile robotics. Of course, as the complexity of a robot increases (e.g., large degree of freedom nonholonomics) and, particularly, as environment dynamics becomes more significant, then the path-planning techniques described earlier become inadequate for grappling with the full scope of the problem. However, for robots moving in largely flat terrain, the mobility decision-making techniques roboticists use often fall under one of the preceding categories.

But a path planner can take into consideration only the environmental obstacles that are known to the robot in *advance*. During path execution the robot's actual sensor values may disagree with expected values due to map inaccuracy or a dynamic environment. Therefore, it is critical that the robot modify its path in real time based on actual sensor values. This is the competence of *obstacle avoidance*, which we discuss next.

6.4   Obstacle avoidance

Local obstacle avoidance focuses on changing the robot's trajectory as informed by its sensors during robot motion. The resulting robot motion is both a function of the robot's current or recent sensor readings *and* its goal position and relative location to the goal position. The obstacle avoidance algorithms presented here depend to varying degrees on the existence of a global map and on the robot's precise knowledge of its location relative to the map. Despite their differences, all of the algorithms can be termed obstacle avoidance algorithms because the robot's local sensor readings play an important role in the robot's future trajectory. We first present the simplest obstacle avoidance systems that are used successfully in mobile robotics. The Bug algorithm represents such a technique in that only the most recent robot sensor values are used, and the robot needs, in addition to current sensor values, only approximate information regarding the direction of the goal. More sophisticated algorithms are presented afterward, taking into account recent sensor history, robot kinematics, and even dynamics.

6.4.1   Bug algorithm

The Bug algorithm [198, 199] is perhaps the simplest obstacle-avoidance algorithm one could imagine. The basic idea is to follow the contour of each obstacle in the robot's way and thus circumnavigate it.

With Bug1, the robot fully circles the object first, then departs from the point with the shortest distance toward the goal (figure 6.14). This approach is, of course, very inefficient but it guarantees that the robot will reach any reachable goal.

With Bug2 the robot begins to follow the object's contour, but departs immediately when it is able to move directly toward the goal. In general this improved Bug algorithm will have significantly shorter total robot travel, as shown in figure 6.15. However, one can still construct situations in which Bug2 is arbitrarily inefficient (i.e., nonoptimal).

A number of variations and extensions of the Bug algorithm exist. We mention one more, the Tangent Bug [161], which adds range sensing and a local environmental representation termed the local tangent graph (LTG). Not only can the robot move more efficiently toward the goal using the LTG, but it can also go along shortcuts when contouring obstacles and switch back to goal seeking earlier. In many simple environments, Tangent Bug approaches globally optimal paths.

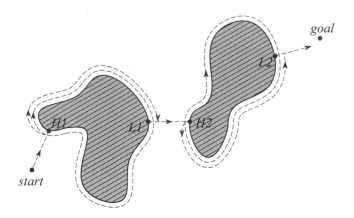

Figure 6.14
Bug1 algorithm with H1, H2, hit points, and L1, L2, leave points [199].

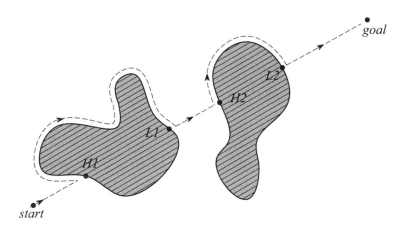

Figure 6.15
Bug2 algorithm with H1, H2, hit points, and L1, L2, leave points [199].

Practical application: example of Bug2. Because of the popularity and simplicity of Bug2, we present a specific example of obstacle avoidance using a variation of this technique. Consider the path taken by the robot in figure 6.15. One can characterize the robot's motion in terms of two states, one that involves moving toward the goal and a second that involves moving around the contour of an obstacle. We will call the former state GOAL-

SEEK and the latter WALLFOLLOW. If we can describe the motion of the robot as a function of its sensor values and the relative direction to the goal for each of these two states, and if we can describe when the robot should switch between them, then we will have a practical implementation of Bug2. The following pseudocode provides the highest-level control code for such a decomposition:

```
public void bug2(position goalPos){
 boolean atGoal = false;

 while( ! atGoal){
   position robotPos = robot.GetPos(&sonars);
   distance goalDist = getDistance(robotPos, goalPos);
   angle goalAngle = Math.atan2(goalPos, robotPos)-robot.GetAngle();
   velocity forwardVel, rotationVel;

   if(goalDist < atGoalThreshold){
     System.out.println("At Goal!");
     forwardVel = 0;
     rotationVel = 0;
     robot.SetState(DONE);
     atGoal = true;
   }
   else{
     forwardVel = ComputeTranslation(&sonars);
     if(robot.GetState() == GOALSEEK){
        rotationVel = ComputeGoalSeekRot(goalAngle);
        if(ObstaclesInWay(goalAngle, &sonars))
          robot.SetState(WALLFOLLOW);
     }
     if(robot.GetState() == WALLFOLLOW){
        rotationVel = ComputeRWFRot(&sonars);
        if( ! ObstaclesInWay(goalAngle, &sonars))
          robot.SetState(GOALSEEK);
     }
   }
   robot.SetVelocity(forwardVel, rotationVel);
 }
}
```

In the ideal case, when encountering an obstacle one would choose between left wall following and right wall following depending on which direction is more promising. In this simple example we have only right wall following, a simplification for didactic purposes that ought not find its way into a real mobile robot program.

Now we consider specifying each remaining function in detail. Consider for our purposes a robot with a ring of sonars placed radially around the robot. This imagined robot will be differential-drive, so that the sonar ring has a clear "front" (aligned with the forward direction of the robot). Furthermore, the robot accepts motion commands of the form shown above, with a rotational velocity parameter and a translational velocity parameter. Mapping these two parameters to individual wheel speeds for each of the two differential-drive chassis' drive wheels is a simple matter.

There is one condition we must define in terms of the robot's sonar readings, ObstaclesInWay(). We define this function to be true whenever any sonar range reading in the direction of the goal (within 45 degrees of the goal direction) is short:

```
private boolean ObstaclesInWay(angle goalAngle, sensorvals sonars) {
    int minSonarValue;
    minSonarValue=MinRange(sonars, goalAngle
                              -(pi/4),goalAngle+(pi/4));
    return (minSonarValue < 200);
} // end ObstaclesInWay() //
```

Note that the function ComputeTranslation() computes translational speed whether the robot is wall-following or heading toward the goal. In this simplified example, we define translation speed as being proportional to the largest range readings in the robot's approximate forward direction:

```
private int ComputeTranslation(sensorvals sonars) {
    int minSonarFront;
    minSonarFront = MinRange(sonars, -pi/4.0, pi/4.0);
    if (minSonarFront < 200) return 0;
    else return (Math.min(500, minSonarFront - 200));
} // end ComputeTranslation() //
```

There is a marked similarity between this approach and the potential field approach described in section 6.3.2. Indeed, some mobile robots implement obstacle avoidance by treating the current range readings of the robot as force vectors, simply carrying out vector addition to determine the direction of travel and speed. Alternatively, many will consider short-range readings to be repulsive forces, again engaging in vector addition to determine an overall motion command for the robot.

When faced with range sensor data, a popular way of determining rotation direction and speed is to simply subtract left and right range readings of the robot. The larger the difference, the faster the robot will turn in the direction of the longer range readings. The following two rotation functions could be used for our Bug2 implementation:

```
private int ComputeGoalSeekRot(angle goalAngle) {
    if (Math.abs(goalAngle) < pi/10) return 0;
    else return (goalAngle * 100);
} // end ComputeGoalSeekRot() //

private int ComputeRWFRot(sensorvals sonars) {
    int minLeft, minRight, desiredTurn;
    minRight = MinRange(sonars, -pi/2, 0);
    minLeft = MinRange(sonars, 0, pi/2);
    if (Math.max(minRight,minLeft) < 200) return (400);
                                        // hard left turn
    else {
        desiredTurn = (400 - minRight) * 2;
        desiredTurn = Math.inttorange(-400, desiredTurn, 400);
        return desiredTurn;
    } // end else
} // end ComputeRWFRot() //
```

Note that the rotation function for the case of right wall following combines a general avoidance of obstacles with a bias to turn right when there is open space on the right, thereby staying close to the obstacle's contour. This solution is certainly not the best solution for implementation of Bug2. For example, the wall follower could do a far better job by mapping the contour locally and using a PID control loop to achieve and maintain a specific distance from the contour during the right wall following action.

Although such simple obstacle avoidance algorithms are often used in simple mobile robots, they have numerous shortcomings. For example, the Bug2 approach does not take into account robot kinematics, which can be especially important with nonholonomic robots. Furthermore, since only the most recent sensor values are used, sensor noise can have a serious impact on real-world performance. The following obstacle avoidance techniques are designed to overcome one or more of these limitations.

6.4.2 Vector field histogram

Borenstein, together with Koren, developed the vector field histogram (VFH) [77]. Their previous work, which was concentrated on potential fields [176], was abandoned due to the method's instability and inability to pass through narrow passages. Later, Borenstein, together with Ulrich, extended the VFH algorithm to yield VFH+ [323] and VFH*[322].

One of the central criticisms of Bug-type algorithms is that the robot's behavior at each instant is generally a function of only its most recent sensor readings. This can lead to undesirable and yet preventable problems in cases where the robot's instantaneous sensor readings do not provide enough information for robust obstacle avoidance. The VFH techniques overcome this limitation by creating a local map of the environment around the robot. This local map is a small occupancy grid, as described in section 5.7 populated only by relatively

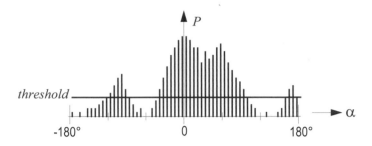

Figure 6.16
Polar histogram [177].

recent sensor range readings. For obstacle avoidance, VFH generates a polar histogram as shown in figure 6.16. The x-axis represents the angle α at which the obstacle was found, and the y-axis represents the probability P that there really is an obstacle in that direction based on the occupancy grid's cell values.

From this histogram a steering direction is calculated. First, all openings large enough for the vehicle to pass through are identified. Then a cost function is applied to every such candidate opening. The passage with the lowest cost is chosen. The cost function G has three terms:

$$G = a \cdot \text{target\_direction} + b \cdot \text{wheel\_orientation} + c \cdot \text{previous\_direction} \qquad (6.15)$$

target_direction = alignment of the robot path with the goal;

wheel_orientation = difference between the new direction and the current wheel orientation;

previous_direction = difference between the previously selected direction and the new direction.

The terms are calculated such that a large deviation from the goal direction leads to a big cost in the term "target direction." The parameters a, b, c in the cost function G tune the behavior of the robot. For instance, a strong goal bias would be expressed with a large value for a. For a complete definition of the cost function, refer to [176].

In the VFH+ improvement, one of the reduction stages takes into account a simplified model of the moving robot's possible trajectories based on its kinematic limitations (e.g., turning radius for an Ackerman vehicle). The robot is modeled to move in arcs or straight lines. An obstacle thus blocks all of the robot's allowable trajectories that pass through the obstacle (figure 6.17a). This results in a masked polar histogram where obstacles are

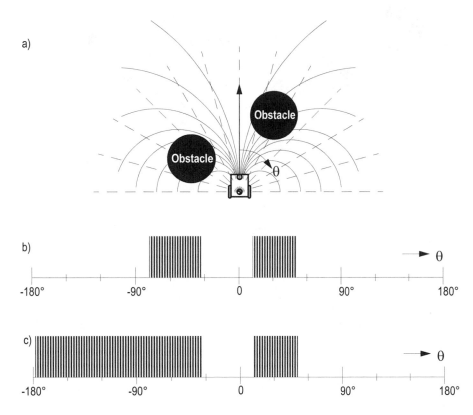

Figure 6.17
Example of blocked directions and resulting polar histograms [54]. (a) Robot and blocking obstacles. (b) Polar histogram. (b) Masked polar histogram.

enlarged so that all kinematically blocked trajectories are properly taken into account (figure 6.17c).

6.4.3 The bubble band technique

This idea is an extension for nonholonomic vehicles of the elastic band concept suggested by Khatib and Quinlan [166]. The original elastic band concept applied only to holonomic vehicles and so we focus only on the bubble band extension made by Khatib, Jaouni, Chatila, and Laumod [165].

A *bubble* is defined as the maximum local subset of the free space around a given configuration of the robot that which can be traveled in any direction without collision. The bubble is generated using a simplified model of the robot in conjunction with range information available in the robot's map. Even with a simplified model of the robot's geometry, it is possible to take into account the actual shape of the robot when calculating the bubble's

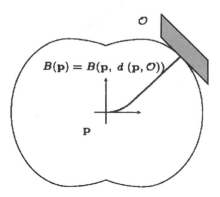

Figure 6.18
Shape of the bubbles around the vehicle. Courtesy of Raja Chatila [165].

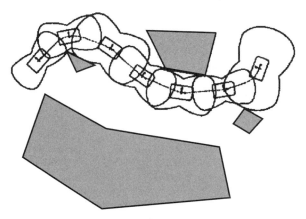

Figure 6.19
A typical bubble band. Courtesy of Raja Chatila [165].

size (figure 6.18). Given such bubbles, a band or string of bubbles can be used along the trajectory from the robot's initial position to its goal position to show the robot's expected free space throughout its path (see figure 6.19).

Clearly, computing the bubble band requires a global map and a global path planner. Once the path planner's initial trajectory has been computed and the bubble band is calculated, then modification of the planned trajectory ensues. The bubble band takes into account forces from modeled objects and internal forces. These internal forces try to minimize the "slack" (energy) between adjacent bubbles. This process, plus a final smoothing operation, makes the trajectory smooth in the sense that the robot's free space will change as smoothly as possible during path execution.

Of course, so far this is more akin to path optimization than obstacle avoidance. The obstacle avoidance aspect of the bubble band strategy comes into play during robot motion. As the robot encounters unforeseen sensor values, the bubble band model is used to deflect the robot from its originally intended path in a way that minimizes bubble band *tension*.

An advantage of the bubble band technique is that one can account for the actual dimensions of the robot. However, the method is most applicable only when the environment configuration is well known ahead of time, just as with offline path-planning techniques.

6.4.4 Curvature velocity techniques

The basic curvature velocity approach. The curvature velocity approach (CVM) from Simmons [291] enables the actual kinematic constraints and even some dynamic constraints of the robot to be taken into account during obstacle avoidance, which is an advantage over more primitive techniques. CVM begins by adding physical constraints from the robot and the environment to a velocity space. The velocity space consists of rotational velocity ω and translational velocity v, thus assuming that the robot travels only along arcs of circles with curvature $c = \omega/v$.

Two types of constraints are identified: those derived from the robot's limitations in acceleration and speed, typically $-v_{max} < v < v_{max}$, $-\omega_{max} < \omega < \omega_{max}$; and, second, the constraints from obstacles blocking certain v and ω values due to their positions. The obstacles begin as objects in a Cartesian grid but are then transformed to the velocity space by calculating the distance from the robot position to the obstacle following some constant curvature robot trajectory, as shown in figure 6.20. Only the curvatures that lie within c_{min} and c_{max} are considered, since that curvature space will contain all legal trajectories.

To achieve real-time performance, the obstacles are approximated by circular objects, and the contours of the objects are divided into few intervals. The distance from an endpoint of an interval to the robot is calculated and in between the endpoints the distance function is assumed to be constant.

The final decision of a new velocity (v and ω) is made by an objective function. This function is only evaluated on that part of the velocity space that fulfills the kinematic and dynamic constraints as well as the constraints due to obstacles. The use of a Cartesian grid for initial obstacle representation enables straightforward sensor fusion if, for instance, a robot is equipped with multiple types of ranging sensors.

CVM takes into consideration the dynamics of the vehicle in useful manner. However a limitation of the method is the circular simplification of obstacle shape. In some environments this is acceptable, while in other environments such a simplification can cause seri-

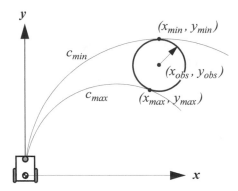

Figure 6.20
Tangent curvatures for an obstacle (from [291]).

ous problems. The CVM method can also suffer from local minima, since no *a priori* knowledge is used by the system.

The lane curvature method.   Ko and Simmons presented an improvement of the CVM that they named the lane curvature method (LCM) [168], based on their experiences with the shortcomings of CVM, which had difficulty guiding the robot through intersections of corridors. The problems stemmed from the approximation that the robot moves only along fixed arcs, whereas in practice the robot can change direction many times before reaching an obstacle.

LCM calculates a set of desired lanes, trading off lane length and lane width to the closest obstacle. The lane with the best properties is chosen using an objective function. The local heading is chosen in such way that the robot will transition to the best lane if it is not in that lane already.

Experimental results have demonstrated better performance as compared to CVM. One caveat is that the parameters in the objective function must be chosen carefully to optimize system behavior.

6.4.5 Dynamic window approaches

Another technique for taking into account robot kinematics constraints is the dynamic window obstacle avoidance method. A simple but very effective dynamic model gives this approach its name. Two such approaches are represented in the literature. The dynamic window approach [130] of Fox, Burgard, and Thrun, and the global dynamic window approach [81] of Brock and Khatib.

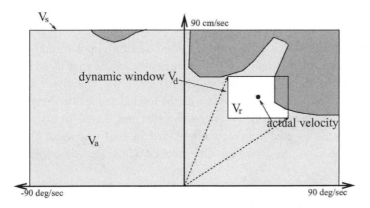

Figure 6.21
The dynamic window approach (courtesy of Dieter Fox [130]). The rectangular window shows the
possible speeds (v, ω) and the overlap with obstacles in configuration space.

The local dynamic window approach. In the local dynamic window approach, the kinematics of the robot is taken into account by searching a well-chosen velocity space. The
velocity space is all possible sets of tuples (v, ω) where v is the velocity and ω is the
angular velocity. The approach assumes that robots move only in circular arcs representing
each such tuple, at least during one timestamp.

Given the current robot speed, the algorithm first selects a *dynamic window* of all tuples
(v, ω) that can be reached within the next sample period, taking into account the acceleration capabilities of the robot and the cycle time. The next step is to reduce the *dynamic
window* by keeping only those tuples that ensure that the vehicle can come to a stop before
hitting an obstacle. The remaining velocities are called admissible velocities. In figure 6.21,
a typical dynamic window is represented. Note that the shape of the dynamic window is
rectangular, which follows from the approximation that the dynamic capabilities for translation and rotation are independent.

A new motion direction is chosen by applying an objective function to all the admissible
velocity tuples in the dynamic window. The objective function prefers fast forward motion,
maintenance of large distances to obstacles and alignment to the goal heading. The objective function O has the form

$$O = a \cdot heading(v, \omega) + b \cdot velocity(v, \omega) + c \cdot dist(v, \omega) \tag{6.16}$$

heading = Measure of progress toward the goal location;

velocity = Forward velocity of the robot $\rightarrow$ encouraging fast movements;

dist = Distance to the closest obstacle in the trajectory.

The global dynamic window approach. The global dynamic window approach adds, as the name suggests, global thinking to the algorithm presented above. This is done by adding *NF1*, or grassfire, to the objective function *O* presented above (see section and figure 6.7). Recall that NF1 labels the cells in the occupancy grid with the total distance L to the goal. To make this faster, the global dynamic window approach calculates the NF1 only on a selected rectangular region that is directed from the robot toward the goal. The width of the region is enlarged and recalculated if the goal cannot be reached within the constraints of this chosen region.

This allows the global dynamic window approach to achieve some of the advantages of global path planning without complete a priori knowledge. The occupancy grid is updated from range measurements as the robot moves in the environment. The NF1 is calculated for every new updated version. If the NF1 cannot be calculated because the robot is surrounded by obstacles, the method degrades to the dynamic window approach. This keeps the robot moving so that a possible way out may be found and NF1 can resume.

The global dynamic window approach promises real-time, dynamic constraints, global thinking, and minimal free obstacle avoidance at high speed. An implementation has been demonstrated with an omnidirectional robot using a 450 MHz on-board PC. This system produced a cycle frequency of about 15 Hz when the occupancy grid was 30×30 m with a 5 cm resolution. Average robot speed in the tests was greater than 1 m/s.

6.4.6 The Schlegel approach to obstacle avoidance

Schlegel [280] presents an approach that considers the dynamics as well as the actual shape of the robot. The approach is adopted for raw laser data measurements and sensor fusion using a Cartesian grid to represent the obstacles in the environment. Real-time performance is achieved by use of precalculated lookup tables.

As with previous methods we have described, the basic assumption is that a robot moves in trajectories built up by circular arcs, defined as curvatures i_c. Given a certain curvature i_c Schlegel calculates the distance l_i to collision between a single obstacle point $[x, y]$ in the Cartesian grid and the robot, depicted in figure 6.22. Since the robot is allowed to be any shape, this calculation is time-consuming, and the result is therefore precalculated and stored in a lookup table.

For example, the search space window V_s is defined for a differential-drive robot to be all the possible speeds of the left and right wheels, v_r, v_l. The dynamic constraints of the robot are taken into account by refining V_s to only those values that are reachable within the next timestep, given the present robot motion. Finally, an objective function chooses the best speed and direction by trading off goal direction, speed, and distance until collision.

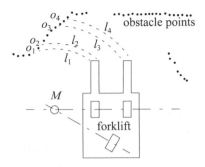

Figure 6.22
Distances l_i resulting from the curvature i_c, when the robot rotates around M (from [280]).

During testing Schlegel used a wavefront path planner. Two robot chassis were used, one with synchro-drive kinematics and one with tricycle kinematics. The tricycle-drive robot is of particular interest because it was a forklift with a complex shape that had a significant impact on obstacle avoidance. Thus the demonstration of reliable obstacle avoidance with the forklift is an impressive result. Of course, a disadvantage of this approach is the potential memory requirements for the lookup table. In their experiments, the authors used lookup tables of up to 2.5 Mb using a 6×6 m Cartesian grid with a resolution of 10 cm and 323 different curvatures.

6.4.7 Nearness diagram
Attempting to close a model fidelity gap in obstacle avoidance methods, the nearness diagram (ND) [222] can be considered to have some similarity to a VFH but solves several of its shortcomings, especially in very cluttered spaces. It was also used in [223] to take into account more precise geometric, kinematic, and dynamic constraints. This was achieved by breaking the problem down into generating the most promising direction of travel with the sole constraint a circular robot, then adapting this to the kinematic and dynamic constraints of the robot, followed by a correction for robot shape if is noncircular (only rectangular shapes were supported in the original publication). Global reasoning was added to the approach and termed the global nearness diagram (GND) in [225], somewhat similar to the GDWA extension to the DWA, but based on a workspace representation (instead of configuration space) and updating free space in addition to obstacle information.

6.4.8 Gradient method
Realizing that current computer technology allows fast recalculation of wavefront propagation techniques, the gradient method [171] formulates a grid global path planning that takes into account closeness to obstacles and allows generating continuous interpolations

of the gradient direction at any given point in the grid. The NF1 is a special case of the proposed algorithm, which calculates a navigation function at each timestep and uses the resulting gradient information to drive the robot toward the goal on a smooth path and not grazing obstacles unless necessary.

6.4.9   Adding dynamic constraints

Attempting to address the lack of dynamic models in most of the obstacle avoidance approaches discussed above, a new kind of space representation was proposed by Minguez, Montano, and Khatib in [224]. The ego-dynamic space is equally applicable to workspace and configuration space methods. It transforms obstacles into distances that depend on the braking constraints and sampling time of the underlying obstacle avoidance method. In combination with the proposed spatial window (PF) to represent acceleration capabilities, the approach was tested in conjunction with the ND and PF methods and gives satisfactory results for circular holonomic robots, with plans to extend it to nonholonomic, noncircular architectures.

6.4.10   Other approaches

The approaches described above are some of the most popularly referenced obstacle avoidance systems. There are, however, a great many additional obstacle avoidance techniques in the mobile robotics community. For example Tzafestas and Tzafestas [321] provide an overview of fuzzy and neurofuzzy approaches to obstacle avoidance. Inspired by nature, Chen and Quinn [98] present a biological approach in which they replicate the neural network of a cockroach. The network is then applied to a model of a four-wheeled vehicle.

The Liapunov functions form a well known theory that can be used to prove stability for nonlinear systems. In Vanualailai, Nakagiri, and Ha [326], the Liapunov functions are used to implement a control strategy for two-point masses moving in a known environment. All obstacles are defined as antitargets with an exact position and a circular shape. The antitargets are then used to build the control laws for the system.

6.4.11   Overview

Table 6.1 gives an overview on the presented obstacle-avoidance approaches.

Table 6.1

Overview of the most popular obstacle-avoidance algorithms

method			model fidelity			view	other requisites			sensors	tested robots	performance		
			shape	kinematics	dynamics		local map	global map	path planner			cycle time	architecture	remarks
Bug	Bug1 [198, 199]	point				local				tactile				very inefficient, robust
	Bug2 [198, 199]	point				local				tactile				inefficient, robust
	Tangent Bug [161]	point				local	local tangent graph			range				efficient in many cases, robust
Vector Field Histogram (VFH)	VFH [77]	simplistic				local	histogram grid			range	synchro-drive (hexagonal)	27 ms	20 MHz, 386 AT	local minima, oscillating trajectories
	VFH+ [176, 323]	circle	basic	simplistic		local	histogram grid			sonars	nonholonomic (GuideCane)	6 ms	66 MHz, 486 PC	local minima
	VFH* [322]	circle	basic	simplistic		essentially local	histogram grid			sonars	nonholonomic (GuideCane)	6 ... 242 ms	66 MHz, 486 PC	fewer local minima

Table 6.1
Overview of the most popular obstacle-avoidance algorithms

method		model fidelity				other requisites					performance	
	shape	kinematics	dynamics	view	local map	global map	path planner	sensors	tested robots	cycle time	architecture	remarks
Dynamic window — Global dynamic window [81]	circle	(holonomic)	basic	global	C-space grid		NF1	180° FOV SCK laser scanner	holonomic (circular)	6.7 ms	450 MHz, PC	turning into corridors
Dynamic window — Dynamic window approach [130]	circle	exact	basic	local	obstacle line field			24 sonars ring, 56 infrared ring, stereo camera	synchro-drive (circular)	250 ms	486 PC	local minima
Curvature velocity — Lane curvature method [168]	circle	exact	basic	local	histogram grid			24 sonars ring, 30° FOV laser	synchro-drive (circular)	125 ms	200 MHz, Pentium	local minima
Curvature velocity — Curvature velocity method [291]	circle	exact	basic	local	histogram grid			24 sonars ring, 30° FOV laser	synchro-drive (circular)	125 ms	66 MHz, 486 PC	local minima, turning into corridors
Bubble band — Bubble band [165]	C-space	exact		local		polygonal	required		various			
Bubble band — Elastic band [166]	C-space			global		polygonal	required		various			

Table 6.1
Overview of the most popular obstacle-avoidance algorithms

	method	model fidelity			view	other requisites			sensors	tested robots	performance		
		shape	kinematics	dynamics		local map	global map	path planner			cycle time	architecture	remarks
Other	Schlegel [280]	polygon	exact	basic	global		grid	wavefront	360° FOV laser scanner	synchrodrive (circular), tricycle (forklift)			allows shape change
	Nearness diagram [222, 223]	circle (but general formulation)	(holonomic)		local				180° FOV SCK laser scanner	holonomic (circular)			local minima
	Global nearness diagram [225]	circle (but general formulation)	(holonomic)		global	grid		NF1	180° FOV SCK laser scanner	holonomic (circular)			
	Gradient method [171]	circle	exact	basic	global	local perceptual space	fused		180° FOV distance sensor	nonholonomic (approx. circle)	100 ms (core algorithm: 10 ms)	266 MHz, Pentium	

6.5 Navigation Architectures

Given techniques for path planning, obstacle avoidance, localization, and perceptual interpretation, how do we combine all of these into one complete robot system for a real-world
application? One way to proceed would be to custom-design an application-specific, monolithic software system that implements everything for a specific purpose. This may be efficient in the case of a trivial mobile robot application with few features and even fewer
planned demonstrations. But for any sophisticated and long-term mobile robot system, the

issue of mobility architecture should be addressed in a principled manner. The study of *navigation architectures* is the study of principled designs for the software modules that constitute a mobile robot navigation system. Using a well-designed navigation architecture has a number of concrete advantages:

6.5.1 Modularity for code reuse and sharing

Basic software engineering principles embrace software modularity, and the same general motivations apply equally to mobile robot applications. But modularity is of even greater importance in mobile robotics because in the course of a single project the mobile robot hardware or its physical environmental characteristics can change dramatically, a challenge most traditional computers do not face. For example, one may introduce a Sick laser rangefinder to a robot that previously used only ultrasonic rangefinders. Or one may test an existing navigator robot in a new environment where there are obstacles that its sensors cannot detect, thereby demanding a new path-planning representation.

We would like to change part of the robot's competence without causing a string of side effects that force us to revisit the functioning of other robot competences. For instance we would like to retain the obstacle avoidance module intact, even as the particular ranging sensor suite changes. In a more extreme example, it would be ideal if the nonholonomic obstacle avoidance module could remain untouched even when the robot's kinematic structure changes from a tricycle chassis to a differential-drive chassis.

6.5.2 Control localization

Localization of robot control is an even more critical issue in mobile robot navigation. The basic reason is that a robot architecture includes multiple types of control functionality (e.g., obstacle avoidance, path planning, path execution, etc.). By localizing each functionality to a specific unit in the architecture, we enable individual testing as well as a principled strategy for control composition. For example, consider collision avoidance. For stability in the face of changing robot software, as well as for focused verification that the obstacle avoidance system is correctly implemented, it is valuable to localize all software related to the robot's obstacle avoidance process. At the other extreme, high-level planning and task decision making are required for robots to perform useful roles in their environment. It is also valuable to localize such high-level decision-making software, enabling it to be tested exhaustively in simulation and thus verified even without a direct connection to the physical robot. A final advantage of localization is associated with learning. Localization of control can enable a specific learning algorithm to be applied to just one aspect of a mobile robot's overall control system. Such targeted learning is likely to be the first strategy that yields successful integration of learning and traditional mobile robotics.

The advantages of localization and modularity provide a compelling case for the use of principled navigation architectures.

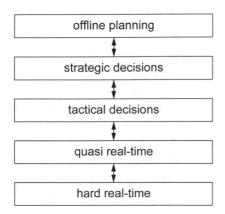

Figure 6.23
Generic temporal decomposition of a navigation architecture.

One way to characterize a particular architecture is by its decomposition of the robot's software. There are many favorite robot architectures, especially when one considers the relationship between artificial intelligence level decision making and lower-level robot control. For descriptions of such high-level architectures, refer to [2] and [39]. Here we concentrate on navigation competence. For this purpose, two decompositions are particularly relevant: temporal decomposition and control decomposition. In section 6.5.3 we define these two types of decomposition, then present an introduction to *behaviors*, which are a general tool for implementing control decomposition. Then, in section 6.5.4 we present three types of navigation architectures, describing for each architecture an implemented mobile robot case study.

6.5.3 Techniques for decomposition

Decompositions identify axes along which we can justify discrimination of robot software into distinct modules. Decompositions also serve as a way to classify various mobile robots into a more quantitative taxonomy. *Temporal decomposition* distinguishes between real-time and non real-time demands on mobile robot operation. *Control decomposition* identifies the way in which various control outputs within the mobile robot architecture combine to yield the mobile robot's physical actions. We will describe each type of decomposition in greater detail.

6.5.3.1 Temporal decomposition

A temporal decomposition of robot software distinguishes between processes that have varying real-time and non-real-time demands. Figure 6.23 depicts a generic temporal decomposition for navigation. In this figure, the most real-time processes are shown at the bottom of the *stack*, with the highest category being occupied by processes with no real-time demands.

The lowest level in this example captures functionality that must proceed with a guaranteed fast cycle time, such as a 40 Hz bandwidth. In contrast, a quasi real-time layer may capture processes that require, for example, 0.1 second response time, with large allowable worst-case individual cycle times. A tactical layer can represent decision making that affects the robot's immediate actions and is therefore subject to some temporal constraints, while a strategic or offline layer represents decisions that affect the robot's behavior over the long term, with few temporal constraints on the module's response time.

Four important, interrelated trends correlate with temporal decomposition. These are not set in stone; there are exceptions. Nevertheless, these general properties of temporal decompositions are enlightening.

Sensor response time. A particular module's sensor response time can be defined as the amount of time between acquisition of a sensor event and a corresponding change in the output of the module. As one moves up the stack in figure 6.23, the sensor response time tends to increase. For the lowest-level modules, the sensor response time is often limited only by the raw processor and sensor speeds. At the highest-level modules, sensor response can be limited by slow and deliberate decision-making processes.

Temporal depth. Temporal depth is a useful concept applying to the temporal window that affects the module's output, both backward and forward in time. *Temporal horizon* describes the amount of look ahead used by the module during the process of choosing an output. *Temporal memory* describes the historical time span of sensor input that is used by the module to determine the next output. Lowest-level modules tend to have very little temporal depth in both directions, whereas the deliberative processes of highest-level modules make use of a large temporal memory and consider actions based on their long-term consequences, making note of large temporal horizons.

Spatial locality. Hand in hand with temporal span, the spatial impact of layers increases dramatically as one moves from low-level modules to high-level modules. Real-time modules tend to control wheel speed and orientation, controlling spatially localized behavior. High-level strategic decision making has little or no bearing on local position, but it informs global position far into the future.

Context specificity. A module makes decisions as a function not only of its immediate inputs but also as a function of the robot's context as captured by other variables, such as the robot's representation of the environment. Lowest-level modules tend to produce outputs directly as a result of immediate sensor inputs, using little context and therefore being

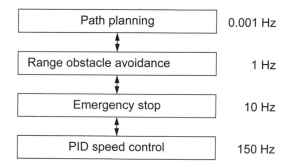

Figure 6.24
Sample four-level temporal decomposition of a simple navigating mobile robot. The column on the right indicates realistic bandwidth values for each module.

relatively context insensitive. Highest-level modules tend to exhibit very high context specificity. For strategic decision making, given the same sensor values, dramatically different outputs are nevertheless conceivable depending on other contextual parameters.

An example demonstrating these trends is depicted in figure 6.24, which shows a temporal decomposition of a simplistic navigation architecture into four modules. At the lowest level, the PID control loop provides feedback to control motor speeds. An emergency stop module uses short-range optical sensors and bumpers to cut current to the motors when it predicts an imminent collision. Knowledge of robot dynamics means that this module by nature has a greater temporal horizon than the PID module. The next module uses longer-range laser rangefinding sensor returns to identify obstacles well ahead of the robot and make minor course deviations. Finally, the path planner module takes the robot's initial and goal positions and produces an initial trajectory for execution, subject to change based on actual obstacles that the robot collects along the way.

Note that the cycle time, or bandwidth, of the modules changes by orders of magnitude between adjacent modules. Such dramatic differences are common in real navigation architectures, and so temporal decomposition tends to capture a significant axis of variation in a mobile robot's navigation architecture.

6.5.3.2 Control decomposition

Whereas temporal decomposition discriminates based on the time behavior of software modules, control decomposition identifies the way in which each module's output contributes to the overall robot control outputs. Presentation of control decomposition requires the evaluator to understand the basic principles of discrete systems representation and analysis. For a lucid introduction to the theory and formalism of discrete systems, see [25, 136].

Consider the robot algorithm and the physical robot instantiation (i.e., the robot form and its environment) to be members of an overall system whose connectivity we wish to

Figure 6.25
Example of a pure serial decomposition.

examine. This overall system S comprises a set M of modules, each module m connected to other modules via inputs and outputs. The system is *closed*, meaning that the input of every module m is the output of one or more modules in M. Each module has precisely one output and one or more inputs. The one output can be connected to any number of other modules inputs.

We further name a special module r in M to represent the physical robot and environment. Usually by r we represent the physical object on which the robot algorithm is intended to have impact, and from which the robot algorithm derives perceptual inputs. The module r contains one input and one output line. The input of r represents the complete action specification for the physical robot. The output of r represents the complete perceptual output to the robot. Of course the physical robot may have many possible degrees of freedom and, equivalently, many discrete sensors. But for this analysis we simply imagine the entire input/output vector, thus simplifying r to just one input and one output. For simplicity, we will refer to the input of r as O and to the robot's sensor readings I. From the point of view of the rest of the control system, the robot's sensor values I are inputs, and the robot's actions O are the outputs, explaining our choice of I and O.

Control decomposition discriminates between different types of control pathways through the portion of this system comprising the robot algorithm. At one extreme, depicted in figure 6.25 we can consider a perfectly linear, or sequential control pathway.

Such a serial system uses the internal state of all associated modules and the value of the robot's percept I in a sequential manner to compute the next robot action O. A pure serial architecture has advantages relating to predictability and verifiability. Since the state and outputs of each module depend entirely on the inputs it receives from the module upstream, the entire system, including the robot, is a single well-formed loop. Therefore, the overall behavior of the system can be evaluated using well-known discrete forward simulation methods.

Figure 6.26 depicts the extreme opposite of pure serial control, a fully parallel control architecture. Because we choose to define r as a module with precisely one input, this parallel system includes a special module n that provides a single output for the consumption of r. Intuitively, the fully parallel system distributes responsibility for the system's control output O across multiple modules, possibly simultaneously. In a pure sequential system, the control flow is a linear sequence through a string of modules. Here, the control flow

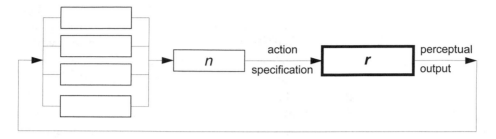

Figure 6.26
Example of a pure parallel decomposition.

contains a *combination* step at which point the result of multiple modules may impact O in arbitrary ways.

Thus parallelization of control leads to an important question: how will the output of each component module inform the overall decision concerning the value of O? One simple combination technique is temporal switching. In this case, called *switched parallel*, the system has a parallel decomposition but at any particular instant in time the output O can be attributed to one specific module. The value of O can of course depend on a different module at each successive time instant, but the instantaneous value of O can always be determined based on the functions of a single module. For instance, suppose that a robot has an obstacle avoidance module and a path-following module. One switched control implementation may involve execution of the path-following recommendation whenever the robot is more than 50 cm from all sensed obstacles and execution of the obstacle-avoidance recommendation when any sensor reports a range closer than 50 cm.

The advantage of such switched control is particularly clear if switching is relatively rare. If the behavior of each module is well understood, then it is easy to characterize the behavior of the switched control robot: it will obstacle avoid at times, and it will path-follow other times. If each module has been tested independently, there is a good chance the switched control system will also perform well. Two important disadvantages must be noted. First, the overall behavior of the robot can become quite poor if the switching is itself a high-frequency event. The robot may be unstable in such cases, switching motion modes so rapidly as to dramatically devolve into behavior that is neither path-following nor obstacle-avoiding. Another disadvantage of switched control is that the robot has no path-following bias when it is obstacle avoiding (and vice versa). Thus in cases where control *ought* to mix recommendations from among multiple modules, the switched control methodology fails.

In contrast, the much more complex *mixed parallel* model allows control at any given time to be shared between multiple modules. For example, the same robot could take the obstacle avoidance module's output at all times, convert it to a velocity vector, and combine it with the path-following module's output using vector addition. Then the output of the robot would never be due to a single module, but would result from the mathematical combination of both modules outputs. Mixed parallel control is more general than switched control, but by that token it is also a more challenging technique to use well. Whereas with switched control most poor behavior arises out of inopportune switching behavior, in mixed control the robot's behavior can be quite poor even more readily. Combining multiple recommendations mathematically does not guarantee an outcome that is globally superior, just as combining multiple vectors when deciding on a swerve direction to avoid an obstacle can result in the very poor decision of going straight ahead. Thus, great care must be taken in mixed parallel control implementations to fashion mixture formulas and individual module specifications that lead to effective mixed results.

Both the switched and mixed parallel architectures are popular in the behavior robotics community. Arkin [2] proposes the *motor-schema* architecture in which *behaviors* (i.e., modules in the earlier discussion) map sensor value vectors to motor value vectors. The output of the robot algorithm is generated, as in mixed parallel systems, using a linear combination of the individual behavior outputs. In contrast, Maes [201, 202] produces a switched parallel architecture by creating a *behavior network* in which a behavior is chosen discretely by comparing and updating activation levels for each behavior. The subsumption architecture of Brooks [82] is another example of a switched parallel architecture, although the active model is chosen via a suppression mechanism rather than activation level. For further discussion, see [2].

One overall disadvantage of parallel control is that verification of robot performance can be extremely difficult. Because such systems often include truly parallel, multithreaded implementations, the intricacies of robot-environment interaction and sensor timing required to represent properly all conceivable module-module interactions can be difficult or impossible to simulate. So, much testing in the parallel control community is performed empirically using physical robots.

An important advantage of parallel control is its biomimetic aspect. Complex organic organisms benefit from large degrees of true parallelism (e.g., the human eye), and one goal of the parallel control community is to understand this biologically common strategy and leverage it to advantage in robotics.

6.5.4   Case studies: tiered robot architectures

We have described temporal and control decompositions of robot architecture, with the common theme that the roboticist is always composing multiple modules together to make up that architecture. Let us turn again toward the overall mobile robot navigation task with this understanding in mind. Clearly, robot behaviors play an important role at the real-time levels of robot control, for example, path-following and obstacle avoidance. At higher temporal levels, more tactical tasks need to modulate the activation of behaviors, or modules,

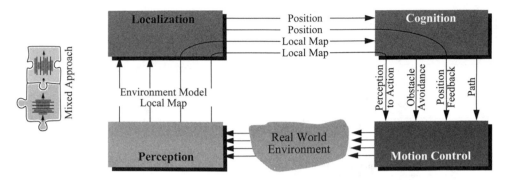

Figure 6.27
The basic architectural example used throughout this text.

in order to achieve robot motion along the intended path. Higher still, a global planner could generate paths to provide tactical tasks with global foresight.

In chapter 1, we introduced a functional decomposition showing such modules of a mobile robot navigator from the perspective of information flow. The relevant figure is shown here again as figure 6.27.

In such a representation, the arcs represent aspects of real-time and non-real-time competence. For instance, obstacle avoidance requires little input from the localization module and consists of fast decisions at the cognition level followed by execution in motion control. In contrast, PID position feedback loops bypass all high-level processing, tying the perception of encoder values directly to lowest-level PID control loops in motion control. The trajectory of arcs through the four software modules provides temporal information in such a representation.

Using the tools of this chapter, we can now present this same architecture from the perspective of a temporal decomposition of functionality. This is particularly useful because we wish to discuss the interaction of strategic, tactical, and real-time processes in a navigation system.

Figure 6.28 depicts a generic tiered architecture based on the approach of Pell and colleagues [256] used in designing an autonomous spacecraft, *Deep Space One*. This figure is similar to figure 6.24 in presenting a temporal decomposition of robot competence. However, the boundaries separating each module from adjacent modules are specific to robot navigation.

Path planning embodies strategic-level decision making for the mobile robot. Path planning uses all available global information in non-real-time to identify the right sequence of local actions for the robot. At the other extreme, *real-time control* represents competences requiring high bandwidth and tight sensor-effector control loops. At its lowest level, this

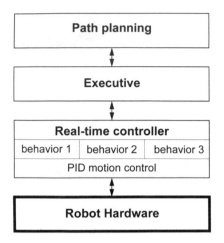

Figure 6.28
A general tiered mobile robot navigation architecture based on a temporal decomposition.

includes motor velocity PID loops. Above those, real-time control also includes low-level behaviors that may form a switch or mixed parallel architecture.

In between the path planner and real-time control tiers sits the *executive*, which is responsible for mediating the interface between planning and execution. The executive is responsible for managing the activation of behaviors based on information it receives from the planner. The executive is also responsible for recognizing failure, saving (placing the robot in a stable state), and even reinitiating the planner as necessary. It is the executive in this architecture that contains all tactical decision making as well as frequent updates of the robot's short-term memory, as is the case for localization and mapping.

It is interesting to note the similarity between this general architecture, used in many specialized forms in mobile robotics today, and the architecture implemented by Shakey, one of the very first mobile robots, in 1969 [242]. Shakey had *LLA* (low-level actions) that formed the lowest architectural tier. The implementation of each LLA included the use of sensor values in a tight loop just as in today's behaviors. Above that, the middle architectural tier included the *ILA* (intermediate-level actions), which would activate and deactivate LLA as required based on perceptual feedback during execution. Finally, the topmost tier for Shakey was STRIPS (Stanford Research Institute Planning System), which provided global look ahead and planning, delivering a series of tasks to the intermediate executive layer for execution.

Although the general architecture shown in figure 6.28 is useful as a model for robot navigation, variant implementations in the robotics community can be quite different. Next, we present three particular versions of the general tiered architecture, describing for each

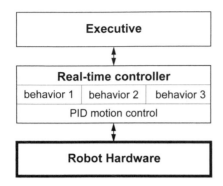

Figure 6.29
A two-tiered architecture for offline planning.

version at least one real-world mobile robot implementation. For broader discussions of various robot architectures, see [39].

6.5.4.1 Offline planning

Certainly the simplest possible integration of planning and execution is no integration at all. Consider figure 6.29, in which there are only two software tiers. In such navigation architectures, the executive does not have a planner at its disposal but must contain a priori all relevant schemes for traveling to desired destinations.

The strategy of leaving out a planner altogether is of course extremely limiting. Moving such a robot to a new environment demands a new instantiation of the navigation system, and so this method is not useful as a general solution to the navigation problem. However such robotic systems do exist, and this method can be useful in two cases.

Static route applications. In mobile robot applications where the robot operates in a completely static environment using a route navigation system, it is conceivable that the number of discrete goal positions is so small that the environmental representation can directly contain paths to all desired goal points. For example, in factory or warehouse settings, a robot may travel a single looping route by following a buried guidewire. In such industrial applications, path-planning systems are sometimes altogether unnecessary when a precompiled set of route solutions can be easily generated by the robot programmers. The Chips mobile robot is an example of a museum robot that also uses this architecture [251]. Chips operates in a unidirectional looping track defined by its colored landmarks. Furthermore, it has only twelve discrete locations at which it is allowed to stop. Due to the simplicity of this environmental model, Chips contains an executive layer that directly caches the path required to reach each goal location rather than a generic map with which a path planner could search for solution paths.

Extreme reliability demands. Not surprisingly, another reason to avoid online planning is to maximize system reliability. Since planning software can be the most sophisticated portion of a mobile robot's software system, and since in theory at least planning can take time exponential to the complexity of the problem, imposing hard temporal constraints on successful planning is difficult if not impossible. By computing all possible solutions offline, the industrial mobile robot can trade versatility for effective constant-time planning (while sacrificing significant memory of course). A real-world example of offline planning for this reason can be seen in the contingency plans designed for space shuttle flights. Instead of requiring astronauts to solve problems online, thousands of conceivable issues are postulated on Earth, and complete conditional plans are designed and published in advance of the shuttle flights. The fundamental goal is to provide an absolute upper limit on the amount of time that passes before the astronauts begin resolving the problem, sacrificing a great deal of ground time and paperwork to achieve this performance guarantee.

6.5.4.2 Episodic planning

The fundamental information-theoretic disadvantage of planning offline is that, during runtime, the robot is sure to encounter perceptual inputs that provide information, and it would be rational to take this additional information into account during subsequent execution. Episodic planning is the most popular method in mobile robot navigation today because it solves this problem in a computationally tractable manner.

As shown in figure 6.30, the structure is three-tiered, as is the general architecture of figure 6.28. The intuition behind the role of the planner is as follows. Planning is computationally intensive, and therefore planning too frequently would have serious disadvantages. But the executive is in an excellent position to identify when it has encountered enough information (e.g., through feature extraction) to warrant a significant change in strategic direction. At such points, the executive will invoke the planner to generate, for example, a new path to the goal.

Perhaps the most obvious condition that triggers replanning is detection of a blockage on the intended travel path. For example, in [281] the path-following behavior returns failure if it fails to make progress for a number of seconds. The executive receives this failure notification, modifies the short-term occupancy grid representation of the robot's surroundings, and launches the path planner in view of this change to the local environment map.

A common technique to delay planning until more information has been acquired is called *deferred planning*. This technique is particularly useful in mobile robots with dynamic maps that become more accurate as the robot moves. For example, the commercially available Cye robot can be given a set of goal locations. Using its grassfire breadth-

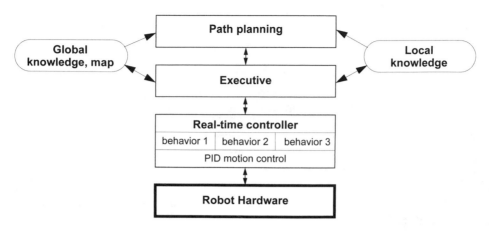

Figure 6.30
A three-tiered episodic planning architecture.

first planning algorithm, this robot will plot a detailed path to the closest goal location only and will execute this plan. Upon reaching this goal location, its map will have changed based on the perceptual information extracted during motion. Only then will Cye's executive trigger the path planner to generate a path from its new location to the next goal location.

The robot Pygmalion implements an episodic planning architecture along with a more sophisticated strategy when encountering unforeseen obstacles in its way [58, 259]. When the lowest-level behavior fails to make progress, the executive attempts to find a way past the obstacle by turning the robot 90 degrees and trying again. This is valuable because the robot is not kinematically symmetric, and so servoing through a particular obstacle course may be easier in one direction than the other.

Pygmalion's environment representation consists of a continuous geometric model as well as an abstract topological network for route planning. Thus, if repeated attempts to clear the obstacle fail, then the robot's executive will temporarily cut the topological connection between the two appropriate nodes and will launch the planner again, generating a new set of waypoints to the goal. Next, using recent laser rangefinding data as a type of local map (see figure 6.30), a geometric path planner will generate a path from the robot's current position to the next waypoint.

In summary, episodic planning architectures are extremely popular in the mobile robot research community. They combine the versatility of responding to environmental changes and new goals with the fast response of a tactical executive tier and behaviors that control real-time robot motion. As shown in figure 6.30, it is common in such systems to have both a short-term local map and a more strategic global map. Part of the executive's job in such

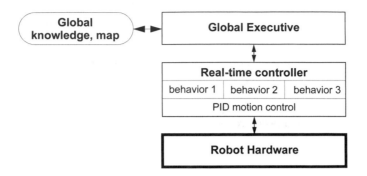

Figure 6.31
An integrated planning and execution architecture in which planning is nothing more than a real-time execution step (behavior).

dual representations is to decide when and if new information integrated into the local map is sufficiently nontransient to be copied into the global knowledge base.

6.5.4.3 Integrated planning and execution
Of course, the architecture of a commercial mobile robot must include more functionality than just navigation. But limiting this discussion to the question of *navigation* architectures leads to what may at first seem a degenerate solution.

The architecture shown in figure 6.31 may look similar to the offline planning architecture of figure 6.29, but in fact it is significantly more advanced. In this case, the planner tier has disappeared because there is no longer a temporal decomposition between the executive and the planner. Planning is simply one small part of the executive's nominal cycle of activities, where the local and global representations are the same. The advantage of this approach is that the robot's actions at every cycle are guided by a global path planner, and they are therefore optimal in view of all of the information the robot has gathered.

Integrated planning and execution has largely been made feasible due to innovations in graph search algorithms (e.g. the D* algorithm described on page page 385) and graph representation (e.g., state lattice graphs, whose edges can be inherently executed by the robotic platform). As a result, formidable real-time implementations devoid of obstacle avoidance modules have emerged: Pivtoraiko et al. [260] showed that graph search on a 3D state lattice (including 2D position and heading) can be as efficient as 2D grid search. In the work of Ferguson et al. [127], an extension to this strategy was successfully employed to navigate an autonomous car in large scale parking lots. Rufli et al. [271, 272] added velocity dimensions to the aforementioned state lattice representation which allowed them to take into account position and velocity information of nearby dynamic obstacles during the planning step. The result is a feasible, globally optimal, time-parametrized path that inherently avoids dynamic obstacles, a task that has traditionally been carried out by local collision avoidance modules.

The described methods naturally face limits of applicability as the size of the environment increases. This issue can be accounted for by applying multiresolution graph approaches, however. Close to the robot, a high-fidelity lattice may be employed, farther away a lower-resolution one. Hundreds of meters distant, the lattice may transition into a road network such as the ones often used in commercial GPS-based navigation systems.

Still, the recent success of integrated planning and execution methods underlines the fact that the designer of a robot navigation architecture must consider not only all aspects of the robot and its environmental task but also the state of processor, GPU, and memory technology. We expect that mobile robot architecture design is sure to remain an active area of innovation for years to come. All forms of technological progress, from robot sensor inventions to processor speed increases, and further parallelization are likely to catalyze new innovations in mobile robot architecture as previously unimaginable tactics become realizable.

6.6   Problems

1. Consider completeness and optimality properties for each of:
    Visibility graph
    Voronoi diagram
    Exact cell decomposition
    Approximate cell decomposition
In the framework of path planning, categorize for each whether it is complete/incomplete, and optimal/ not guaranteed-optimal for path planning.

2. Consider an Ackerman steering 4-wheel high speed Martian rover. Consider all the obstacle avoidance techniques described in 6.2. Explain for every option its advantage or disadvantage, in one sentence each, for this specific application. Specifically do so for: Schlegel, local dynamic window, LCM, CVM, VFH, Bubble band, Bug.

3. Consider an autonomous driving robot for highway driving. Propose a temporal decomposition, as in figure 6.24, with at least five levels, describing control frequency at each level and the specific driving skills/behaviors incorporated at that level.

4. Challenge Question.
    Consider a robot that navigates with range sensors that have a limited useful range r. Propose a path-planning method based on the ones in 6.3 that is complete and maintains a safe distance from objects while also staying within distance r from objects whenever possible.

Bibliography

Books

[1]   Adams, M.D., *Sensor Modelling: Design and Data Processing for Autonomous Navigation*. World Scientific Series in Robotics and Intelligent Systems. Singapore, World Scientific Publishing, 1999.

[2]   Arkin, R.C., *Behavioral Robotics*. Cambridge MA, MIT Press, 1998.

[3]   Bar-Shalom, Y., Li, X.-R., *Estimation and Tracking: Principles, Techniques, and Software*. Norwood, MA, Artech House, 1993.

[4]   Benosman, R., Kang, S. B., *Panoramic Vision: Sensors, Theory, and Applications*, New York, Springer-Verlag, 2001.

[5]   Borenstein, J., Everett, H.R., Feng, L., *Navigating Mobile Robots: Systems and Techniques*. Natick, MA, A.K. Peters, Ltd., 1996.

[6]   Borenstein, J., Everett, H.R., Feng, L., *Where Am I? Sensors and Methods for Mobile Robot Positioning*. Technical report, Ann Arbor, University of Michigan, 1996. Available at http://www-personal.engin.umich.edu/~johannb/position.htm.

[7]   Bradski, G., Kaehler, A., *Learning OpenCV: Computer Vision with the OpenCV Library*, Sebastopol, CA, O'Reilly Media, Inc., 1st edition, 2008.

[8]   Breipohl, A.M., *Probabilistic Systems Analysis: An Introduction to Probabilistic Models, Decisions, and Applications of Random Processes*. New York, John Wiley & Sons, 1970.

[9]   Bundy, A. (editor), *Artificial Intelligence Techniques: A Comprehensive Catalogue*. New York, Springer-Verlag, 1997.

[10]   Canudas de Wit, C., Siciliano, B., and Bastin G. (editors), *Theory of Robot Control*. New York, Springer, 1996.

[11]   Carroll, R.J., Ruppert, D., *Transformation and Weighting in Regression*. New York, Chapman and Hall, 1988.

[12]   Cox, I.J., Wilfong, G.T. (editors), *Autonomous Robot Vehicles*. New York, Springer-Verlag, 1990.

[13]   Craig, J.J., *Introduction to Robotics: Mechanics and Control*. 2nd edition. Boston, Addison-Wesley, 1989.

[14]   De Silva, C.W., *Control Sensors and Actuators*. Upper Saddle River, NJ, Prentice-Hall, 1989.

[15]   Daniilidis, K., Klette, R., *Imaging Beyond the Pinhole Camera*. New York, Springer, 2006.

[16] Dietrich, C.F., *Uncertainty, Calibration and Probability*. Bristol, UK, Adam Hilger, 1991.

[17] Draper, N.R., Smith, H., *Applied Regression Analysis*. 3rd edition. New York, John Wiley & Sons, 1988.

[18] Duda, R.O., Hart, P.E., Stork, D.G., *Pattern Classification*. New York, Wiley, 2001.

[19] Duda, R. O., Hart, P.E. *Pattern Classification and Scene Analysis*. New York, John Wiley & Sons, 1973.

[20] Everett, H.R., *Sensors for Mobile Robots: Theory and Applications*. New York, Natick, MA, A.K. Peters, Ltd., 1995.

[21] Faugeras, O., *Three-Dimensional Computer Vision: A Geometric Viewpoint*. Cambridge, MA, MIT Press, 1993.

[22] Faugeras, O., Luong, Q.T., *The Geometry of Multiple Images*. Cambridge, MA, MIT Press, 2001.

[23] Floreano, D., Zufferey, J.C., Srinivasan, M.V., Ellington, C., *Flying Insects and Robots*, Springer, 2009.

[24] Forsyth, D. A., Ponce, J., *Computer Vision: A Modern Approach*. Upper Saddle River, NJ, Prentice Hall, 2003.

[25] Genesereth, M.R., Nilsson, N.J., *Logical Foundations of Artificial Intelligence*. Palo Alto, CA, Morgan Kaufmann, 1987.

[26] Gonzalez, R., Woods, R., *Digital Image Processing*. 3rd edition. New York, Pearson Prentice Hall, 2008.

[27] Hammond, J. H., *The Camera Obscura: A Chronicle*. Bristol, UK, Adam Hilger, 1981.

[28] Haralick, R.M., Shapiro, L.G., *Computer and Robot Vision, 1+2*. Boston, Addison-Wesley, 1993.

[29] Hartley, R.I., Zisserman, A. *Multiple View Geometry*. Cambridge, UK, Cambridge University Press, 2004.

[30] Jones, J., Flynn, A., *Mobile Robots, Inspiration to Implementation*. Natick, MA, A.K. Peters, Ltd., 1993.

[31] Kortenkamp, D., Bonasso, R.P., Murphy, R.R. (editors), *Artificial Intelligence and Mobile Robots; Case Studies of Successful Robot Systems*. Cambridge, MA, AAAI Press / MIT Press, 1998.

[32] Latombe, J.C., *Robot Motion Planning*. Norwood, MA, Kluwer Academic, 1991.

[33] LaValle, S.M. *Planning Algorithms*, Cambridge, UK, Cambridge University Press, 2006.

[34] Lee, D., *The Map-Building and Exploration Strategies of a Simple Sonar-Equipped Mobile Robot*. Cambridge, UK, Cambridge University Press, 1996.

[35] Leonard, J.E., Durrant-Whyte, H.F., *Directed Sonar Sensing for Mobile Robot Navigation*. Norwood, MA, Kluwer Academic, 1992.

[36] Ma, Y., S. Soatto, S., Kosecka, J., Sastry, S., *An Invitation to 3-D Vision: From Images to Geometric Models*. New York, Springer-Verlag, 2003.

[37] Manyika, J., Durrant-Whyte, H.F., *Data Fusion and Sensor Management: A Decentralized Information-Theoretic Approach*. Palo Alto, CA, Ellis Horwood, 1994.

[38] Mason, M., *Mechanics of Robotics Manipulation*. Cambridge, MA, MIT Press, 2001.

[39]   Murphy, R.R., *Introduction to AI Robotics.* Cambridge, MA, MIT Press, 2000.

[40]   Nourbakhsh, I., *Interleaving Planning and Execution for Autonomous Robots.* Norwood, MA, Kluwer Academic, 1997.

[41]   Papoulis, A. *Probability, Random Variables, and Stochastic Processes*, 4th edition. New York, McGraw-Hill, 2001.

[42]   Raibert, M.H., *Legged Robots That Balance.* Cambridge, MA, MIT Press, 1986.

[43]   Ritter, G.X., Wilson, J.N., *Handbook of Computer Vision Algorithms in Image Algebra.* Boca Raton, FL, CRC Press, 1996.

[44]   Russell, S., Norvig, P., *Artificial Intelligence: A Modern Approach.* 3rd edition. New York, Prentice Hall International, 2010.

[45]   Schraft, R.D., Schmierer, G., *Service Roboter.* Natick, MA, A.K. Peters, Ltd, 2000.

[46]   Sciavicco, L., Siciliano, B., *Modeling and Control of Robot Manipulators.* New York, McGraw-Hill, 1996.

[47]   Siciliano, B., Khatib, O., *Springer Handbook of Robotics*, Springer, 2008.

[48]   Slama, C.C., *Manual of Photogrammetry.* 4th edition. Falls Church VA, American Society of Photogrammetry,1980.

[49]   Szeliski, R., *Computer Vision: Algorithms and Applications*, New York, Springer, 2010.

[50]   Tennekes, H., *The Simple Science of Flight: From Insects to Jumbo Jets.* Cambridge, MA, MIT Press, 1996.

[51]   Thrun, S., Burgard, W., Fox, D., *Probabilistic Robotics.* Cambridge, MA, MIT Press, 2005.

[52]   Todd, D.J, *Walking Machines: An Introduction to Legged Robots.* London, Kogan Page Ltd, 1985.

[53]   Trucco, E., Verri, A., *Introductory Techniques for 3-D Computer Vision.* New York, Prentice Hall, 1998.

[54]   Zufferey, J.C., *Bio-inspired Flying Robots: Experimental Synthesis of Autonomous Indoor Flyers*, EPFL Press, 2008.

Papers

[55]   Aho, A.V., "Algorithms for finding patterns in strings," in J. van Leeuwen (editor), *Handbook of Theoretical Computer Science*, Cambridge, MA, MIT Press, 1990, Volume A, chapter 5, 255–300.

[56]   Angeli, A., Filliat, D., Doncieux, S., Meyer, J.A., "Fast and incremental method for loop-closure detection using bags of visual words," *IEEE Transactions on Robotics*, 24(5): 1027–1037, October, 2008.

[57]   Arras, K.O., Castellanos, J.A., Siegwart, R., "Feature multi-hypothesis localization and tracking for mobile robots using geometric constraints," in *Proceedings of the IEEE International Conference on Robotics and Automation (ICRA'2002)*, Washington, DC, May , 2002.

[58]   Arras, K.O., Persson, J., Tomatis, N., Siegwart, R., "Real-time obstacle avoidance for polygonal robots with a reduced dynamic window," in *Proceedings of the IEEE International Conference on Robotics and Automation (ICRA 2002)*, Washington, DC, May, 2002.

[59]   Arras, K.O., Siegwart, R.Y., "Feature extraction and scene interpretation for map navigation and map building," in *Proceedings of SPIE, Mobile Robotics XII*, 1997.

[60]   Arras, K.O., Tomatis, N., "Improving robustness and precision in mobile robot localization by using laser range finding and monocular vision," in *Proceedings of the Third European Workshop on Advanced Mobile Robots (Eurobot 99)*, Zurich, September, 1999.

[61]   Astolfi, A., "Exponential stabilization of a mobile robot," *in Proceedings of 3rd European Control Conference*, Rome, September, 1995.

[62]   Bailey, T., Durrant-Whyte, H., "Simultaneous localization and mapping: Part II," *IEEE Robotics and Automation Magazine*, 108–117, 2006.

[63]   Bailey, T., "Mobile robot localisation and mapping in extensive outdoor environments," Ph.D. thesis, University of Sydney, 2002.

[64]   Baker, S., Nayar, S., "A theory of single-viewpoint catadioptric image formation," *International Journal of Computer Vision* 35, no. 2: 175–196, 1999.

[65]   Barnard, K., Cardei V., Funt, B., "A comparison of computational color constancy algorithms," *IEEE Transactions on Image Processing* 11: 972–984, 2002.

[66]   Barreto, J. P., Araujo, H., "Issues on the geometry of central catadioptric image formation. *International Conference on Computer Vision and Pattern Recognition* (CVPR), 2001.

[67]   Barreto, J. P., Araujo, H., "Fitting conics to paracatadioptric projection of lines," *Computer Vision and Image Understanding* 101(3): 151–165. March, 2006.

[68]   Barreto, J. P., Araujo, H., "Geometric properties of central catadioptric line images and their application in calibration," *IEEE Transactions on Pattern Analysis and Machine Intelligence*, 27(8): 1237–1333, August 2005.

[69]   Barron, J.L., Fleet, D.J., Beauchemin, S.S., "Performance of optical flow techniques," *International Journal of Computer Vision*, 12: 43–77, 1994.

[70]   Batavia, P., Nourbakhsh, I., "Path planning for the cye robot," *in Proceedings of the IEEE/RSJ International Conference on Intelligent Robots and Systems (IROS'00)*, Takamatsu, Japan, November 2000.

[71]   Bay, H., Ess, A., Tuytelaars, T., Van Gool, L., "Speeded-up robust features (SURF)," *International Journal on Computer Vision and Image Understanding* 110, no. 3: 346–359, 2008.

[72]   Besl, P., McKay, N., "A method for registration of 3-D shapes," *IEEE Transactions on Pattern Analysis and Machine Intelligence* (PAMI) 14, no. 2: 239–256, February 1992.

[73]   Bicchi, A., Marigo, A., Piccoli, B., "On the reachability of quantized control systems," *IEEE Transactions on Automatic Control,*. 4, no. 47: 546–563, 2002.

[74]   Biederman, I., "Recognition-by-components: A theory of human image understanding," *Psychological Review*, 2, no. 94: 115–147, 1987.

[75]   Blackwell, D., "Conditional expectation and unbiased sequential estimation," *Annals of Mathematical Statistics* 18: 105–110, 1947.

[76]   Blösch, M., Weiss, S., Scaramuzza, D., Siegwart, R., "Vision based MAV navigation in unknown and unstructured environments," *IEEE International Conference on Robotics and Automation* (ICRA 2010), Anchorage, Alaska, May 2010.

[77]   Borenstein, J., Koren, Y., "The vector field histogram – fast obstacle avoidance for mobile robots." *IEEE Journal of Robotics and Automation* 7: 278–288, 1991.

[78] Borges, G. A., Aldon, M.-J., "Line Extraction in 2D Range Images for Mobile Robotics," *Journal of Intelligent and Robotic Systems* 40: 267–297, 2004.

[79] Bosse, M., Newman, P., Leonard, J., Teller, S., "Simultaneous localization and map building in large-scale cyclic environments using the Atlas framework," *International Journal of Robotics Research* 23, no. 12: 1113–1139, 2004.

[80] Bosse, M., Rikoski, R., Leonard, J., Teller, S., "Vanishing points and 3d lines from omnidirectional video," *International Conference on Image Processing*, 2002.

[81] Brock, O., Khatib, O., "High-speed navigation using the global dynamic window approach," *in Proceeding of the IEEE International Conference on Robotics and Automation*, Detroit, May 1999.

[82] Brooks, R., "A robust layered control system for a mobile robot," *IEEE Transactions of Robotics and Automation*, RA-2:14–23, March 1986.

[83] Brown, H.B., Zeglin, G.Z., "The bow leg hopping robot", *in Proceedings of the IEEE International Conference on Robotics and Automation*, Leuwen, Belgium, May 1998.

[84] Bruce, J., Balch,T., and Veloso, M., "Fast and inexpensive color image segmentation for interactive robots," *in Proceedings of the IEEE/RSJ International Conference on Intelligent Robots and Systems (IROS'00)*, Takamatsu, Japan, 2000.

[85] Burgard,W., Cremers, A., Fox, D., Hahnel, D., Lakemeyer, G., Schulz, D., Steiner, W., Thrun, S., "Experiences with an interactive museum tour-guide robot," *Artificial Intelligence* 114: 1–53, 2000.

[86] Burgard, W., Derr, A., Fox, D., Cremers, A., "Integrating Global Position Estimation and Position Tracking for Mobile Robots: The Dynamic Markov Localization Approach," *in Proceedings of the 1998 IEEE/RSJ International Conference of Intelligent Robots and Systems (IROS'98)*, Victoria, Canada, October 1998.

[87] Burgard, W., Fox, D., Henning, D., "Fast grid-based position tracking for mobile robots," *in Proceedings of the 21th German Conference on Artificial Intelligence (KI97)*, Freiburg, Germany, Springer-Verlag, 1997.

[88] Burgard, W., Fox, D., Jans, H., Matenar, C., Thrun, S., "sonar mapping of large-scale mobile robot environments using EM," *in Proceedings of the International Conference on Machine Learning*, Bled, Slovenia, 1999.

[89] Cabani, C., Mac Lean, W. J., "Implementation of an affine-covariant feature detector in field-programmable gate arrays," in *Proceedings of the International Conference on Computer Vision Systems*, 2007.

[90] Campion, G., Bastin, G., D'Andréa-Novel, B., "Structural properties and classification of kinematic and dynamic models of wheeled mobile robots." *IEEE Transactions on Robotics and Automation* 12, no. 1: 47–62, 1996.

[91] Canny, J. F., "A computational approach to edge detection," *IEEE Transactions on Pattern Analysis and Machine Intelligence*, 679–698, 1986.

[92] Canudas de Wit, C., Sordalen, O.J., "Exponential stabilization of mobile robots with nonholonomic constraints." *IEEE Transactions on Robotics and Automation* 37: 1791–1797, 1993.

[93] Caprari, G., Estier, T., Siegwart, R., "Fascination of down scaling–alice the sugar cube robot." *Journal of Micro-Mechatronics* 1: 177–189, 2002.

[94] Caprile, B., Torre, V., "Using vanishing points for camera calibration." *International Journal of Computer Vision*. 4: 127–140, 1990.

[95] Castellanos, J.A., Tardos, J.D., Schmidt, G., "Building a global map of the environment of a mobile robot: The importance of correlations," in *Proceedings of the 1997 IEEE Conference on Robotics and Automation*, Albuquerque, NM, April 1997.

[96] Castellanos, J.A., Tardos, J.D., "Laser-based segmentation and localization for a mobile robot," in *Robotics and Manufacturing: Recent Trends in Research and Applications*, volume 6. ASME Press, 1996.

[97] Censi, A., Carpin, S., "HSM3D: Feature-less global 6DOF scan-matching in the hough/radon domain," *IEEE International Conference on Robotics and Automation* (ICRA), 2009.

[98] Chen, C.T., Quinn, R.D., "A crash avoidance system based upon the cockroach escape response circuit," in *Proceedings of the IEEE International Conference on Robotics and Automation*, Albuquerque, NM, April 1997.

[99] Chenavier, F., Crowley, J.L., "Position estimation for a mobile robot using vision and odometry," in *Proceedings of the IEEE International Conference on Robotics and Automation*, Nice, France, May 1992.

[100] Cheeseman, P., Smith, P. "On the representation and estimation of spatial uncertainty," *International Journal of Robotics* 5: 56–68, 1986.

[101] Chomat, O., Colin deVerdiere, V., Hall, D., Crowley, J., "Local scale selection for gaussian based description techniques," in *Proceedings of the European Conference on Computer Vision*, Dublin, Ireland, 117–133, 2000.

[102] Chong, K.S., Kleeman, L., "Accurate odometry and error modelling for a mobile robot," in *Proceedings of the IEEE International Conference on Robotics and Automation*, Albuquerque, NM, April 1997.

[103] Choset, H., Walker, S., Eiamsa-Ard, K., Burdick, J., "Sensor exploration: Incremental construction of the hierarchical generalized voronoi graph." *The International Journal of Robotics Research* 19: 126–148, 2000.

[104] Collins, A. Ruina, R. Tedrake, M. Wisse, "Efficient bipedal robots based on passive-dynamic walkers," *Science* 307, no. 5712: 1082 - 1085, 2005.

[105] Csorba, M. "Simultaneous localisation and map building," *Ph.D. thesis*, University of Oxford, Oxford, 1997.

[106] Cox, I.J., Leonard, J.J., "Modeling a dynamic environment using a bayesian multiple hypothesis approach," *Artificial Intelligence* 66: 311–44, 1994.

[107] Corke, P.I., Strelow, D., Singh, S., "Omnidirectional visual odometry for a planetary rover," *IEEE/RSJ International Conference on Intelligent Robots and Systems*, 2004.

[108] Cummins, M., Newman, P., "FAB-MAP: Probabilistic localization and mapping in the space of appearance," *The International Journal of Robotics Research* 27(6): 647–665, 2008.

[109] Cummins, M., Newman, P., "Highly scalable appearance-only SLAM – FAB-MAP 2.0," *In Robotics Science and Systems (RSS)*, Seattle, USA, June 2009.

[110] Davison, A.J., "Real-time simultaneous localisation and mapping with a single camera," *International Conference on Computer Vision*, 2003.

[111] Davison, A.J. "Active search for real-time vision," *In International Conference on Computer Vision*, 2005.

[112] Davison, A. J., Reid, I., Molton, N., Stasse, O., "MonoSLAM: Real-time single camera SLAM," *IEEE Transactions on Pattern Analysis and Machine Intelligence* 29, no. 6, June, 2007.

[113] Dellaert, F. "Square root SAM," *Proceedings of the Robotics Science and Systems Conference*, 2005.

[114] Dijkstra, E.W. "A note on two problems in connexion with graphs," *Numerische Mathematik* 1: 269–271, 1959.

[115] Dowlingn, K., Guzikowski, R., Ladd, J., Pangels, H., Singh, S., Whittaker, W.L., "NAVLAB: An autonomous navigation testbed," *Technical report CMU-RI-TR-87-24, Robotics Institute*, Pittsburgh, Carnegie Mellon University, November 1987.

[116] Duckett, T., Marsland, S.,Shapiro, J. "Learning globally consistent maps by relaxation," *IEEE International Conference on Robotics and Automation*, 2000.

[117] Duckett, T., Marsland, S.,Shapiro, J. "Fast, on-line learning of globally consistent maps," *Autonomous Robots* 12, no. 3: 287–300, 2002.

[118] Dudek, G., Jenkin, M., "Inertial sensors, GPS, and odometry," *Springer Handbook of Robotics*, Springer, 2008.

[119] Dugan, B., "Vagabond: A demonstration of autonomous, robust outdoor navigation," *in Video Proceedings of the IEEE International Conference on Robotics and Automation*, Atlanta, GA, May 1993.

[120] Durrant-Whyte, H., Bailey, T., "Simultaneous localization and mapping: Part I," *IEEE Robotics and Automation Magazine*, 99–108, 2006.

[121] Einsele, T., "Real-time self-localization in unknown indoor environments using a panorama laser range finder," in *Proceedings of the IEEE/RSJ International Conference on Intelligent Robots and Systems*, 697–702, 1997.

[122] Elfes, A., "Sonar real world mapping and navigation," in [12].

[123] Ens, J., Lawrence, P., "An investigation of methods for determining depth from focus." *IEEE Transactions on Pattern Analysis and Machine Intelligence* 15: 97–108, 1993.

[124] Espenschied, K.S., Quinn, R.D., "Biologically-inspired hexapod robot design and simulation," *in AIAA Conference on Intelligent Robots in Field, Factory, Service and Space*, Houston, Texas, March, 1994.

[125] Falcone, E., Gockley, R., Porter, E., Nourbakhsh, I., "The personal rover project: the comprehensive design of a domestic personal robot," *Robotics and Autonomous Systems, Special Issue on Socially Interactive Robots* 42: 245–258, 2003.

[126] Feder, H.J.S., Slotine, J-J.E., "Real-time path planning using harmonic potentials in dynamic environments," in *Proceedings of the IEEE International Conference on Robotics and Automation*, Albuquerque, NM, April 1997.

[127] Ferguson, D., Howard, T., Likhachev, M., "Motion planning in urban environments: Part II," *Proceedings of the IEEE/RSJ International Conference on Intelligent Robots and Systems* (IROS), 2008.

[128] Fischler, M. A., Bolles, R. C. "RANSAC random sampling concensus: A paradigm for model fitting with applications to Image analysis and automated cartography,". *Communications of ACM* 26: 381–395, 1981.

[129] Fox, D., "KLD-sampling: Adaptive particle filters and mobile robot localization," *Advances in Neural Information Processing Systems 14*. MIT Press, 2001.

[130] Fox, D., Burgard,W., Thrun, S., "The dynamic window approach to collision avoidance," *IEEE Robotics and Automation Magazine* 4: 23–33, 1997.

[131] Fraundorfer, F., Engels, C., Nister, D., "Topological mapping, localization and navigation using image collections," *IEEE/RSJ Conference on Intelligent Robots and Systems* 1, 2007.

[132] Freedman, B., Shpunt, A., Machline, M., Arieli, Y., "Depth mapping using projected patterns," *US Patent no. US20100118123A1*, May 13, 2010. http://www.freepatentsonline.com/20100118123.pdf

[133] Fusiello, A., Trucco, E., Verri, A., "A compact algorithm for rectification of stereo pairs," *Machine Vision and Applications*, 12(1): 16–22, 2000.

[134] Gächter, S., Harati, A., Siegwart, R., "Incremental object part detection toward object classification in a sequence of noisy range images," *Proceedings of the IEEE International Conference on Robotics and Automation (ICRA 2008)*, Pasadena, USA, May 2008.

[135] Gander,W., Golub, G.H., Strebel, R., "Least-squares fitting of circles and ellipses," *BIT Numerical Mathematics* 34, no. 4: 558–578, December 1994.

[136] Genesereth, M.R. "Deliberate agents," *Technical Report Logic-87-2*. Stanford, CA, Stanford University, Logic Group, 1987.

[137] Geyer, C., Daniilidis, K., "A unifying theory for central panoramic systems and practical applications," *European Conference on Computer Vision* (ECCV), 2000.

[138] Goedeme, T., Nuttin, M., Tuytelaars, T., Van Gool, L., "Markerless computer vision based localization using automatically generated topological maps," *European Navigation Conference* GNSS, Rotterdam, 2004.

[139] Golfarelli, M., Maio, D., Rizzi, S. "Elastic correction of dead-reckoning errors in map building," *IEEE/RSJ International Conference on Intelligent Robots and Systems*, 1998.

[140] Golub, G., Kahan,W., "Calculating the singular values and pseudo-inverse of a matrix." *Journal SIAM Numerical Analysis* 2: 205–223, 1965.

[141] Grisetti, G., Stachniss, C., Grzonka, S., Burgard, W., "A tree parameterization for efficiently computing maximum likelihood maps using gradient descent," *Robotics Science and Systems* (RSS), 2007.

[142] Grzonka, S., Grisetti, G., Burgard, W. "Towards a navigation system for autonomous indoor flying," *IEEE International Conference on Robotics and Automation*, 2009.

[143] Gutmann, J.S., Burgard, W., Fox, D., Konolige, K., "An experimental comparison of localization methods," in *Proceedings of the 1998 IEEE/RSJ International. Conference of Intelligent Robots and Systems* (IROS'98), Victoria, Canada, October 1998.

[144] Guttman, J.S., Konolige, K., "Incremental mapping of large cyclic environments," in P*roceedings of the IEEE International Symposium on Computational Intelligence in Robotics and Automation (CIRA)*, Monterey, November 1999.

[145] Hähnel, D., Fox, D., Burgard, W., Thrun, S. "A highly efficient FastSLAM algorithm for generating cyclic maps of large-scale environments from raw laser range measurements," *Proceedings of the Conference on Intelligent Robots and Systems*, 2003.

[146] Harris, C., Stephens, M., "A combined corner and edge detector," *Proceedings of the 4th Alvey Vision Conference*, 1988.

[147] Hart, P. E., Nilsson, N. J., Raphael, B. "A formal basis for the heuristic determination of minimum cost paths," *IEEE Transactions on Systems Science and Cybernetics* 4, no. 2: 100–107, 1968.

[148] Hashimoto, S., "Humanoid robots in Waseda University—Hadaly-2 and WABIAN," *in IARP First International Workshop on Humanoid and Human Friendly Robotics*, Tsukuba, Japan, October 1998.

[149] Heale, A., Kleeman, L.: "A real time DSP sonar echo processor," in *Proceedings of the IEEE/RSJ International Conference on Intelligent Robots and Systems (IROS'00)*, Takamatsu, Japan, 2000.

[150] Heymann, S., Maller, K., Smolic, A., Froehlich, B., Wiegand, T., "SIFT implementation and optimization for general-purpose GPU," *in Proceedings of the International Conference in Central Europe on Computer Graphics, Visualization and Computer Vision*, 2007.

[151] Horn, B.K.P., Schunck, B.G., "Determining optical flow," *Artificial Intelligence*, 17: 185–203, 1981.

[152] Horswill, I., "Visual collision avoidance by segmentation," in *Proceedings of IEEE International Conference on Robotics and Automation*, 902–909, 1995, IEEE Press, Munich, November 1994.

[153] Hoyt, D.F., Taylor, C.R, "Gait and the energetics of locomotion in horses," *Nature* 292: 239–240, 1981.

[154] Jacobs, R. and Canny, J., "Planning smooth paths for mobile robots," in *Proceeding. of the IEEE Conference on Robotics and Automation*, IEEE Press, 2–7, 1989.

[155] Jeffreys, H. and Jeffreys, B. S. "Methods of mathematical physics," *Cambridge, Cambridge University Press*, 305-306, 1988.

[156] Jennings, J., Kirkwood-Watts, C., Tanis, C., "Distributed map-making and navigation in dynamic environments," in *Proceedings of the 1998 IEEE/RSJ International Conference on Intelligent Robots and Systems (IROS'98)*, Victoria, Canada, October 1998.

[157] Jensfelt, P., Austin, D., Wijk, O., Andersson, M., "Feature based condensation for mobile robot localization," in *Proceedings of the IEEE International Conference on Robotics and Automation*, San Francisco, May 24–28, 2000.

[158] Jensfelt, P., Christensen, H., "Laser based position acquisition and tracking in an indoor environment," in *Proceedings of the IEEE International Symposium on Robotics and Automation* 1, 1998.

[159] Jogan, M., Leonardis, A. "Robust localization using panoramic viewbased recognition," in *Proceedings of ICPR00* 4: 136–139, 2000.

[160] Jung, I., Lacroix, S., "Simultaneous localization and mapping with stereovision," in *Proceedings of the 11th International Symposium Robotics Research*, Siena, Italy, 2005.

[161] Kamon, I., Rivlin, E., Rimon, E., "A new range-sensor based globally convergent navigation algorithm for mobile robots," in *Proceedings of the IEEE International Conference on Robotics and Automation*, Minneapolis, April 1996.

[162] Kelly, A., "Pose determination and tracking in image mosaic based vehicle position estimation," in *Proceeding of the IEEE/RSJ International Conference on Intelligent Robots and Systems (IROS'00)*, Takamatsu, Japan, 2000.

[163] Khatib, O., Real-time obstacle avoidance for manipulators and mobile robots, *International Journal of Robotics Research* 5, no. 1, 1986.

[164] Khatib, M., Chatila, R., "An extended potential field approach for mobile robot sensor motions," in *Proceedings of the Intelligent Autonomous Systems IAS-4*, IOS Press, Karlsruhe, Germany, March 1995, 490–496.

[165] Khatib, M., Jaouni, H., Chatila, R., Laumod, J.P., "Dynamic path modification for car-like nonholonomic mobile robots," in *Proceedings of IEEE International Conference on Robotics and Automation*, Albuquerque, NM, April 1997.

[166] Khatib, O., Quinlan, S., "Elastic bands: connecting, path planning and control," in *Proceedings of IEEE International Conference on Robotics and Automation*, Atlanta, GA, May 1993.

[167] Klein, G., Murray, D., "Parallel Tracking and Mapping for Small AR Workspaces," *Proceedings of the International Symposium on Mixed and Augmented Reality* (ISMAR'07), Nara, Japan, 2007.

[168] Ko, N.Y., Simmons, R., "The lane-curvature method for local obstacle avoidance," in *Proceedings of the 1998 IEEE/RSJ International Conference on Intelligent Robots and Systems (IROS'98)*, Victoria, Canada, October 1998.

[169] Koenig, S., Simmons, R., "Xavier: A robot navigation architecture based on partially observable markov decision process models," in [31].

[170] Koenig, S., Likhachev, M., "Fast replanning for navigation in unknown terrain," *IEEE Transactions on Robotics* 21(3): 354–363, 2005.

[171] Konolige, K.,. "A gradient method for realtime robot control," in *Proceedings of the IEEE/RSJ Conference on Intelligent Robots and Systems*, Takamatsu, Japan, 2000.

[172] Konolige, K., "Small vision systems: Hardware and implementation," in *Proceedings of Eighth International Symposium on Robotics Research*, Hayama, Japan, October 1997.

[173] Konolige, K., "Large-scale map-making," *AAAI National Conference on Artificial Intelligence*, 2004.

[174] Konolige, K., Agrawal, M., Solà, J., "Large scale visual odometry for rough terrain," *International Symposium on Research in Robotics* (ISRR), November, 2007.

[175] Koperski, K., Adhikary, J., Han, J., "Spatial data mining: Progress and challenges survey paper," in *Proceedings of the ACM SIGMOD Workshop on Research Issues on Data Mining and Knowledge Discovery*, Montreal, June 1996.

[176] Koren, Y., Borenstein, J., "High-speed obstacle avoidance for mobile robotics," in *Proceedings of the IEEE Symposium on Intelligent Control* 382–384, Arlington, VA, August 1988.

[177] Koren, Y., Borenstein, J., "Real-time obstacle avoidance for fast mobile robots in cluttered environments," in *Proceedings of the IEEE International Conference on Robotics and Automation*, Los Alamitos, CA, May 1990.

[178] Kruppa, E., "Zur ermittlung eines objektes aus zwei perspektiven mit innerer orientierung," *Sitzungsberichte Österreichische Akademie der Wissenschaften, Mathematisch-naturwissenschaftliche Klasse, Abteilung II a*, volume 122: 1939-1948, 1913.

[179] Kuipers, B., Byun, Y.T., "A robot exploration and mapping strategy based on a semantic hierarchy of spatial representations," *Journal of Robotics and Autonomous Systems*, 8: 47–63, 1991.

[180] Kuo, A., "Choosing your steps carfully," *Robotics & Automation Magazine*, 2007.

[181] Lacroix, S., Mallet, A., Chatila, R., Gallo, L., "Rover self localization in planetary-like environments," *in Proc. Int. Symp. Artic. Intell., Robot., Autom. Space* (i-SAIRAS), Noordwijk, The Netherlands, 1999.

[182] Lamon, P., Nourbakhsh, I., Jensen, B., Siegwar,t R., "Deriving and matching image fingerprint sequences for mobile robot localization," in *Proceedings of the 2001 IEEE International Conference on Robotics and Automation*, Seoul, Korea, May 2001.

[183] Latombe, J.C., Barraquand, J., "Robot motion planning: A distributed presentation approach." *International Journal of Robotics Research*, 10: 628–649, 1991.

[184] Lauria, M., Estier, T., Siegwart, R.: "An innovative space rover with extended climbing abilities," in *Video Proceedings of the 2000 IEEE International Conference on Robotics and Automation*, San Francisco, May 2000.

[185] LaValle, S. M., "Rapidly-exploring random trees: A new tool for path planning," *Technical Report, Computer Science Dept.*, Iowa State University, October 1998.

[186] Lavalle, S. M.: "Rapidly-exploring random trees: Progress and prospects," In *Algorithmic and Computational Robotics: New Directions*, pp. 293-308, 2000.

[187] Lazanas, A., Latombe, J.C., "Landmark robot navigation," in *Proceedings of the Tenth National Conference on AI*. San Jose, CA, July 1992.

[188] Lazanas, A. Latombe, J.C., "Motion planning with uncertainty: A landmark approach." *Artificial Intelligence*, 76: 285–317, 1995.

[189] Lee, S.-O., Cho, Y.-J., Hwang-Bo, M., You, B.-J., Oh, S.-R.: "A stabile target-tracking control for unicycle mobile robots," in *Proceedings of the 2000 IEEE/RSJ International Conference on Intelligent Robots and Systems*, Takamatsu, Japan, 2000.

[190] Leonard, J.J., Rikoski, R.J., Newman, P.M., Bosse, M., "Mapping partially observable features from multiple uncertain vantage points," *International Journal of Robotics Research* 21, no. 10: 943–975, 2002.

[191] Likhachev, M., Gordon, G., Thrun, S. "ARA*: Anytime A* with provable bounds on sub-optimality," *Advances in Neural Information Processing Systems* (NIPS), 2003.

[192] Likhachev, M., Ferguson, D., Gordon, G., Stentz, A., Thrun, S., "Anytime dynamic A*: An anytime, replanning algorithm," *Proceedings of the International Conference on Automated Planning and Scheduling* (ICAPS), 2005.

[193] Lindeberg, T., "Feature detection with automatic scale selection," *International Journal of Computer Vision* 30, no. 2: 79-116, 1998.

[194] Longuet-Higgins, H.C., "A computer algorithm for reconstructing a scene from two projections," *Nature* 293: 133–135, September, 1981.

[195] Louste, C. and Liegois, A., Path planning for non-holonomic vehicles: a potential viscous fluid method, Robotica 20: 291–298, 2002.

[196] Lowe, David G., "Object recognition from local scale-invariant features," *Proceedings of the International Conference on Computer Vision,* 1999.

[197] Lowe, D. G., "Distinctive image features from scale-invariant keypoints," *International Journal of Computer Vision* 60 (2): 91-110, 2004.

[198] Lumelsky, V., Skewis, T., "Incorporating range sensing in the robot navigation function," *IEEE Transactions on Systems, Man, and Cybernetics* 20: 1058–1068, 1990.

[199] Lumelsky, V., Stepanov, A., "Path-planning strategies for a point mobile automaton moving amidst unknown obstacles of arbitrary shape," in [12].

[200] Lu, F., Milios, E. "Globally consistent range scan alignment for environment mapping," *Autonomous Robots* 4: 333–349,1997.

[201] Maes, P., "The dynamics of action selection," in *Proceedings of the Eleventh International Joint Conference on Artificial Intelligence*, Detroit, 1989.

[202] Maes, P., "Situated Agents Can Have Goals," *Robotics and Autonomous Systems*, 6: 49–70. 1990.

[203] Maimone, M., Cheng, Y., Matthies, L., "Two years of visual odometry on the mars exploration rovers," *Journal on Field Robotics* 24, no. 3: 169–186, 2007.

[204] Makadia, A., Patterson, A., Daniilidis, K., "Fully automatic registration of 3D point clouds," *IEEE Conference on Computer Vision and Pattern Recognition*, New York, June 2006.

[205] Martinelli, A., Siegwart, R., "Estimating the odometry error of a mobile robot during navigation," in *Proceedings of the European Conference on Mobile Robots (ECMR 2003)*, Warsaw,   September 4–6, 2003.

[206] Masoud, S.A., Masoud, A.A., "Motion planning in the presence of directional and regional avoidance constraints unsing nonlinear, anisotropic, harmonic potential fields: a physical metaphor," *IEEE Transactions on Systems, Man and Cybernetics* 32, no. 6: 705–723, 2002.

[207] Masoud, S.A., Masoud, A.A., "Kinodynamic motion planning: a novel type of nonlinear, passive damping forces and advantages," *IEEE Robotics Automation Magazine* 17, no. 1: 85–99, 2010.

[208] Matsumoto, Y., Inaba, M., Inoue, H., "Visual navigation using viewsequenced route representation," *IEEE International Conference on Robotics and Automation*, 1996.

[209] Maybeck,P.S., "The Kalman filter: An introduction to concepts," in [12].

[210] Matas, J., Chum, O., Urban, M., Pajdla, T., "Robust wide-baseline stereo from maximally stable extremal regions," in *Proceedings of the British Machine Vision Conference*, 384–393, 2002.

[211] McGeer, T., "Passive dynamic walking," *International Journal of Robotics Research* 9, no. 2: 62–82, 1990.

[212] Mei, C., Rives, P., "Single view point omnidirectional camera calibration from planar grids," *IEEE International Conference on Robotics and Automation* (ICRA), 2007.

[213] Menegatti, E., Maedab, T., Ishiguro, H., "Image-based memory for robot navigation using properties of omnidirectional images," *Robotics and Autonomous System* 47, no. 4: 251–267, July, 2004.

[214] Meng, M., Kak, A.C.. "Mobile robot navigation using neural networks and nonmetrical environmental models," *IEEE Control Systems Magazine*, 13(5): 30–39, October 1993.

[215] Metropolis, N., Ulam, S. "The Monte Carlo method," *Journal of the American Statistical Association* 44, no. 247: 335–341, 1949.

[216] Mikolajczyk, K., C. Schmid, "Indexing based on scale-invariant interest points," *in Proceedings of the International Conference on Computer Vision*, 525–531, Vancouver, Canada, 2001.

[217] Mikolajczyk, K., Schmid, C., "Scale and affine invariant interest point detectors," *International Journal of Computer Vision* 1, no. 60: 63–86, 2004.

[218] Mikolajcyk, K. and Schmid, C., "An affine invariant interest point detector," *in Proceedings of the 7th European Conference on Computer Vision*, Denmark, 2002.

[219] Mikolajczyk, K., "Scale and Affine Invariant Interest Point Detectors," PhD thesis, INRIA Grenoble, 2002.

[220] Mikolajczyk, K., Tuytelaars, T., Schmid, C., Zisserman, A., Matas, J., Schaffalitzky, F., Kadir, T.,Van Gool, L. "A comparison of affine region detectors," *International Journal of Computer Vision*, 65(1-2): 43–72, 2005.

[221] Minetti, A.E. ,Ardigò, L.P., Reinach, E., Saibene, F., "The relationship between mechanical work and energy expenditure of locomotion in horses," *Journal of Experimental Biology* 202, no. 17, 1999.

[222] Minguez, J., Montano, L., "Nearness diagram navigation (ND): A new real time collision avoidance approach," in *Proceedings of the IEEE/RSJ International Conference on Intelligent Robots and Systems*, Takamatsu, Japan, October 2000.

[223] Minguez, J., Montano, L., "Robot navigation in very complex, dense, and cluttered indoor / outdoor environments," in *Proceeding of International Federation of Automatic Control (IFAC2002)*, Barcelona, April 2002.

[224] Minguez, J., Montano, L., Khatib, O., "Reactive collision avoidance for navigation with dynamic constraints," in *Proceedings of the 2002 IEEE/RSJ International Conference on Intelligent Robots and Systems*, 2002.

[225] Minguez, J., Montano, L., Simeon, T., Alami, R., "Global nearness diagram navigation (GND)," in *Proceedings of the 2001 IEEE International Conference on Robotics and Automation*, 2001.

[226] Mondada, F., Bonani, M., Raemy, X., Pugh, J., Cianci, C., Klaptocz, A., Magnenat, S., Zufferey, J.-C., Floreano, D. and Martinoli, A. "The e-puck, a robot designed for education in engineering," *The 9th Conference on Autonomous Robot Systems and Competitions*, 2009.

[227] Montiel, J.M.M. , Civera, J., Davison, A.J., "Unified inverse depth parametrization for monocular SLAM," *Proc. of the Robotics Science and Systems Conference*, 2006.

[228] Moutarlier, P., Chatila, R., "An experimental system for incremental environment modeling by an autonomous mobile robot," *1st International Symposium on Experimental Robotics*, 1989.

[229] Moutarlier, P., Chatila, R. "Stochastic multisensory data fusion for mobile robot location and environment modeling," *5th Int. Symposium on Robotics Research*, 1989.

[230] Montano, L., Asensio, J.R., "Real-time robot navigation in unstructured environments using a 3D laser range finder," in *Proceedings of the IEEE/RSJ International Conference on Intelligent Robot and Systems*, IROS 97, September 1997.

[231] Montemerlo, M., Thrun, S., Koller, D., Wegbreit, B. "FastSLAM: A factored solution to the simultaneous localization and mapping problem," *Proceedings of the AAAI National Conference on Artificial Intelligence*, 2002.

[232] Montemerlo, M., Thrun, S., Koller, D., Wegbreit, B. "Fast-SLAM 2.0: An improved particle filtering algorithm for simultaneous localization and mapping that provably converges," *International Joint Conference on Artificial Intelligence*, 2003.

[233] Moravec, H. and Elfes, A.E., "High Resolution Maps from Wide Angle Sonar," in *Proceedings of the 1985 IEEE International Conference on Robotics and Automation*, March 1985.

[234] Moravec, H. P., "Towards automatic visual obstacle avoidance," *Proceedings of the 5th International Joint Conference on Artificial Intelligence*, 1977.

[235] Moravec, H. P., "Visual mapping by a robot rover," *International Joint Conference on Artificial Intelligence*, 1979.

[236] Moravec, H., "Obstacle avoidance and navigation in the real world by a seeing robot rover," *PhD thesis*, Stanford University, 1980.

[237] Moutarlier, P., Chatila, R., "Stochastic multisensory data fusion for mobile robot location and environment modelling," in *Proceedings of the 5th International Symposium of Robotics Research*, Tokyo, 1989.

[238] Murillo, A.C., Kosecka, J., "Experiments in Place Recognition using Gist Panoramas," *Proceedings of the International Workshop on Omnidirectional Vision* (OMNIVIS'09), 2009.

[239] Murphy, K., Russell, S. "Rao-Blackwellized particle filtering for dynamic Bayesian networks," *In Sequential Monte Carlo Methods in Practice*, ed. by A. Doucet, N. de Freitas, N. Gordon, 499–516, Springer, 2001.

[240] Nayar, S.K., "Catadioptric omnidirectional camera." *IEEE CVPR*, 482–488, 1997.

[241] Nayar, S., Watanabe, M., and Noguchi, M., "Real-time focus range sensor." *In Fifth International Conference on Computer Vision*, 995–1001, Cambridge, Massachusetts, 1995.

[242] Nilsson, N.J., "Shakey the robot." *SRI, International, Technical Note*, Menlo Park, CA, 1984, No. 325.

[243] Nistér, D. Stewénius, H., "Scalable recognition with a vocabulary tree," *IEEE International Conference on Computer Vision and Pattern Recognition*, 2006.

[244] Nistér, D., Naroditsky, O., Bergen, J., "Visual odometry for ground vehicle applications," *Journal of Field Robotics* 23, no. 1: 3–20, 2006.

[245] Nistér, D., Naroditsky, O., Bergen, J., "Visual odometry," *IEEE International Conference on Computer Vision and Pattern Recognition*, 2004.

[246] Nistér, D., "An efficient solution to the five-point relative pose problem," *IEEE Transactions on Pattern Analysis and Machine Intelligence* (PAMI), 26(6): 756-770, June 2004.

[247] Nguyen, V., Martinelli, A., Tomatis, N., Siegwart, R. "A comparison of line extraction algorithms using 2D laser rangefinder for indoor mobile robotics," *IEEE/RSJ Intenational Conference on Intelligent Robots and Systems*, IROS, 2005.

[248] Noth, André, "Design of solar powered airplanes for continuous flight," *Ph.D. thesis, Autonomous Systems Lab, ETH Zurich*, Switzerland, December 2008.

[249] Nourbakhsh, I.R., "Dervish: An office-navigation robot," in [31].

[250] Nourbakhsh, I.R., Andre. D., Tomasi, C., Genesereth, M.R., "Mobile robot obstacle avoidance via depth from focus," *Robotics and Autonomous Systems*, 22: 151–158, 1997.

[251] Nourbakhsh, I.R., Bobenage, J., Grange, S., Lutz, R., Meyer, R, Soto, A., "An affective mobile educator with a full-time job," *Artificial Intelligence*, 114: 95–124, 1999.

[252] Nourbakhsh, I.R., Powers, R., Birchfield, S., "DERVISH, an office-navigation robot." *AI Magazine*, 16: 39–51, summer 1995.

[253] Oliva, A., Torralba, A., "Modeling the shape of the scene: A holistic representation of the spatial envelope," International Journal of Computer Vision, 42(3):145–175, 2001.

[254] Oliva, A., Torralba, A., "Building the gist of a scene: The role of global image features in recognition," in *Visual Perception, Progress in Brain Research*, 155:23–36, Elsevier, 2006.

[255] Omer, A.M.M., Ghorbani, R., Hun-ok Lim, Takanishi, A., "Semi-passive dynamic walking for biped walking robot using controllable joint stiffness based on dynamic simulation," *IEEE/ASME International Conference on Advanced Intelligent Mechatronics*, Singapore, 2009.

[256] Pell, B., Bernard, D., Chien, S., Gat, E., Muscettola, N., Nayak, P., Wagner, M., Williams, B., "An autonomous spacecraft agent prototype," *Autonomous Robots* 5: 1–27, 1998.

[257] Pavlidis, T., Horowitz, S. L. "Segmentation of plane curves," *IEEE Transactions on Computers* C-23(8): 860–870, 1974.

[258] Pentland, A.P., "A new sense for depth of field," *IEEE Transactions on Pattern Analysis and Machine Intelligence (PAMI)*, 9: 523–531, 1987.

[259] Philippsen, R., Siegwart, R., "Smooth and efficient obstacle avoidance for a tour guide robot," in *Proceedings of the IEEE International Conference on Robotics and Automation (ICRA 2003)*, Taipei, Taiwan, 2003.

[260] Pivtoraiko, M., Knepper, R., A., Kelly, A. "Differentially constrained mobile robot motion planning in state lattices," *Journal of Field Robotics* 26, no. 1: 308–333, 2009.

[261] Pfister, S. T., Roumeliotis, S. I., Burdick, J. W. "Weighted line fitting algorithms for mobile robot map building and efficient data representation," in *Proceedings of the IEEE International Conference on Robotics and Automation*, 2003.

[262] Pratt, J., Pratt, G., "Intuitive control of a planar bipedal walking robot," in *Proceedings of the IEEE International Conference on Robotics and Automation (ICRA '98)*, Leuven, Belgium, May 1998.

[263] Rao, C.R."Information and accuracy obtainable in estimation of statistical parameters," *Bulletin of the Calcutta Mathematical Society* 37: 81–91, 1945.

[264] Raibert, M. H., Brown, H. B., Jr., Chepponis, M., "Experiments in balance with a 3D one-legged hopping machine," *International Journal of Robotics Research*, 3: 75–92, 1984.

[265] Remy, C., Buffinton, K., Siegwart, R., "Stability analysis of passive dynamic walking of quadrupeds," *International Journal of Robotics Research*, 2009.

[266] Ringrose, R., "Self-stabilizing running," in *Proceedings of the IEEE International Conference on Robotics and Automation (ICRA '97)*, Albuquerque, NM, April 1997.

[267] Rosten, E., Drummond, T., "Fusing points and lines for high performance tracking," in *Proceedings of the International Conference on Computer Vision*, 1508–1511, 2005.

[268] Rosten, E., Drummond, T., "Machine learning for high-speed corner detection," in *Proceedings of the European Conference on Computer Vision*, 430-443, 2006.

[269] Rowe, A., Rosenberg, C., Nourbakhsh, I., "A simple low cost color vision system," in *Proceedings of Tech Sketches for CVPR 2001*, Kuaii, Hawaii, December 2001.

[270] Rubner, Y., Tomasi, C., Guibas, L., "The earth mover's distance as a metric for image retrieval," *STAN-CS-TN-98-86, Stanford University*, 1998.

[271] Rufli, M., Ferguson, D., Siegwart, R., "Smooth path planning in constrained environments," *Proceedings of the IEEE International Conference on Robotics and Automation* (ICRA), 2009.

[272] Rufli, M., Siegwart, R., "On the application of the D* search algorithm to time based planning on lattice graphs," *Proceedings of the European Conference on Mobile Robots* (ECMR), 2009.

[273] Scaramuzza, D., "Omnidirectional vision: from calibration to robot motion estimation,", *PhD thesis n. 17635, ETH Zurich*, February 2008.

[274] Scaramuzza, D., Martinelli, A., Siegwart, R., "A flexible technique for accurate omnidirectional camera calibration and structure from motion," *IEEE International Conference on Computer Vision Systems (ICVS 2006)*, New York, January 2006.

[275] Scaramuzza, D., Martinelli, A. Siegwart, R., "A toolbox for easily calibrating omnidirectional cameras," *IEEE/RSJ International Conference on Intelligent Robots and Systems* (IROS 2006), Beijing, China, October 2006.

[276] Scaramuzza, D., Fraundorfer, F., Pollefeys, M., "Closing the loop in appearance-guided omnidirectional visual odometry by using vocabulary trees," *Robotics and Autonomous System Journal (Elsevier)*, 2010.

[277] Scaramuzza, D., Fraundorfer, F., Pollefeys, M., and Siegwart, R., "Absolute scale in structure from motion from a single vehicle mounted camera by exploiting nonholonomic constraints," *IEEE International Conference on Computer Vision* (ICCV 2009), Kyoto, October, 2009.

[278] Scaramuzza, D., Fraundorfer, F., and Siegwart, R., Real-time monocular visual odometry for on-road vehicles with 1-point RANSAC, *IEEE International Conference on Robotics and Automation* (ICRA 2009), Kobe, Japan, May 2009.

[279] Scaramuzza, D., Siegwart, R., "Appearance guided monocular omnidirectional visual odometry for outdoor ground vehicles," *IEEE Transactions on Robotics* 24, no. 5, October 2008.

[280] Schlegel, C., "Fast local obstacle under kinematic and dynamic constraints," in *Proceedings of the IEEE International Conference on Intelligent Robot and Systems (IROS 98)*, Victoria, Canada 1998.

[281] Schultz, A., Adams, W., "Continuous localization using evidence grids," in *Proceedings of the IEEE International Conference on Robotics and Automation (ICRA '98)*, May 1998.

[282] Schweitzer, G., Werder, M., "ROBOTRAC – a mobile manipulator platform for rough terrain," in *Proceedings of the International Symposium on Advanced Robot Technology (ISART)*, Tokyo, Japan, March, 1991.

[283] Shi, J., Malik, J., "Normalized cuts and image segmentation," *IEEE Transactions on Pattern Analysis and Machine Intelligence (PAMI)* 82: 888–905, 2000.

[284] Shi, J., Tomasi, C., "Good features to track," *IEEE Conference on Computer Vision and Pattern Recognition*, 1994.

[285] Schmid, C., Mohr, R., Bauckhage, C., "Evaluation of interest point detectors," *International Journal of Computer Vision* 37, no. 2: 151–172, 2000.

[286] Se, S., Barfoot, T., Jasiobedzki, P., "Visual motion estimation and terrain modeling for planetary rovers," *Proceedings of the International Symposium on Artificial Intelligence for Robotics and Automation in Space*, 2005.

[287] Siadat, A., Kaske, A., Klausmann, S., Dufaut, M., Husson, R. "An optimized segmentation method for a 2D laser-scanner applied to mobile robot navigation," *Proceedings of the 3rd IFAC Symposium on Intelligent Components and Instruments for Control Applications*, 1997.

[288] Siegwart R., Arras, K., Bouabdallah, S., Burnier, D., Froidevaux, G., Greppin, X., Jensen, B., Lorotte, A., Mayor, L., Meisser, M., Philippsen, R., Piguet, R., Ramel, G., Terrien, G., Tomatis, N., "Robox at Expo.02: A large scale installation of personal robots," *Journal of Robotics and Autonomous Systems* 42: 203–222, 2003.

[289] Siegwart, R., Lamon, P., Estier, T., Lauria, M, Piguet, R., "Innovative design for wheeled locomotion in rough terrain," *Journal of Robotics and Autonomous Systems* 40: 151–162, 2002.

[290] Simhon, S., Dudek, G., "A global topological map formed by local metric maps," *Proceedings of the 1998 IEEE/RSJ International Conference on Intelligent Robots and Systems (IROS'98)*, Victoria, Canada, October 1998.

[291] Simmons, R., "The curvature velocity method for local obstacle avoidance," *Proceedings of the IEEE International Conference on Robotics and Automation*, Minneapolis, April 1996.

[292] Sinha, S. N., Frahm, J. M., Pollefeys, M., Genc, Y., "GPU video feature tracking and matching," *in EDGE, Workshop on Edge Computing Using New Commodity Architectures*, 2006.

[293] Sivic, J. and Zisserman, A., "Video Google: A text retrieval approach to object matching in videos," *Proceedings of the International Conference on Computer Vision*, 2003.

[294] Smith, R., Self, M., Cheeseman, P., "Estimating uncertain spatial relationships in robotics," *Autonomous Robot Vehicles,* I. J. Cox and G. T. Wilfong (editors), Springer-Verlag, 167–193, 1990.

[295] Smith, R.C. , Cheeseman, P., "On the representation and estimation of spatial uncertainty, *International Journal of Robotics Research* 5, no. 4: 56–68, 1986.

[296] Smith, S. M., Brady, J. M., "SUSAN - A new approach to low level image processing," *International Journal of Computer Vision* 23, no. 34: 45–78, 1997.

[297] Snavely, N., Seitz, S.M., Szeliski, R., "Photo Tourism: Exploring photo collections in 3D," *ACM Transactions on Graphics*, 25(3), August 2006.

[298] Snavely, N., Seitz, S.M., Szeliski, R., "Modeling the World from Internet Photo Collections," *International Journal of Computer Vision*, 2007

[299] Soatto, S., Brockett, R., "Optimal structure from motion: Local ambiguities and global estimates,", *International Conference on Computer Vision and Pattern Recognition*, 1998.

[300] Sordalen, O.J., Canudas de Wi,t C., "Exponential control law for a mobile robot: extension to path following," *IEEE Transactions on Robotics and Automation*, 9: 837–842, 1993.

[301] Sorg, H.W., "From serson to draper – two centuries of gyroscopic development," *Navigation* 23: 313–324, 1976.

[302] Steinmetz, B.M., Arbter, K., Brunner, B., Landzettel, K., "Autonomous vision navigation of the nanokhod rover," *Proceedings of i-SAIRAS 6th International Symposium on Artificial Intelligence, Robotics and Automation in Space*, 2001.

[303] Stentz, A., "The focussed D* algorithm for real-time replanning," in *Proceedings of IJCAI-95*, August 1995.

[304] Stentz, A., "Optimal and efficient path planning for partially-known environments," *Proceedings of the International Conference on Robotics and Automation*, 1994.

[305] Stevens, B.S., Clavel, R., Rey, L., "The DELTA parallel structured robot, yet more performant through direct drive," *Proceedings of the 23rd International Symposium on Industrial Robots*, 1992.

[306] Takeda, H., Facchinetti, C., Latombe, J.C., "Planning the motions of a mobile robot in a sensory uncertainty field," *IEEE Transactions on Pattern Analysis and Machine Intelligence*, 16: 1002–1017, 1994.

[307] Tardif, J., Pavlidis, Y., Daniilidis, K., "Monocular visual odometry in urban environments using an omnidirectional camera," *IEEE/RSJ International Confrence on Intelligent Robots and Systems*, 2008.

[308] Taylor, R., Probert, P., "Range finding and feature extraction by segmentation of images for mobile robot navigation," *Proceedings of the IEEE International Conference on Robotics and Automation*, ICRA, 1996.

[309] Thrun, S., Burgard, W., Fox, D., "A probabilistic approach to concurrent mapping and localization for mobile robots." *Autonomous Robots* 31: 1–25. 1998.

[310] Thrun, S., et al., "Minerva: A second generation museum tour-guide robot," *Proceedings of the IEEE International Conference on Robotics and Automation (ICRA'99)*, Detroit, May 1999.

[311] Thrun, S., Fox, D., Burgard, W., Dellaert, F., "Robust Monte Carlo localization for mobile robots," *Artificial Intelligence*, 128: 99–141, 2001.

[312] Thrun, S. "A probabilistic online mapping algorithm for teams of mobile robots," *International Journal of Robotics Research* 20, no. 5: 335–363, 2001.

[313] Thrun, S. "Simultaneous localization and mapping," *Springer Tracts in Advanced Robotics* 38, no. 5: 13–41, 2008.

[314] Thrun, S., Gutmann, J.-S., Fox, D., Burgard, W., Kuipers, B., "Integrating topological and metric maps for mobile robot navigation: A statistical approach," *Proceedings of the National Conference on Artificial Intelligence (AAAI)*, 1998.

[315] Thrun, S., Thayer, S., Whittaker, W., Baker, C., Burgard, W., Ferguson, D., Hähnel, D., Montemerlo, M., Morris, A., Omohundro, Z., Reverte, C., Whittaker, W. "Autonomous exploration and mapping of abandoned mines," *IEEE Robotics and Automation Magazine* 11, no. 4: 79–91, 2004.

[316] Tomasi, C., Shi, J., "Image deformations are better than optical flow," *Mathematical and Computer Modelling* 24: 165–175, 1996.

[317] Tomatis, N., Nourbakhsh, I., Siegwart, R., "Hybrid simultaneous localization and map building: A natural integration of topological and metric," *Robotics and Autonomous Systems* 44, 3–14, 2003.

[318] Triggs, B., McLauchlan, P., Hartley, R., Fitzgibbon, A., "Bundle adjustment — a modern synthesis," *International Conference on Computer Vision*, 1999.

[319] Tsai, R. "A versatile camera calibration technique for high-accuracy 3D machine vision metrology using off-the-shelf TV cameras and lenses," *IEEE Journal of Robotics and Automation* 3, no. 4: 323–344, August 1987.

[320] Tuytelaars, T., Mikolajczyk, K., "Local invariant feature detectors: a survey," *Source, Foundations and Trends in Computer Graphics and Vision* 3 , no. 3, 2007.

[321] Tzafestas, C.S., Tzafestas, S.G., "Recent algorithms for fuzzy and neurofuzzy path planning and navigation of autonomous mobile robots," *Systems-Science* 25: 25–39, 1999.

[322] Ulrich, I., Borenstein, J., "VFH*: Local obstacle avoidance with look-ahead verification," *Proceedings of the IEEE International Conference on Robotics and Automation*, San Francisco, May 2000.

[323] Ulrich, I., Borenstein, J., "VFH+: Reliable obstacle avoidance for fast mobile robots," *Proceedings of the International Conference on Robotics and Automation (ICRA'98)*, Leuven, Belgium, May 1998.

[324] Ulrich, I., Nourbakhsh, I., "Appearance obstacle detection with monocular color vision," *the Proceedings of the AAAI National Conference on Artificial Intelligence*. Austin, TX. August 2000.

[325] Ulrich, I., Nourbakhsh, I., "Appearance-based place recognition for topological localization," *Proceedings of t he IEEE International Conference on Robotics and Automation*, San Francisco, 1023–1029, April 2000.

[326] Vanualailai, J., Nakagiri, S., Ha, J-H., "Collision avoidance in a two-point system via Liapunov's second method," *Mathematics and Simulation* 39: 125–141, 1995.

[327] Van Winnendael, M., Visenti G., Bertrand, R., Rieder, R., "Nanokhod microrover heading towards Mars," *Proceedings of the Fifth International Symposium on Artificial Intelligence, Robotics and Automation in Space* (ESA SP-440), Noordwijk, Netherlands, 1999.

[328] Vandorpe, J., Brussel, H. V., Xu, H. "Exact dynamic map building for a mobile robot using geometrical primitives produced by a 2D range finder," *Proceedings of the IEEE International Conference on Robotics and Automation*, ICRA, 901–908, 1996.

[329] Weiss, G., Wetzler, C., Puttkamer, E., "Keeping track of position and orientation of moving indoor systems by correlation of range-finder scans," *Proceedings of the IEEE/RSJ International Conference on Intelligent Robots and Systems (IROS'94)*, Munich, September 1994.

[330] Weingarten, J., Gruener, G. and Siegwart, R., "A state-of-the-art 3D sensor for robot navigation," *Proceedings of IROS*, Sendai, September 2004.

[331] Weingarten, J. and Siegwart, R., "3D SLAM using planar segments," Proceedings of IROS, Beijing, October 2006.

[332] Wullschleger, F.H., Arra,s K.O., Vestli, S.J., "A flexible exploration framework for map building," *Proceedings of the Third European Workshop on Advanced Mobile Robots (Eurobot 99)*, Zurich, September 1999.

[333] Yagi, Y., Kawato, S.,"Panorama scene analysis with conic projection," *Proceedings of the IEEE International Conference on Intelligent Robots and Systems (IROS), Workshop on Towards a New Frontier of Applications*, 1990.

[334] Yamauchi, B., Schultz, A., Adams, W., "Mobile robot exploration and map-building with continuous localization," *Proceedings of the IEEE International Conference on Robotics and Automation (ICRA'98)*, Leuven, Belgium, May 1998.

[335] Ying, X., Hu, Z., "Can we consider central catadioptric cameras and fisheye cameras within a unified imaging model?," *European Conference on Computer Vision* (ECCV), Lecture Notes in Computer Science, Springer Verlag, May 2004.

[336] Zhang, L., Ghosh, B. K., "Line segment based map building and localization using 2D laser rangefinder," *Proceedings of the IEEE International Conference on Robotics and Automation*, 2000.

[337] Zhang, Z., "A flexible new technique for camera calibration," *Microsoft Research Technical Report 98-71*, December 1998
see also http://research.microsoft.com/~zhang.

Referenced Webpages

[338] Fisher, R.B. (editor), "CVonline: On-line Compendium of Computer Vision," Available at www.dai.ed.ac.uk/CVonline.

[339] The Intel Image Processing Library/Integrated Performance Primitives (Intel IPP): http://software.intel.com/en-us/intel-ipp.

[340] Source code release site: www.cs.cmu.edu/~jbruce/cmvision.

[341] Newton Labs website: www.newtonlabs.com.

[342] For probotics: http://www.personalrobots.com.

[343] OpenCV, the Open Source Computer Vision library: http://opencv.willowgarage.com/wiki.

[344] Passive walking: www-personal.umich.edu/~artkuo/Passive_Walk/passive_walking.html.

[345] Passive walking, the Cornell Ranger: http://ruina.tam.cornell.edu/research/topics/locomotion_and_robotics/ranger/ranger2008.php.

[346] Computer Vision industry: http://www.cs.ubc.ca/spider/lowe/vision.html.

[347] Camera Calibration Toolbox for Matlab: http://www.vision.caltech.edu/bouguetj/calib_doc.

[348] List of camera calibration softwares: http://www.vision.caltech.edu/bouguetj/calib_doc/htmls/links.html.

[349] Omnidirectional camera calibration toolbox from Christopher Mei http://www.robots.ox.ac.uk/~cmei/Toolbox.html.

[350] Omnidirectional camera calibration toolbox from Joao Barreto http://www.isr.uc.pt/~jpbar/CatPack/pag1.htm.

[351] Omnidirectional camera calibration toolbox from Davide Scaramuzza: google "ocamcalib" or go to http://robotics.ethz.ch/~scaramuzza/Davide_Scaramuzza_files/Research/OcamCalib_Tutorial.htm.

[352] Open source software for SLAM and loop-closing: http://openslam.org.

[353] Open source software for multi-view structure from motion: http://phototour.cs.washington.edu/bundler

[354] Microsoft Photosynth: http://photosynth.net

[355] Photo Tourism: http://phototour.cs.washington.edu/

[356] Voodoo Camera Tracker: A tool for the integration of virtual and real scenes http://www.digilab.uni-hannover.de/docs/manual.html

[357] Augmented-reality toolkit (ARToolkit): http://www.hitl.washington.edu/artoolkit

[358] Parallel Tracking and Mapping (PTAM): http://www.robots.ox.ac.uk/~gk/PTAM

Index

Wellman, Michael P., Amy Greenwald, and Peter Stone, *Autonomus Bidding Agents: Strategies and Lessons from Trading Agent Competition*

Floreano, Dario, and Claudio Mattiusi, *Bio-Inspired Artificial Intelligence: Theories, Methods, and Technologies*

Sterling, Leon S., and Kuldar Taveter, *The Art of Agent-Oriented Modeling*

Stoy, Kasper, David Brandt, and David J. Christensen, *An Introduction to Self-Reconfigurable Robots*

Siegwart, Roland, Illah R. Nourbakhsh, and Davide Scaramuzza, *Introduction to Autonomous Mobile Robots*, second edition